Springer Monographs in Mathematics

This series publishes advanced monographs giving well-written presentations of the "state-of-the-art" in fields of mathematical research that have acquired the maturity needed for such a treatment. They are sufficiently self-contained to be accessible to more than just the intimate specialists of the subject, and sufficiently comprehensive to remain valuable references for many years. Besides the current state of knowledge in its field, an SMM volume should ideally describe its relevance to and interaction with neighbouring fields of mathematics, and give pointers to future directions of research.

Dorina Mitrea · Irina Mitrea · Marius Mitrea ·
Joel H. Shapiro

Fredholm Theory in Topological Vector Spaces

Springer

Dorina Mitrea
Department of Mathematics
Baylor University
Waco, USA

Irina Mitrea
Department of Mathematics
Texas A&M University
Philadelphia, USA

Marius Mitrea
Department of Mathematics
Baylor University
Waco, USA

Joel H. Shapiro
Fariborz Maseeh Department
of Mathematics and Statistics
Portland State University
Portland, USA

ISSN 1439-7382 ISSN 2196-9922 (electronic)
Springer Monographs in Mathematics
ISBN 978-3-032-24737-7 ISBN 978-3-032-24738-4 (eBook)
https://doi.org/10.1007/978-3-032-24738-4

Mathematics Subject Classification: 46A16, 46A32, 46A50, 47A53, 47B06, 47A55

This Springer imprint is published by the registered company Springer Nature Switzerland AG
The registered company address is: Gewerbestrasse 11, 6330 Cham, Switzerland

Dedicated to the memory of our
colleague, collaborator, and friend,
Nigel Kalton

Preface

Our principal aim is to explore the extent to which Riesz and Fredholm theories of operators continue to function within the context of topological vector spaces. The essence of the book is its focus on extending classical operator theories from Banach and Hilbert spaces to the more general setting of topological vector spaces, and the fact that it provides a comprehensive framework for understanding Fredholm and Riesz theories in this broader context. The main takeaway is that local convexity and the existence of continuous linear functionals are not as necessary for these theories as the current literature would seem to suggest. Overall, the work undertaken here addresses the challenges and highlights the success of doing functional analysis in spaces that are not locally convex, providing novel insights into the resilience and robustness of the aforementioned theories.

The starting point is a review of some key concepts and results pertaining to topological spaces, such as basic definitions and properties of open and closed sets, neighborhoods, and compactness. Next, metrics, norms, and the notion of completeness are introduced. This is followed by definitions and properties of vector spaces, subspaces, and linear topological spaces. Subsequently, a detailed discussion about inner products, orthogonality, and the geometry of Hilbert spaces, takes place. At the heart of the matter one finds the subject of Riesz theory in topological vector spaces. The treatment here is jump-started by the study of compact operators, eigenvalues, and eigensubspaces, then continues on to more complex results, including spectral theory (where characterizations of compact operators and their spectral properties play a basic role). At this stage, the door is open to begin in earnest building a Fredholm theory in rather inclusive topological settings. Adopting this level of generality requires re-working things from scratch, starting with producing suitable definitions and then developing basic properties of Fredholm operators, including a detailed analysis of the Fredholm index, its stability, and applications.

The book includes numerous exercises to reinforce the theoretical concepts and provide practical applications. These exercises range from basic proofs to more complex problems involving the application of Fredholm and Riesz theories.

The authors gratefully acknowledge support from the Simons Foundation.

Dorina Mitrea, Waco, TX, USA
Irina Mitrea, Philadelphia, PA, USA
Marius Mitrea, Waco, TX, USA
Joel H. Shapiro, Portland, OR, USA

Introduction

It is well-known that the Riesz theory of compact operators and the Fredholm theory of bounded operators work effectively in the context of Banach spaces as they do in the setting of Hilbert spaces. What is not very well-known is that the Riesz theory of compact operators survives completely intact in arbitrary Hausdorff topological vector spaces.

The landscape of Fredholm theory is more nuanced. While a substantial portion of the theory carries over to Hausdorff topological vector spaces with minimal disruption, certain results may fail unless the ambient space is more accommodating. Chief among these is the continuity of the Fredholm index - a delicate issue that necessitates additional hypotheses (see Exercise 3.4.16 for an example demonstrating that some extra hypothesis is really needed here). To safeguard the validity of this aspect, we transition from the general framework of Hausdorff topological vector spaces to the more restricted one of p-Banach spaces. Remarkably, Fredholm theory remains intact in this, and other more general, 'dual-less' settings. In fact, local convexity and the attendant existence of continuous linear functionals is not required for either theory, while assumptions about metrizability and completeness need not appear until one is ready to discuss continuity of the Fredholm index.

Even better, once some very basic facts about linear topological spaces have been put in place, the arguments that establish the general Riesz and Fredholm theories in the context of linear topological spaces are not essentially more difficult than those needed for Hilbert space case. In fact, we claim that the absence of miracles associated with completeness, duality, and Hilbert-space geometry in the current theory actually *clarifies* what is going on.

To place matters into a proper perspective, let us record a theorem at the heart of Fredholm theory in *Banach spaces*. As a preamble, we need some basic definitions. Let X, Y be two Banach spaces and denote by $\mathcal{L}(X \to Y)$ the collection of all linear bounded operators from X into Y. For each $T \in \mathcal{L}(X \to Y)$ indicate its range by the symbol $\operatorname{ran} T$ and its kernel by the symbol $\ker T$. Define the space of Fredholm operators from X to Y as

$$\Phi(X \to Y) := \left\{T \in \mathcal{L}(X \to Y) : \operatorname{ran} T \text{ closed and } \dim(\ker T),\ \operatorname{codim}(\operatorname{ran} T) < \infty\right\}. \tag{0.0.1}$$

Also, consider the index function

$$\begin{aligned} i : \Phi(X \to Y) &\longrightarrow \mathbb{Z} \ \text{ defined by} \\ i(T) := \dim(\ker T) - \operatorname{codim}(\operatorname{ran} T) &\ \text{ for each } \ T \in \Phi(X \to Y). \end{aligned} \tag{0.0.2}$$

In addition, introduce the class of left Fredholm operators on X as

$$\Phi_+(X \to Y) := \left\{T \in \mathcal{L}(X \to Y) : \ \operatorname{ran} T \ \text{closed and} \ \dim(\ker T) < \infty\right\}, \tag{0.0.3}$$

and the class of right Fredholm operators on X as

$$\Phi_-(X \to Y) := \left\{T \in \mathcal{L}(X \to Y) : \ \operatorname{codim}(\operatorname{ran} T) < \infty\right\}. \tag{0.0.4}$$

The collection of semi-Fredholm operators from X to Y is then defined as the union $\Phi_-(X \to Y) \cup \Phi_+(X \to Y)$. The index function (0.0.2) may then be extended to the set of all semi-Fredholm operators by setting

$$\begin{gathered} i : \Phi_-(X \to Y) \cup \Phi_+(X \to Y) \longrightarrow \mathbb{Z} \cup \{\pm\infty\} \\ \text{defined as } \ i(T) := \dim(\ker T) - \operatorname{codim}(\operatorname{ran} T) \\ \text{for each operator } \ T \in \Phi_-(X \to Y) \cup \Phi_+(X \to Y). \end{gathered} \tag{0.0.5}$$

The following theorem summarizes various standard properties of Fredholm and semi-Fredholm operators in *Banach spaces*. In this functional analytic setting, the literature is enormous; see, e.g., [3, Chs. 1, 3], [8], [16, Chs. XI, XVII], [17, Vol. I, Sects. IV.6, IV.10], [24], [27, Sect. I.3], [34, Ch. 2], [35, Ch. III], [40, Chs. 5, 7], [46], and the references therein.

Theorem 0.0.1. *The following assertions hold for every Banach spaces X, Y, Z:*

(1) Any semi-Fredholm operator $T : X \to Y$ has closed range.
(2) If $T \in \Phi_\pm(X \to Y)$ and $S \in \Phi_\pm(Y \to Z)$ then $ST \in \Phi_\pm(X \to Z)$ and

$$i(ST) = i(S) + i(T). \tag{0.0.6}$$

In particular, for any operators $T \in \Phi(X \to Y)$ and $S \in \Phi(Y \to Z)$ it follows that $ST \in \Phi(X \to Z)$ and that (0.0.6) *holds.*

(3) If $T \in \Phi_\pm(X \to Y)$ then $T^* \in \Phi_\mp(Y^* \to X^*)$ and $i(T^*) = -i(T)$. In particular, for any operator $T \in \Phi(X \to Y)$ one has $T^* \in \Phi(Y^* \to X^*)$ and $i(T^*) = -i(T)$.
(4) Given $T \in \mathcal{L}(X \to Y)$, one has $T \in \Phi_+(X \to Y)$ if and only if there exist a Banach space Z, a compact operator $K : X \to Z$, and a constant $C \in (0, \infty)$ such that

$$\|x\|_X \leq C\|Tx\|_Y + \|Kx\|_Z \ \textit{ for all } \ x \in X. \tag{0.0.7}$$

In particular, $\Phi_+(X \to Y)$ *is both open in* $\mathcal{L}(X \to Y)$ *and stable under addition of compact operators. Moreover, the index is unaffected by addition of compact operators. Another consequence of* (0.0.7) *is the fact that*[1]

$$\text{if } X_0 \text{ is a closed subspace of } X \text{ and } T \in \Phi_+(X \to X) \text{ with } TX_0 \subseteq X_0, \text{ then also } T|_{X_0} \in \Phi_+(X_0 \to X_0). \tag{0.0.8}$$

(5) The set $\Phi_-(X \to Y)$ *is open in* $\mathcal{L}(X \to Y)$ *and* $\Phi_-(X \to Y)$ *is stable under addition of compact operators. Furthermore, the index is unaffected by addition of compact operators.*

(6) Under addition of compact operators, the set $\Phi(X \to Y)$ *is stable and the index is unaffected. Also,*

$$\Phi(X \to Y) = \Phi_-(X \to Y) \cap \Phi_+(X \to Y). \tag{0.0.9}$$

(7) Given any operator $T \in \mathcal{L}(X \to Y)$*, one has* $T \in \Phi(X \to Y)$ *if and only if there exist* $S_1, S_2 \in \mathcal{L}(Y \to X)$ *and* $K_1 : Y \to Y$*,* $K_2 : X \to X$ *compact operators such that*

$$TS_1 = I_Y + K_1 \text{ and } S_2T = I_X + K_2 \tag{0.0.10}$$

where I_X, I_Y *are the identity operators on X and Y, respectively. In fact, one may take* $S_1 = S_2 \in \Phi(X \to Y)$*, i.e., T is Fredholm if and only if it is invertible modulo compact operators.*

(8) The index function (0.0.5) *is continuous, hence locally constant.*

The role played by duality in the proof of this theorem is prefigured by item *(3)* which shows that duality acts as a portal, allowing one to seamlessly move between the class of lower semi-Fredholm operators $\Phi_+(X \to Y)$, and the class of upper semi-Fredholm operators $\Phi_-(X \to Y)$, while retaining a transparent measure of control for the index.

When dealing with generic topological linear spaces, one of the main challenges is to avoid the use of duality, and develop a "dual-less" approach to Fredholm theory. The price to pay is giving up on the classical Hahn-Banach Theorem[2], and its countless consequences. In this vein, the reader is referred to Exercise 6 in §1.9 where it is shown the Hahn-Banach Theorem fails for the p-Banach space ℓ^p with $0 < p < 1$, even though it has enough continuous linear functionals to separate points (cf. Exercise 3 in §1.9). The reader should also be mindful of the fact that while every Banach space becomes a p-Banach space, with $0 < p < 1$, when its original norm is replaced by its p-th power, the space ℓ^p is not isomorphic to a Banach space if $0 < p < 1$; see Exercise 6 in §1.9. In addition, Exercise 7 in §1.9 shows that for $0 < p < 1$ the space $L^p([0, 1])$ doesn't just fail to be a Banach space, it actually has a trivial dual! Our main aim is to show that Fredholm theory survives this (and other more general) "dual-less" settings. In the bigger picture, the present work falls

---

[1] by contrast, generally speaking, Fredholmness does not survive restriction to an invariant subspace; see Exercise 3.1.4 on this topic

[2] asserting that, for a normed space X, any continuous linear functional on a subspace of X extends to an element in the topological dual X^*

under the umbrella of *functional analysis on non-locally convex topological vector spaces*, where the powerful Hahn-Banach Theorem (for normed vector spaces) and its myriad of corollaries, cease to be effective, thus rendering arguments based on duality unsuitable. Early work in this regard is due to Hyers [21] and Williamson [45]. More timely expositions may be found in the monographs [23] and [33].

One of the goals of this monograph is to make this beautiful and surprising body of results available to a reader who is not necessarily an expert in functional analysis. In particular, no previous experience with topological vector spaces is required and, in fact, this material should also serve as a (hopefully) smooth introduction to this subject. The presentation proceeds in a manner which allows us to take whatever additional steps seem appropriate in order to further customize the development. For example, if one cares only about *metrizable* linear topological spaces, then one should ignore the material on nets and subnets – all the arguments will still work after replacing the term "net" by "sequence" whenever it appears. If one wishes to assume, in addition, that the metrizable spaces are *complete*, then any proofs (or assumptions) of relative openness for particular classes of continuous linear maps with closed range become superfluous: in this context the Open Mapping Theorem provides the desired property without further ado. Finally, for the quickest tour possible, one should concentrate on the case of Hilbert spaces. Then, in addition to the shortcuts mentioned above, one may safely ignore any acrobatics associated with the subject of complementation.

The emerging theory is neat and satisfies the innate curiosity regarding the natural boundaries of Fredholm theory. Yet the most compelling reason for pursuing this line of research stems from its relevance in other fields, most notably in the context of Partial Differential Equations. To offer an example, work with the scale of Hardy spaces in the whole Euclidean space $\mathbb{R}^n$. For $p \in (0, \infty)$, $H^p(\mathbb{R}^n)$ consists of tempered distributions f in $\mathbb{R}^n$ with the property that their radial maximal function, defined as $(\mathcal{M}_{\rm rad} f)(x) := \sup_{t>0} |(\Phi_t * f)(x)|$ for each $x \in \mathbb{R}^n$ (where Φ is a fixed background Schwartz function in $\mathbb{R}^n$ with $\int_{\mathbb{R}^n} \Phi \neq 0$ and $\Phi_t(x) := t^{-n}\Phi(x/t)$ for each $t > 0$ and $x \in \mathbb{R}^n$), satisfies

$$\|f\|_{H^p(\mathbb{R}^n)} := \|\mathcal{M}_{\rm rad} f\|_{L^p(\mathbb{R}^n)} < +\infty. \tag{0.0.11}$$

It is then well known that

$$H^p(\mathbb{R}^n) = L^p(\mathbb{R}^n) \ \text{ if } \ 1 < p < \infty, \tag{0.0.12}$$

but $H^p(\mathbb{R}^n)$ fails to be Banach in the regime $p \in (0, 1)$, although it is quasi-Banach whose dual separates points. This invites the question of identifying the Banach envelope (see [23]) of the quasi-Banach space $H^p(\mathbb{R}^n)$ with $p \in (0, 1)$, i.e., its "minimal enlargement" to a Banach space $\widehat{H^p}(\mathbb{R}^n)$. In this regard, it has been proved in [26] that

$$\widehat{H^p}(\mathbb{R}^n) = B^{1,1}_{-n\left(\frac{1}{p}-1\right)}(\mathbb{R}^n) \ \text{ for each } \ p \in (0, 1), \tag{0.0.13}$$

where $B^{p,q}_s(\mathbb{R}^n)$ is the scale of Besov spaces in $\mathbb{R}^n$. This can be used to show that $H^p(\mathbb{R}^n)$ fails to be locally convex in the regime $p \in (0, 1)$. A similar phenomenon occurs for the brand of Hardy space $\mathcal{H}^p(\partial\mathbb{D})$ with $0 < p < 1$, of traces of holomorphic functions subject to a uniform p-th power integrability condition on concentric circles in the unit disk $\mathbb{D}$ in the complex plane. Indeed, as shown by Duren, Romberg, and Shields in [11], there is a spectacular failure of the classical Hahn-Banach Theorem in this context (specifically, this space has proper, closed, weakly dense subspaces).

Mindful of these insights and nuances, let us now inject Fredholm theory into the proceedings by turning to the task of solving the Neumann Problem for the Laplace operator $\Delta := \sum_{j=1}^n \partial_j^2$ in a bounded domain $\Omega \subseteq \mathbb{R}^n$, with boundary data prescribed arbitrarily in the Hardy space $H^p(\partial\Omega)$ with $p \in (\frac{n-1}{n}, 1)$ (as defined in [7], [29, §4]):

$$\begin{cases} u \in \mathscr{C}^\infty(\Omega), \\ \Delta u = 0 \ \text{ in } \ \Omega, \\ \mathcal{N}(\nabla u) \in L^p(\partial\Omega), \\ \partial_\nu u = f \in H^p(\partial\Omega). \end{cases} \tag{0.0.14}$$

Above, ∂_ν denotes the normal derivative on $\partial\Omega$, while $\mathcal{N}$ stands for the nontangential maximal operator (see the discussion in [28, §8.2]). Looking for a candidate expressed as a single layer potential operator acting on a density $g \in H^p(\partial\Omega)$ of our choice then permits reducing the entire Neumann Problem to the issue of solving the boundary integral equation

$$\left(-\tfrac{1}{2}I + K\right)g = f. \tag{0.0.15}$$

Here, I is the identity, and K is the principal-value singular integral operator

$$Kg(x) = \lim_{\varepsilon \to 0^+} \frac{1}{\omega_{n-1}} \int\limits_{\substack{y \in \partial\Omega \\ |x-y|>\varepsilon}} \frac{\langle \nu(x), x-y \rangle}{|x-y|^n} g(y)\, d\sigma(y), \qquad x \in \partial\Omega, \tag{0.0.16}$$

where ω_{n-1} is the area of the unit sphere in $\mathbb{R}^{n-1}$ and σ is the surface measure on $\partial\Omega$. If $\partial\Omega$ is $\mathscr{C}^2$-smooth, then the integral kernel of K in (0.0.16) is only weakly singular, and this renders the operator in question compact on the Hardy space $H^p(\partial\Omega)$. In turn, this opens the door for employing Fredholm theory in dealing with the all-important boundary integral equation (0.0.15), since $-\frac{1}{2}I + K$ is a compact perturbation of $-\frac{1}{2}I$ (which is obviously invertible). This being said, the Hardy space $H^p(\partial\Omega)$ is equipped with

$$\|f\|_{H^p(\partial\Omega)} := \|f^\#\|_{L^p(\partial\Omega)} \tag{0.0.17}$$

where $f^\#$ is the Fefferman-Stein grand maximal function of f (cf. [13], [25], [29, §4]).

The failure of $\|\cdot\|_{L^p(\partial\Omega)}$ with $p < 1$ to satisfy the triangle inequality, is then inherited by $\|\cdot\|_{H^p(\partial\Omega)}$, so the Hardy space $H^p(\partial\Omega)$ is *not* Banach in the regime $p \in (\frac{n-1}{n}, 1)$. In fact, much as its counterpart in the whole Euclidean space, it fails to be locally convex. This takes us out of the zone of comfort for Fredholm theory as we knew it, namely forces us to work outside the class of Banach spaces. A lot is riding on this, since frankly there aren't a whole lot of other options to deal with the Neumann Problem (0.0.14), especially as this further surges to new heights by allowing more complicated elliptic systems to replace the Laplacian and the underlying domain is allowed to be considerably rougher than $\mathscr{C}^2$. One is also very much interested in prescribing boundary data in other types of function spaces besides the Hardy scale, such as in Besov and Triebel-Lizorkin spaces, both of which are not necessarily locally convex (see [12], [15], [26]). For more on this topic, the interested reader is referred to [5], [9], [22] [28]-[32], and the references included there.

Here is another application[3] illustrating the necessity and importance of generalizing the classical Fredholm theory beyond the confines of Banach spaces. Work in the complex plane $\mathbb{C} \equiv \mathbb{R}^2$. Consider the unit disk $\mathbb{D} := \{z \in \mathbb{C} : |z| < 1\}$, whose boundary is the torus

$$\mathbb{T} := \partial\mathbb{D} = \{z \in \mathbb{C} : |z| = 1\}. \tag{0.0.18}$$

Bring in the Triebel-Lizorkin scale $F^{p,q}_s(\mathbb{T})$, indexed by $0 < p, q \leq \infty$ and $s \in \mathbb{R}$, and set

$$h^p(\mathbb{T}) := F^{p,2}_0(\mathbb{T}), \quad 0 < p < \infty. \tag{0.0.19}$$

As is well known, $h^p(\mathbb{T})$ simply becomes the Lebesgue space $L^p(\mathbb{T})$ in the regime $p \in (1, \infty)$, and becomes a (real) local Hardy space on the torus for each integrability exponent $p \in (0, 1)$. Several equivalent descriptions of the later scale are available. For example, the local Hardy space $h^p(\mathbb{T})$ corresponding to $p \in (\frac{1}{2}, 1)$ may be characterized as the ℓ^p-span of local p-atoms, or as the space of distributions on the torus for which the Fefferman-Stein grand maximal function belongs to $L^p(\mathbb{T})$ (see, e.g., [1], [18], and the references therein). Henceforth, restrict attention to $h^p(\mathbb{T})$ with $p \in (\frac{1}{2}, 1)$. Classically, the Cauchy singular integral operator C on $\mathbb{T}$ (which may be regarded as a pseudodifferential operator of order zero) is bounded on the Triebel-Lizorkin scale on the torus, so it induces a linear and bounded mapping from the local Hardy space $h^p(\mathbb{T})$ into itself. In addition, in this context one has

$$C^2 = \tfrac{1}{4} I \text{ on } h^p(\mathbb{T}), \tag{0.0.20}$$

where I is the identity operator on $h^p(\mathbb{T})$. As a consequence,

$$P := \tfrac{1}{2} I + C \text{ is a projection on } h^p(\mathbb{T}), \tag{0.0.21}$$

i.e., P is linear, bounded, and satisfies $P^2 = P$ on $h^p(\mathbb{T})$. Use this to define the following closed subspace of $h^p(\mathbb{T})$:

$$\mathscr{H}^p(\mathbb{T}) := \operatorname{Ran} P = \{Pf : f \in h^p(\mathbb{T})\}. \tag{0.0.22}$$

For each given function $\phi \in \mathscr{C}^\infty(\mathbb{T})$, denote by M_ϕ the operator of multiplication by ϕ and introduce the Toeplitz operator with symbol ϕ by setting

[3] We are grateful to one of our referees for suggesting this.

$$T_\phi := PM_\phi. \tag{0.0.23}$$

Since the Hardy space $h^p(\mathbb{T})$ is stable under multiplication by smooth functions (in a quantitative fashion), it follows that the Toeplitz operator T_ϕ induces a well-defined linear and bounded mapping

$$T_\phi : \mathscr{H}^p(\mathbb{T}) \longrightarrow \mathscr{H}^p(\mathbb{T}) \text{ for each } \phi \in \mathscr{C}^\infty(\mathbb{T}). \tag{0.0.24}$$

Simple algebra shows that

$$T_\phi T_\psi - T_{\phi\psi} = P[M_\phi, P]M_\psi \text{ for each } \phi, \psi \in \mathscr{C}^\infty(\mathbb{T}), \tag{0.0.25}$$

where $[\cdot, \cdot]$ is the usual commutator bracket. Note that $[M_\phi, P] = [M_\phi, C]$ and the latter commutator is a smoothing operator when acting on Triebel-Lizorkin spaces on the torus, for each $\phi \in \mathscr{C}^\infty(\mathbb{T})$. In particular, thanks to classical embedding results, said commutator becomes a compact operator on $h^p(\mathbb{T})$. Bearing this in mind, we see from (0.0.25) that

$$\begin{gathered} T_\phi T_\psi - T_{\phi\psi} \text{ is a compact operator on } \mathscr{H}^p(\mathbb{T}) \\ \text{for each } \phi, \psi \in \mathscr{C}^\infty(\mathbb{T}). \end{gathered} \tag{0.0.26}$$

Specialize next (0.0.26) to the case when $\phi \in \mathscr{C}^\infty(\mathbb{T})$ never vanishes and $\psi := 1/\phi$. Then ψ belongs to $\mathscr{C}^\infty(\mathbb{T})$ and $T_{\phi\psi} = I$, the identity operator on $\mathscr{H}^p(\mathbb{T})$. Consequently, (0.0.26) implies

$$\begin{gathered} T_\phi T_{1/\phi} - I \text{ and } T_{1/\phi} T_\phi - I \text{ are compact operators on } \mathscr{H}^p(\mathbb{T}) \\ \text{for each } \phi \in \mathscr{C}^\infty(\mathbb{T}) \text{ which never vanishes.} \end{gathered} \tag{0.0.27}$$

Hence,

$$\begin{gathered} T_\phi \text{ is invertible, modulo compact operators, on } \mathscr{H}^p(\mathbb{T}) \\ \text{for each } \phi \in \mathscr{C}^\infty(\mathbb{T}) \text{ which never vanishes.} \end{gathered} \tag{0.0.28}$$

According to Corollary 3.2.1 (see §3.2 for a discussion) the property singled out in (0.0.28) is equivalent to saying that

$$\begin{gathered} T_\phi \text{ is a Fredholm operator on } \mathscr{H}^p(\mathbb{T}) \\ \text{for each } \phi \in \mathscr{C}^\infty(\mathbb{T}) \text{ which never vanishes.} \end{gathered} \tag{0.0.29}$$

As such, this opens the door for studying the Toeplitz operator T_ϕ from the perspective of Fredholm theory when this is considered on the scale of quasi-Banach spaces[4] $\mathscr{H}^p(\mathbb{T})$ with $p \in (\frac{1}{2}, 1)$. Among other things, given any function $\phi \in \mathscr{C}^\infty(\mathbb{T})$ which never vanishes, the result established in (0.0.29) invites the question of relating the index of the Fredholm operator T_ϕ, acting on $\mathscr{H}^p(\mathbb{T})$ with $p \in (\frac{1}{2}, 1)$, to the winding number of $\phi : \mathbb{T} \to \mathbb{C}$ about the point $0 \in \mathbb{C}$.

[4] the spaces in question fail to be genuinely Banach

When one carries out a similar program in relation to Lebesgue spaces $L^p(\mathbb{T})$ with $p \in (1, \infty)$, in place of local Hardy spaces on the torus, the aforementioned issues (i.e., Fredholmness and the relationship of the index to the winding number of the symbol) have been amply discussed in the literature. See, for example, [4] and the references therein. While the Lebesgue scale $L^p(\mathbb{T})$ in the regime $p \in (1, \infty)$ consists entirely of Banach spaces, our present work paves the way for studying these basic questions for a more general scale, which also allows the inclusion of spaces which are only quasi-Banach.

Besides its aesthetic and intrinsic values, the work in this monograph thus plays a direct role in addressing the aforementioned issues, by producing a more potent brand of Fredholm theory capable of dealing with such functional analytic settings. Yet the presentation has a light touch, permitting a non-specialist to become familiar with an ample repertoire of basic tools and results along the way, which have independent intrinsic value. The presentation is fairly self-contained and is punctuated by a multitude of exercises designed to add new nuances and degrees of understanding to the material. This also makes this monograph particularly well-suited for self-study, as well as for teaching a one (or two) semester course on Fredholm theory. We wish to mention that material relevant to Fredholm theory (or Fredholm-type results) in settings that go beyond locally convex spaces may be found in a number of sources, including [20], [37], [36], [43], and [44], though none of those sources is directly concerned with developing a Fredholm theory in general topological vector spaces in a systematic fashion.

The layout of the monograph is as follows. In Chapter 1 we discuss topologic, algebraic, and functional analytic tools providing foundational material necessary for understanding the more complex theories and applications appearing later. Topics here include: topological spaces (open and closed sets, neighborhoods, Hausdorff spaces, continuity, homeomorphisms, and the concept of a topology), metric and normed spaces (convergence, Cauchy sequences, and completeness), vector spaces and linear topological spaces (fundamental properties and examples, the concept of basis for a vector space, linear independence, span, dimension, local bases for vector topologies, locally bounded spaces, linear topological completeness), Hilbert spaces (inner products, orthogonality, the Pythagorean Theorem, and weak boundedness), p-Banach spaces, quasi-Banach spaces, the Aoki-Rolewicz Theorem in topological vector spaces, and the role of local convexity in ensuring a plentiful supply of continuous linear functionals on the space in question.

Chapter 2 is reserved for developing a brand of Riesz theory in topological vector spaces. This is divided into what is called "elementary Riesz theory" where the focus is on the eigenvalues and eigensubspaces of compact operators, and "advanced Riesz theory" on linear topological spaces, where the main result is that compact perturbations of the identity behave like linear operators on finite-dimensional vector spaces. These are separated by what we refer to as an algebraic interlude, where the range and null-space sequences of operators are analyzed from a purely algebraic point of view. For more on this topic, the reader is referred to [42].

In Chapter 3 we then turn in earnest to Fredholm theory in topological vector spaces. First, Fredholm operators are properly introduced in the setting of topological vector spaces, and connections with the Calkin algebra are presented. We then explore the properties of the Fredholm index, including its continuity in the context of p-Banach spaces, and link the Fredholm theory to spectra. While our treatment so far has been limited to the "diagonal" case, i.e., when a linear operator acts from a linear space into itself, we indicate how the theory developed up to this point can be upgraded to become applicable to the "off-diagonal" case, allowing an operator to act between two (possibly) different linear spaces. In the final section we specialize matters to the category of quasi-Banach spaces (which, as a class, is identical to the category of all p-Banach spaces with $p \in (0, 1]$ arbitrary), thus obtaining a satisfactory Fredholm theory in this general functional analytic setting.

Contents

Chapter 1
Topologic, Algebraic, and Functional Analytic Tools

Abstract Overall, this chapter lays the groundwork and provides foundational material necessary for understanding the more complex theories and applications discussed later in the book, by providing essential tools, concepts, and results in topology, algebra, and functional analysis.

There are eleven sections in this chapter, covering the following topics: §1.1 deals with topological spaces, relative topology, the concept of neighborhood, Hausdorff topological spaces, continuous functions acting between topological spaces, the product topology, the notion of local base at a point, interior and closure, compact topological spaces, relative compactness, limit points, the concept of net (and subnet), limits of nets. In §1.2 we discuss distance and metric spaces, Cauchy sequences and completeness in metric spaces. In §1.3 we introduce vector spaces, linear combinations, linear independence, the notion of basis in a vector space, the dimension of a vector space, linear transformations (operators/mappings), the concept of kernel and range for a linear operator, linear isomorphisms, algebraic complement, the concept of codimension, quotient spaces. In §1.4 we introduce normed vector spaces, the notion of p-norm and p-Banach space, semi-norms, quasi-norms, and the topology induced by a norm or a quasi-norm.

In §1.5 we consider inner products, Hilbert spaces, orthogonality. In §1.6 we initiate the study of linear topological spaces. The material is organized into eight subsections, addressing the following topics: local bases for vector topologies in §1.6.1, characterizations of local bases in §1.6.2, closed neighborhoods in §1.6.3, Hausdorff linear topological spaces in §1.6.4, the notion of boundedness for linear topological spaces in §1.6.5, weak boundedness in Hilbert spaces in §1.6.6, locally bounded linear topological spaces in §1.6.7, the notion of distance to a subspace of a Hilbert space in §1.6.8. Additional exercises pertaining to this material are contained in §1.7.

Next, §1.8 which is concerned with finite-dimensional issues, is divided into seven subsections, presenting: linear topological completeness in §1.8.1, dissecting and

D. Mitrea et al., *Fredholm Theory in Topological Vector Spaces*, Springer Monographs in Mathematics, https://doi.org/10.1007/978-3-032-24738-4_1

reassembling linear topological spaces in §1.8.2, projections in §1.8.3, subspaces of codimension one in §1.8.4, a finite-dimensional denouement in §1.8.5, finite-dimensional extensions in §1.8.6, and a converse to the Heine-Borel Theorem in §1.8.7. Once again, additional exercises concerning this material are collected in §1.9. Subsequently, the Aoki-Rolewicz Theorem in topological vector spaces is discussed in §1.10, and we revisit quasi-Banach spaces in §1.11. The last section in this chapter, §1.12, is a foray into locally convex topological vector spaces. The goal here is to indicate how local convexity is an essential structural ingredient guaranteeing the existence of an abundance of continuous linear functional on the space in question.

1.1 Topological Spaces

In this section, we review fundamental definitions and results related to abstract topological spaces, including the notion of a topology on a set, the properties of open and closed sets, neighborhoods, Hausdorff spaces, continuity, homeomorphisms, compactness, nets, and convergence, among others. Although this material may serve as a quick introduction to point-set topology, the selection of topics here is guided primarily by their relevance to the more advanced concepts developed later in this monograph. The same philosophy applies to the material in §1.2 dealing with metric spaces, §1.3 dealing with vector spaces, and §1.4 dealing with normed spaces.

Before beginning in earnest, we briefly elaborate on some standard notation and conventions used throughout the monograph (with the understanding that more specialized notation will be introduced in the text as needed). As it is usually the case, we shall let $\mathbb{N}$ stand for the set of positive natural numbers $\{1, 2, \dots\}$, $\mathbb{R}$ stand for the set of real numbers, $\mathbb{C}$ stand for the set of complex numbers, and $\varnothing$ for the empty set. Another convention followed is the use of the symbol ":=" in place of the usual "=" to indicate a definition (namely, that the term in the left side is defined as being the expression in the right side).

Definition 1.1.1. *A* `topology` *on a set X is a collection τ of subsets of X satisfying the following properties:*

(1) $\varnothing \in \tau$ and $X \in \tau$;
(2) τ is closed under union, i.e., the union of the elements of every sub-collection of τ belongs τ;
(3) τ is closed under finite intersections, i.e., the intersection of the elements of every finite sub-collection of τ belongs to τ.

The ordered pair (X, τ) is called a `topological space`*. One shall often omit specific mention of τ if no confusion will arise. The sets of the topology τ are called* `open sets` *of X, and the elements of the topological space (X, τ) are called* `points`*.*

It turns out that a given topology induces, in a hereditary fashion, canonical topologies on all subsets of the ambient.

Exercise 1.1.2. Given a topological space (X, τ) and a set $Y \subseteq X$, prove that

$$\tau\big|_Y := \{Y \cap O : O \in \tau\} \tag{1.1.1}$$

is a topology on Y.

In the above context, $\tau\big|_Y$ is called the `relative topology` induced by X on Y.

Definition 1.1.3. *Let (X, τ) be a topological space. Then*

(1) $A \subseteq X$ is called a `closed set` *of X provided its complement is open, i.e., $X \setminus A \in \tau$;*
(2) If $x \in X$, then $A \subseteq X$ is called a `neighborhood` *of x provided there exists $O \in \tau$ such that $x \in O$ and $O \subseteq A$ (note that neighborhoods need not be open). If X is a vector space, a neighborhood of the origin 0 will also be referred to as a* `zero-neighborhood`.
(3) For each $x \in X$ let $\mathcal{N}(x; \tau)$ denote the family of neighborhoods in X of the point x, relative to the topology τ, i.e.,

$$\mathcal{N}(x; \tau) := \{A \subseteq X : A \text{ is a neighborhood of } x\}. \tag{1.1.2}$$

Something that comes up frequently, is that every superset of a neighborhood of a point is itself a neighborhood of said point.

We recall next the following elementary result to the effect that a set is open if and only if it is a neighborhood all of its points.

Proposition 1.1.4. *Let (X, τ) be a topological space and $A \subseteq X$. Then the following holds:*

$$A \in \tau \iff A \in \mathcal{N}(a; \tau), \quad \forall\, a \in A. \tag{1.1.3}$$

Proof. The direct implication is trivial. Conversely, suppose that, for each $a \in A$, there exists an $O_a \in \tau$ such that $a \in O_a \subseteq A$. Then the set $\bigcup_{a \in A} O_a \in \tau$ by part *(2)* of Definition 1.1.1, and coincides with the set A. □

Definition 1.1.5. *A topological space X is called* `Hausdorff` *if for each pair of points $x_1, x_2 \in X$ such that $x_1 \neq x_2$ there exist neighborhoods $N_j \in \mathcal{N}(x_j; \tau)$ with $j = 1, 2$, such that $N_1 \cap N_2 = \varnothing$.*

Being Hausdorff is a hereditary property:

Exercise 1.1.6. Given a Hausdorff topological space (X, τ) and a set $Y \subseteq X$, prove that $(Y, \tau\big|_Y)$ is also Hausdorff.

We next introduce the concept of continuous function, and related notions.

Definition 1.1.7. *Assume that (X, τ_X) and (Y, τ_Y) are topological spaces. Then*

(1) A function $f : (X, \tau_X) \longrightarrow (Y, \tau_Y)$ *is said to be* τ_X-to-τ_Y continuous at $x_o \in X$*, or simply* continuous at $x_o \in X$*, provided*[1]

$$\forall\, N \in \mathcal{N}(f(x_o); \tau_Y)\ \exists\ V \in \mathcal{N}(x_o; \tau_X)\ \text{ such that }\ f(V) \subseteq N. \tag{1.1.4}$$

(2) A function $f : (X, \tau_X) \longrightarrow (Y, \tau_Y)$ *is said to be* τ_X-to-τ_Y continuous*, or simply* continuous*, if it is continuous at each point of* X*.*
(3) A function $f : (X, \tau_X) \longrightarrow (Y, \tau_Y)$ *is called* homeomorphism *of the topological space* (X, τ_X) *onto the topological space* (Y, τ_Y) *provided this is a bijection of* X *onto* Y *which is* τ_X*-to-*τ_Y *continuous and has a* τ_Y*-to-*τ_X *continuous inverse.*

A particularly versatile characterization of continuity, via pre-images, is contained in the proposition below and the subsequent exercise.

Proposition 1.1.8. *Suppose that* (X, τ_X) *and* (Y, τ_Y) *are two topological spaces, and consider a function* $f : (X, \tau_X) \longrightarrow (Y, \tau_Y)$*. Then*

$$f \text{ is continuous} \iff f^{-1}(O) \in \tau_X, \quad \text{whenever } \ O \in \tau_Y. \tag{1.1.5}$$

Proof. For the direct implication suppose that f is continuous and let $O \in \tau_Y$. For every $x \in f^{-1}(O)$ we have $O \in \mathcal{N}(f(x); \tau_Y)$ and hence, by the continuity of f at x, there exists $V \in \mathcal{N}(x; \tau_X)$ such that $f(V) \subseteq O$. Thus $x \in V \subseteq f^{-1}(O)$, which yields that $f^{-1}(O) \in \mathcal{N}(x; \tau_X)$. Since x was arbitrarily chosen in $f^{-1}(O)$, by appealing to Proposition 1.1.4 it follows that $f^{-1}(O) \in \tau_X$ as desired.

Reciprocally, let $x_o \in X$ and assume that $N \in \mathcal{N}(f(x_o); \tau_Y)$. Then there exists some $O \in \tau_Y$ such that $f(x_o) \in O \subseteq N$. Since $f^{-1}(O)$ is open in X (by hypothesis) and contains x_o, the set $V := f^{-1}(O)$ is a neighborhood of x_o. Also clearly, we have $f(V) \subseteq O \subseteq N$. As a result, the function f is continuous at every point of X. □

Exercise 1.1.9. Suppose (X, τ_X) and (Y, τ_Y) are topological spaces, and consider some function $f : (X, \tau_X) \longrightarrow (Y, \tau_Y)$. Also, fix a point $x_o \in X$. Show that f is continuous at x_o if and only if

$$f^{-1}(W) \in \mathcal{N}(x_o; \tau_X) \ \text{ for every } \ W \in \mathcal{N}(f(x_o); \tau_Y). \tag{1.1.6}$$

Hint: This can be seen by re-wording item *(1)* in Definition 1.1.7; alternatively, use Proposition 1.1.8.

Here is a companion result to Proposition 1.1.8:

Exercise 1.1.10. Assume (X, τ_X) and (Y, τ_Y) are two topological spaces, and consider a function $f : (X, \tau_X) \longrightarrow (Y, \tau_Y)$. Then f is continuous if and only if the pre-image under f of every closed set in (Y, τ_Y) is a closed set in (X, τ_X).

Hint: Use Proposition 1.1.8 and the fact that pre-imaging commutes with complementation.

---

[1] as is customary, for each set $A \subseteq X$ we define $f(A) := \{y \in Y : \text{ there is } a \in A \text{ such that } y = f(a)\}$

A very useful fact is that *continuity is preserved under restrictions*, as made precise in the following statement.

Exercise 1.1.11. Let (X, τ_X) and (Y, τ_Y) be topological spaces and consider a continuous function $f : (X, \tau_X) \longrightarrow (Y, \tau_Y)$. Also, assume that $A \subseteq X$ and $B \subseteq Y$ are such that $f(A) \subseteq B$. Then the function

$$g : (A, \tau_X\big|_A) \longrightarrow (B, \tau_Y\big|_B) \text{ given by } g(x) := f(x) \text{ for each } x \in A \tag{1.1.7}$$

is continuous.

Hint: Prove that $g^{-1}(O \cap B) = A \cap f^{-1}(O)$ for each $O \subseteq Y$.

Exercise 1.1.12. Given a topological space (X, τ) and a set $Y \subseteq X$, prove that the canonical inclusion map $\iota : (Y, \tau\big|_Y) \to (X, \tau)$, defined by $\iota(x) := x$ for each $x \in Y$, is a continuous function.

Hint: Use Proposition 1.1.8 and Exercise 1.1.2.

Another essential fact is that *continuity is preserved under compositions*, in the precise sense indicated below.

Exercise 1.1.13. Let (X, τ_X), (Y, τ_Y), and (Z, τ_Z) be three topological spaces, and consider two continuous functions $f : (X, \tau_X) \to (Y, \tau_Y)$ and $g : (Y, \tau_Y) \to (Z, \tau_Z)$. Show that the composition $g \circ f : (X, \tau_X) \to (Z, \tau_Z)$ is itself a continuous function.

Hint: Use Proposition 1.1.8, or, alternatively, Exercise 1.1.9.

Definition 1.1.14. *Let (X, τ) be a topological space. Call a* `basis` *for the topology τ any family $\mathcal{B} \subseteq \tau$ such that every open subset of X may be written as the union of sets from $\mathcal{B}$, i.e.,*

$$\forall\, O \in \tau \;\; \exists\, \mathcal{F} \subseteq \mathcal{B} \;\text{ such that }\; O = \bigcup_{W \in \mathcal{F}} W. \tag{1.1.8}$$

An important observation, which is apparent from (1.1.8), is that:

$$\begin{array}{c}\text{in the context of Definition 1.1.14, whenever}\\ x \in O \in \tau \text{ there exists } W \in \mathcal{B} \text{ such that } x \in W \subseteq O.\end{array} \tag{1.1.9}$$

It is of importance to place a topological structure on the Cartesian product of two sets, each of which equipped with a certain topology, of the sort indicated in the next definition.

Definition 1.1.15. *Given two topological spaces, (X_0, τ_0) and (X_1, τ_1), the* `product topology` *$\tau_0 \times \tau_1$ on $X_0 \times X_1$ is the collection of all unions (be they finite or infinite) of sets of the form $O_0 \times O_1$ with $O_0 \in \tau_0$ and $O_1 \in \tau_1$.*

In the context of the above definition,

$$\begin{gathered}\text{the set } \{O_0 \times O_1 : O_0 \in \tau_0 \text{ and } O_1 \in \tau_1\} \\ \text{is a basis for the topology } \tau_0 \times \tau_1.\end{gathered} \tag{1.1.10}$$

Proposition 1.1.16. *Suppose (X_0, τ_0) and (X_1, τ_1) are two topological spaces. Then for any two sets $Y_0 \subseteq X_0$ and $Y_1 \subseteq X_1$ it follows that*

$$(\tau_0 \times \tau_1)\big|_{Y_0 \times Y_1} = (\tau_0|_{Y_0}) \times (\tau_1|_{Y_1}). \tag{1.1.11}$$

Proof. For an arbitrary set $O \subseteq Y_0 \times Y_1$ we have $O \in (\tau_0 \times \tau_1)\big|_{Y_0 \times Y_1}$ if and only if there exists some $W \in \tau_0 \times \tau_1$ such that $O = W \cap (Y_0 \times Y_1)$ (cf. Exercise 1.1.2), if and only if there exist $\{O_j^\alpha\}_{\alpha \in A} \subseteq \tau_j$ for $j = 0, 1$ such that $O = \big(\bigcup_{\alpha \in A}(O_0^\alpha \times O_1^\alpha)\big) \cap (Y_0 \times Y_1)$ (cf. Definition 1.1.15), if and only if there exist $\{O_j^\alpha\}_{\alpha \in A} \subseteq \tau_j$ for $j = 0, 1$ such that $O = \bigcup_{\alpha \in A}(O_0^\alpha \cap Y_0) \times (O_1^\alpha \cap Y_1)$, if and only if $O \in (\tau_0|_{Y_0}) \times (\tau_1|_{Y_1})$ (see Definition 1.1.15 in this regard). □

Exercise 1.1.17. Let (X_j, τ_j), where $j = 1, 2$, be two topological spaces, and suppose $Y_j \subseteq X_j$, with $j = 1, 2$, are two closed sets. Show that the set $Y_1 \times Y_2$ is itself closed in the product topological space $(X_1 \times X_2, \tau_1 \times \tau_2)$.

Hint: Use the fact that

$$(X_1 \times X_2) \setminus (Y_1 \times Y_2) = [(X_1 \setminus Y_1) \times X_2] \cup [X_1 \times (X_2 \setminus Y_2)]. \tag{1.1.12}$$

Exercise 1.1.18. Suppose (X_j, τ_{X_j}) and (Y_j, τ_{Y_j}), $j = 1, 2$, are two pairs of topological spaces. Also, suppose $f_j : (X_j, \tau_{X_j}) \to (Y_j, \tau_{Y_j})$, $j = 1, 2$, are two given functions, and consider

$$\begin{gathered} f : (X_1 \times X_2, \tau_{X_1} \times \tau_{X_2}) \longrightarrow (Y_1 \times Y_2, \tau_{Y_1} \times \tau_{Y_2}). \\ \text{defined by } f(x_1, x_2) := (f_1(x_1), f_2(x_2)) \text{ for each } (x_1, x_2) \in X_1 \times X_2. \end{gathered} \tag{1.1.13}$$

Show that f is continuous if and only if its components, f_1 and f_2, are continuous.

Hint: Use (1.1.10) (cf. also (1.1.9)), and item *(1)* in Definition 1.1.7.

When doing analysis in a topological space, instead of looking at every neighborhood of a given point, a more economical (yet equivalent) way is to merely consider a sub-collection of the entire family of neighborhoods of that point which exhibits a certain feature.

Definition 1.1.19. *Given a nonempty topological space (X, τ), a* `local base at` *$x \in X$ for the topology τ is a collection of sets $\mathcal{U} \subseteq \mathcal{N}(x; \tau)$ with the property that*

$$\forall\, V \in \mathcal{N}(x; \tau) \;\; \exists\, U \in \mathcal{U} \;\; \textit{such that} \;\; U \subseteq V. \tag{1.1.14}$$

Exercise 1.1.20. Let (X, τ) be a topological space and consider a collection $\mathcal{B} \subseteq \tau$. Show that $\mathcal{B}$ is a basis for the topology τ (in the sense of Definition 1.1.14) if and only if for each $x \in X$ there exists $\mathcal{U}_x \subseteq \mathcal{B}$ a local base at x for the topology τ (in the sense of Definition 1.1.19).

Hint: Assuming $\mathcal{B}$ is a basis for τ, show that for each point $x \in X$ the collection $\mathcal{U}_x := \{O \in \mathcal{B} : x \in O\}$ is a local base at x for the topology τ. In the other direction, given any $O \in \tau$, for each $x \in O$ we may find $V_x \in \mathcal{U}_x \subseteq \mathcal{B}$ such that $x \in V_x \subseteq O$. Hence, $O = \bigcup_{x \in O} V_x$.

We continue by recalling the notions of interior and closure for an arbitrary subset of a topological space.

Definition 1.1.21. *Suppose that (X, τ) is a topological space and that $A \subseteq X$. Then*

(1) A point x is said to be an `interior point` *of A whenever $A \in \mathcal{N}(x; \tau)$. The set of interior points of A is called the* `interior` *of A and is denoted by* $\mathrm{Int}(A)$.
(2) The `closure` *of A in (X, τ) is the set of all points $x \in X$ such that every neighborhood of x intersects A, and is denoted by $\overline{A}$. Also, write $\overline{A}^{(X,\tau)}$ when one wishes to emphasize the dependence of the closure on the topological context.*

It follows that, if A is a subset of a topological space (X, τ), $\mathrm{Int}(A)$ is the union of all the open sets contained in A, and hence is the largest (with respect to $\subseteq$) open set contained in A. Also $\overline{A}$ is the intersection of all closed sets in (X, τ) which contain A, and hence is the smallest (again, with respect to $\subseteq$) closed set containing A. In particular, we have the following result:

Proposition 1.1.22. *In a topological space, a set is open if and only if it coincides with its interior. Also, a set is closed if and only if it coincides with its closure.*

Exercise 1.1.23. Given a topological space (X, τ) and a set $A \subseteq X$, show that a point $x \in X$ belongs to $\overline{A}$ if and only if $O \cap A \neq \varnothing$ whenever $x \in O \in \tau$.

The next proposition (whose proof follows immediately by unraveling definitions) shows that every point which is not in the closure of a subset of a topological space must be interior to its complement, and conversely.

Proposition 1.1.24. *Let A be a subset of a topological space (X, τ). Then the following hold:*

$$X \setminus \overline{A} = \mathrm{Int}(X \setminus A) \quad \text{and} \quad X \setminus \mathrm{Int}(A) = \overline{X \setminus A}. \tag{1.1.15}$$

Proposition 1.1.25. *Let (X, τ) be a topological space, and a consider a closed subset Y of X. Then, given any $Z \subseteq Y$, it follows that (recall the piece of notation introduced in item (2) of Definition 1.1.21)*

$$\overline{Z}^{(X,\tau)} = \overline{Z}^{(Y,\tau|_Y)}. \tag{1.1.16}$$

Proof. We have the following sequence of equivalences:

$$
\begin{aligned}
x \in \overline{Z}^{(X,\tau)} &\Longleftrightarrow x \in X \text{ and } O \cap Z \neq \varnothing \text{ for all } O \in \tau \text{ with } x \in O \\
&\Longleftrightarrow x \in Y \text{ and } (O \cap Y) \cap Z \neq \varnothing \text{ for all } O \in \tau \text{ with } x \in O \cap Y \\
&\Longleftrightarrow x \in \overline{Z}^{(Y,\tau|_Y)}.
\end{aligned}
\tag{1.1.17}
$$

The first and last equivalences are based on Exercise 1.1.23 and (1.1.1). For the left-to-right implication in the middle equivalence, observe that if its hypotheses are satisfied then

$$
\varnothing \neq O \cap Z \subseteq O \cap Y \Longrightarrow O \cap Y \neq \varnothing \Longrightarrow x \in \overline{Y}^{(X,\tau)} = Y, \tag{1.1.18}
$$

where the first inclusion uses the fact that Z is a subset of Y, and the last equality is a consequence of the fact that Y is closed in (X, τ). The right-to-left implication in the middle equivalence is clear, and this finishes the proof. □

Definition 1.1.26. *A family $\mathcal{C}$ of subsets of a topological space (X,τ) is said to* `cover` *a given set $A \subseteq X$, or to form a* `covering` *of A, if the union of the sets of $\mathcal{C}$ contains A. That is*

$$
A \subseteq \bigcup_{C \in \mathcal{C}} C. \tag{1.1.19}
$$

If in addition to (1.1.19) *one has $\mathcal{C} \subseteq \tau$ then $\mathcal{C}$ is called an* `open covering` *of A. Moreover, any family $\mathcal{S} \subseteq \mathcal{C}$ which covers A is called a* `subcovering` (*of $\mathcal{C}$*) *of the set A.*

Definition 1.1.27. *A subset A of a topological space (X,τ) is called* `compact` *provided every open covering of A contains a finite subcovering, i.e., a subcovering of finite cardinality. If X is compact in (X,τ) then (X,τ) is called a* `compact topological space`*. Call the topological space (X,τ)* `locally compact` *if every point in X has a compact neighborhood.*

Compactness is a hereditary property, in the precise sense described below.

Exercise 1.1.28. Let (X,τ) be a topological space. Suppose Y is a subset of X, and K is a subset of Y. Then K is compact in (X,τ) if and only if K is compact in $(Y, \tau|_Y)$ (cf. (1.1.1)).

Exercise 1.1.29. Given two topological spaces, (X_0, τ_0) and (X_1, τ_1), show that the set $K_0 \times K_1$ is compact in $(X_0 \times X_1, \tau_0 \times \tau_1)$ whenever K_0 is compact in (X_0, τ_0) and K_1 is compact in (X_1, τ_1).

Hint: Use (1.1.10).

Theorem 1.1.30. *Let (X,τ) be a topological space. Then the following hold.*

(1) If X is compact (i.e., if (X,τ) is a compact topological space) then every closed set $A \subset X$ is compact in X.

(2) If the topological space (X, τ) is Hausdorff then, if A is compact then A is closed in (X, τ).

Proof. To treat part *(1)*, let (X, τ) be a compact topological space and assume that $A \subseteq X$ is closed. Given a open covering C of A, the collection $\mathcal{D} := C \cup \{X \setminus A\}$ forms an open covering of X. By compactness of X, some finite subcollection of $\mathcal{D}$ must cover X. Thus, upon discarding $X \setminus A$ from this subcollection, if possible, the resulting family is a finite subcovering of A.

Turning attention to part *(2)*, consider (X, τ) a Hausdorff topological space and $A \subseteq X$ a compact set. We shall prove that $X \setminus A \in \tau$ by making use of Proposition 1.1.4. To this end, fix $x \in X \setminus A$ and notice that the Hausdorff property immediately gives that there exist open neighborhoods $N_a \in \mathcal{N}(x; \tau)$ and $V_a \in \mathcal{N}(a; \tau)$ such that

$$N_a \cap V_a = \varnothing \qquad \forall\, a \in A. \tag{1.1.20}$$

In particular, the family $\{V_a : a \in A\} \subseteq \tau$ is an open covering of the compact A, and as a consequence there exist $a_1, \dots, a_n \in A$ such that

$$A \subseteq \bigcup_{i=1}^{n} V_{a_i} =: V. \tag{1.1.21}$$

In addition, if we set $N := \bigcap_{i=1}^{n} N_{a_i}$, then $N \in \tau \cap \mathcal{N}(x; \tau)$ and, using (1.1.20)-(1.1.21),

$$N \subseteq X \setminus V \subseteq X \setminus A. \tag{1.1.22}$$

This completes the proof of the fact that $X \setminus A \in \tau$, establishing part *(2)* in Theorem 1.1.30. □

Theorem 1.1.31. *Let (X, τ_X) and (Y, τ_Y) be topological spaces such that X is compact, and suppose $f : (X, \tau_X) \longrightarrow (Y, \tau_Y)$ is a continuous function. Then $f(X)$ is a compact subset of (Y, τ_Y).*

Proof. Let $C \subseteq \tau_Y$ be a covering of the set $f(X)$. Since f is continuous, by Proposition 1.1.8, the family $\{f^{-1}(C) : C \in C\}$ satisfies $\{f^{-1}(C) : C \in C\} \subseteq \tau_X$. Since in addition this is also a covering of the compact set X there exist $C_1, \dots, C_n \in C$ such that the family $\{f^{-1}(C_i) : i = 1, \dots, n\}$ is a subcovering of X. Then the family $\{C_i : i = 1, \dots, n\} \subseteq C$ is a finite subcover of $f(X)$. This finishes the proof of the fact that $f(X)$ is compact in (Y, τ_Y). □

Exercise 1.1.32. Let (X, τ_X) and (Y, τ_Y) be topological spaces and consider a continuous function $f : (X, \tau_X) \to (Y, \tau_Y)$. Then $f(K)$ is a compact subset of (Y, τ_Y) whenever K is a compact subset of (X, τ_X).

Hint: Reason as in the proof of Theorem 1.1.31. Alternatively, use that $(K, \tau_X|_K)$ is a compact topological space and invoke Theorem 1.1.31.

Here is a companion to Definition 1.1.27.

Definition 1.1.33. *Call a subset of a topological space* `relatively compact` *if it has a compact closure.*

Proposition 1.1.34. *Let (X, τ) be a Hausdorff topological space, and consider a set $A \subseteq X$. Then A is relatively compact if and only if A is contained in a compact subset of X.*

Proof. In one direction, if A is relatively compact then $\overline{A}$ is compact and contains A. In the opposite direction, assume K is a compact subset of X containing A. Then K is closed (since X is Hausdorff), hence K contains $\overline{A}$. As a closed subset of a compact set, then $\overline{A}$ is compact. □

Much like compactness, relative compactness is also hereditary, as noted below.

Proposition 1.1.35. *Let (X, τ) be a topological space. Then every subset of a relatively compact set in (X, τ) is itself relatively compact in (X, τ). Also, if Y is a closed subset of (X, τ) and Z is a subset of Y, then Z is relatively compact in (X, τ) if and only if Z is relatively compact in $(Y, \tau\big|_Y)$ (cf.* (1.1.1)*).*

Proof. The first claim readily follows based on item *(1)* in Theorem 1.1.30. To deal with the last claim, assume first that Z is relatively compact in (X, τ), so $\overline{Z}^{(X,\tau)}$ is a compact set in (X, τ). Since Z is a subset of Y which is closed in (X, τ), it follows that $\overline{Z}^{(X,\tau)}$ is contained in Y. Granted these, we may invoke Exercise 1.1.28 to conclude that $\overline{Z}^{(X,\tau)}$ is compact in $(Y, \tau|_Y)$. However, as a consequence of Proposition 1.1.25, we have $\overline{Z}^{(X,\tau)} = \overline{Z}^{(Y,\tau|_Y)}$. As such, $\overline{Z}^{(Y,\tau|_Y)}$ is compact in $(Y, \tau|_Y)$, making Z relatively compact in $(Y, \tau\big|_Y)$.

In the converse direction, assume now Z is a relatively compact set in $(Y, \tau\big|_Y)$. Again by Proposition 1.1.25, we have $\overline{Z}^{(X,\tau)} = \overline{Z}^{(Y,\tau|_Y)}$, and the latter set is, by assumption, compact in $(Y, \tau\big|_Y)$. Since the set in question is also contained in Y (since Z is a subset of the closed set Y), Exercise 1.1.28 applies and gives that $\overline{Z}^{(X,\tau)}$ is compact in (X, τ). Thus, Z is relatively compact in (X, τ), finishing the proof. □

Exercise 1.1.36. Suppose (X, τ_X) is a topological space, and (Y, τ_Y) is a Hausdorff topological space. Let $f : (X, \tau_X) \to (Y, \tau_Y)$ be a continuous function. Show that if $A \subseteq X$ is relatively compact in τ_X then $f(A) \subseteq Y$ is relatively compact in τ_Y.

Hint: Since $f(A) \subseteq f(\overline{A})$, and $f(\overline{A}) \subseteq Y$ is compact in τ_Y (given that $\overline{A} \subseteq X$ compact in τ_X and f is continuous; cf. Exercise 1.1.32), use Proposition 1.1.34 to conclude that $f(A) \subseteq Y$ is relatively compact in τ_Y.

Another important notion is that of limit point, formally introduced below.

Definition 1.1.37. *Let (X, τ) be a topological space and consider $A \subseteq X$ and $x \in A$. Then x is called a* `limit point of the set` *A provided $x \in \overline{A \setminus \{x\}}$. The set of all limit points of A is denoted by A'.*

An immediate observation is that if (X, τ) is a topological space, $A \subseteq X$ and $x \in A$, then

$$x \in A' \iff \forall N \in \mathcal{N}(x; \tau) \; \exists \, y_N \in N \cap A \; \text{ such that } \; y_N \neq x. \tag{1.1.23}$$

Theorem 1.1.38. *If A is a subset of a topological space* (X, τ)*, then*

$$\overline{A} = A \cup A'. \tag{1.1.24}$$

Proof. We start by showing first the left-to-right inclusion in (1.1.24) and to this end let $x \in \overline{A}$. If $x \in A$, then clearly $x \in A \cup A'$. Otherwise $x \in \overline{A} \setminus A$ and consequently

$$N \cap A \neq \varnothing, \qquad \forall N \in \mathcal{N}(x; \tau), \tag{1.1.25}$$

which, in turn, guarantees that the right hand-side of the equivalence (1.1.23) is satisfied. Thus $x \in A'$ and clearly $x \in A \cup A'$.

To show the reverse inclusion in (1.1.24) note that $A' \subseteq \overline{A}$ follows readily from definitions, whereas $A \subseteq \overline{A}$ is a trivial consequence of the last part of the discussion immediately following Definition 1.1.21. □

Theorem 1.1.39. *Let* (X, τ) *be a compact topological space. Then each infinite subset* $A \subseteq X$ *has a limit point, i.e.,* $A' \neq \varnothing$.

Proof. By contradiction, suppose that $A \subseteq X$ is an infinite set and $A' = \varnothing$. Then, by Theorem 1.1.38, we have $A = \overline{A}$ which, in turn, implies that A is a closed set of X (cf. the second statement in Proposition 1.1.22). Furthermore, this yields $A = \overline{A} \setminus A'$ which combined with (1.1.23) allows us to conclude that for each $a \in A$ there exists $N_a \in \mathcal{N}(a, \tau) \cap \tau$ such that

$$N_a \cap A = \{a\}. \tag{1.1.26}$$

Since X is compact and the family $\{\{N_a\}_{a \in A}, X \setminus A\}$ is an open covering of X, we may conclude that X is covered by finitely many of these open sets. However, in view of the fact that $(X \setminus A) \cap A = \varnothing$, the set A must be covered by $\{N_{a_1}, \dots, N_{a_n}\}$, for some finitely many points $a_1, \dots, a_n \in A$. In conjunction with (1.1.26), this entails that A must be finite, leading to a contradiction. □

With the piece of terminology introduced in Definition 1.1.33, Theorem 1.1.39 yields:

Theorem 1.1.40. *Let* (X, τ) *be a topological space, and let A be a relatively compact subset of X. Then every infinite subset of A must have a limit point in X.*

Proof. Apply Theorem 1.1.39 for the compact topological space $(\overline{A}, \tau_{\overline{A}})$, where

$$\tau_{\overline{A}} := \{O \cap \overline{A} : O \in \tau\} \tag{1.1.27}$$

is the relative topology induced by τ on $\overline{A}$ (cf. (1.1.1)). □

The remainder of this section is essentially reserved for discussing notions of *convergence* in topological spaces. First, for ordinary sequences[2], we have the following definition.

Definition 1.1.41. *Suppose (X, τ) is a topological space and assume $\{x_j\}_{j\in\mathbb{N}}$ is a sequence of points in X. Call this sequence* `convergent to` *$x_* \in X$ (in the topology τ) if for every $V \in \mathcal{N}(x_*; \tau)$ there exists some $j_V \in \mathbb{N}$ with the property that $x_j \in V$ whenever $j \in \mathbb{N}$ satisfies $j \geq j_V$. Refer to x_* as the* `limit` *of the sequence in question. Simply say that $\{x_j\}_{j\in\mathbb{N}}$ is* `convergent` *(in the topology τ, or simply in X) if it has a limit in X.*

Succinctly put, a sequence of points in a topological space (X, τ) converges to some $x_* \in X$ if every neighborhood of x_* contains all but finitely many terms of the sequence. If (X, τ) is Hausdorff, then such a limit point is unique.

Exercise 1.1.42. Let (X, τ_X) and (Y, τ_Y) be two topological spaces and consider a continuous function $f : (X, \tau_X) \to (Y, \tau_Y)$. Show that f is `sequentially continuous`, i.e., for each point $x_* \in X$ and each sequence $\{x_j\}_{j\in\mathbb{N}} \subseteq X$ which converges to x_* in τ_X it follows that the sequence of images $\{f(x_j)\}_{j\in\mathbb{N}} \subseteq Y$ converges to $f(x_*)$ in τ_Y.

Working with sequences is particularly useful for topologies induced by metrics (a topic we will address later). In generic topological spaces, there is a more inclusive notion of "sequence," called *net*, that allows the terms to be labeled by partially ordered sets, thus more general than $\mathbb{N}$. This concept is essential because, in general topological spaces, nets replace sequences as the primary tool for characterizing continuity and compactness. We elaborate on this point of view below.

Definition 1.1.43. *A* `directed set` *is a set D equipped with a binary relation $\preceq$ satisfying the following properties:*

(1) $\alpha \preceq \alpha$ for each $\alpha \in D$.
(2) For each $\alpha, \beta, \gamma \in D$ the following implication holds: $\alpha \preceq \beta$ and $\beta \preceq \gamma \Longrightarrow \alpha \preceq \gamma$.
(3) For each $\alpha, \beta \in D$ there exists $\gamma \in D$ such that $\alpha \preceq \gamma$ and $\beta \preceq \gamma$.

If $\alpha \preceq \beta$, this shall also occasionally be denoted by $\beta \succeq \alpha$.

The prototypical example of a directed set is the set of natural numbers, $\mathbb{N}$, with $\preceq := \leq$. Another example, that is of fundamental importance in topology, is the collection $\mathcal{N}(x; \tau)$ of all neighborhoods of a point x in a topological space (X, τ), with $\preceq := \supseteq$.

Definition 1.1.44. *A* `net` *in a set X is a mapping from a directed set D into X, i.e.,*

$$D \ni \alpha \mapsto x_\alpha \in X. \tag{1.1.28}$$

[2] The reader is alerted to the fact that we will interchangeably use curly brackets $\{x_j\}_{j\in\mathbb{N}}$, and round parentheses $(x_j)_{j\in\mathbb{N}}$, to denote sequences. Also, we will occasionally identify a sequence $\{x_j\}_{j\in\mathbb{N}}$ with the set consisting of all distinct elements in the sequence, i.e., $\{x_j : j \in \mathbb{N}\}$ with the tacit understanding that repetitions are naturally discarded.

Such a mapping shall be denoted in the sequel by $\{x_\alpha\}_{\alpha\in D}$, *or simply by* $\{x_\alpha\}$ *(alternatively,* $(x_\alpha)_{\alpha\in D}$, *or even* (x_α) *when appropriate) if the directed set D over which the net is indexed is understood from the context.*

Definition 1.1.45. *Let* (X,τ) *be a topological space, D a directed set, and consider a net* $\{x_\alpha\}_{\alpha\in D} \subseteq X$ *in X indexed over D. A point* $x \in X$ *is called a* `limit` *of the net* $\{x_\alpha\}_{\alpha\in D}$ *if*

$$\forall N \in \mathcal{N}(x;\tau) \;\; \exists \alpha_N \in D \;\; \textit{such that} \;\; x_\alpha \in N \;\; \textit{whenever} \;\; \alpha \geq \alpha_N. \tag{1.1.29}$$

If the net $\{x_\alpha\}_{\alpha\in D}$ *has* $x \in X$ *as a limit, then the net* $\{x_\alpha\}_{\alpha\in D}$ *is said to be* `convergent` *to x and this shall be denoted by*

$$x_\alpha \to x \;\; \textit{or} \;\; \lim x_\alpha = x. \tag{1.1.30}$$

It is immediate to see that convergent nets in a Hausdorff topological space have unique limits. Also, if (X,τ) is a topological space, $x \in X$ is fixed, and D is a directed set then the constant net

$$\{x_\alpha\}_{\alpha\in D} \;\; \text{where} \;\; x_\alpha := x \quad \forall\, \alpha \in D, \tag{1.1.31}$$

can be easily seen to be convergent to x. A more relevant example is offered by the lemma below.

Lemma 1.1.46. *Let* (X,τ) *be a topological space and fix some point* $x \in X$. *For each* $\omega \in \mathcal{N}(x;\tau)$ *select a point* $x_\omega \in \omega$. *After directing the set of neighborhoods of x by reverse inclusion, this gives rise to the net* (x_ω), *indexed by all neighborhoods in* $\mathcal{N}(x;\tau)$. *Then*

$$x_\omega \to x. \tag{1.1.32}$$

Proof. Having fixed a neighborhood $N \in \mathcal{N}(x;\tau)$, it follows that if $\omega \in \mathcal{N}(x;\tau)$ is such that $\omega \subseteq N$ we necessarily have $x_\omega \in \omega \subseteq N$, from which we see that $x_\omega \in N$. Thus, the net (x_ω) is indeed convergent to x in (X,τ). □

Our next result highlights the deep connection between continuity and nets. Unlike sequences, which may fail to capture the full structure of general topological spaces, nets provide a complete characterization of continuity. We now present this "net-ential" characterization of continuity:

Theorem 1.1.47. *Let* (X,τ_X) *and* (Y,τ_Y) *be topological spaces, and let* $f : X \to Y$ *be a function. Then, for every* $x \in X$,

$$f \;\; \textit{is continuous at} \;\; x \iff f(x_\alpha) \to f(x) \;\; \textit{for every net} \;\; x_\alpha \to x. \tag{1.1.33}$$

Proof. To deal with the right-pointing implication, assume f is continuous at some point $x \in X$. This implies

$$\forall\, U \in \mathcal{N}(f(x);\tau_Y) \;\; \text{there exists} \;\; V \in \mathcal{N}(x;\tau_X) \;\; \text{such that} \;\; f(V) \subseteq U. \tag{1.1.34}$$

Let (x_α) be a net in X such that

$$x_\alpha \to x. \tag{1.1.35}$$

This means

$$\forall\, V \in \mathcal{N}(x;\tau_X) \ \text{ there exists } \alpha_0 \text{ such that } x_\alpha \in V \text{ whenever } \alpha \geq \alpha_0. \tag{1.1.36}$$

Take any neighborhood $U \in \mathcal{N}(f(x);\tau_Y)$. By (1.1.34), there exists $V \in \mathcal{N}(x;\tau_X)$ with the property that $f(V) \subseteq U$. By (1.1.36), there exists α_0 such that $x_\alpha \in V$ whenever $\alpha \geq \alpha_0$. Hence,

$$\alpha \geq \alpha_0 \implies f(x_\alpha) \in f(V) \subseteq U. \tag{1.1.37}$$

Thus, $(f(x_\alpha))$ converges to $f(x)$ in (Y, τ_Y), as desired.

Consider next the left-pointing implication in (1.1.33). Now the working assumption is that $f(x_\alpha) \to f(x)$ for every net (x_α) convergent to x. Seeking a contradiction, suppose f is not continuous at x. Then there exists $U \in \mathcal{N}(f(x);\tau_Y)$ for which

$$\forall\, V \in \mathcal{N}(x;\tau_X) \implies f(V) \nsubseteq U. \tag{1.1.38}$$

As such, for each $V \in \mathcal{N}(x;\tau_X)$ it is possible to choose a point $x_V \in V$ with the property that $f(x_V) \notin U$. Direct the set of neighborhoods of x by reverse inclusion, and define the net (x_V) indexed by all neighborhoods in $\mathcal{N}(x;\tau_X)$. Lemma 1.1.46 guarantees that

$$x_V \to x. \tag{1.1.39}$$

However,

$$f(x_V) \notin U, \quad \forall\, V \in \mathcal{N}(x;\tau_X), \tag{1.1.40}$$

so the net $(f(x_V))$ does not converge to $f(x)$. This contradicts our current working assumption, therefore f must be continuous at x. The proof of the theorem is now complete. □

We continue by including a characterization of the limit points of a set in a topological space using nets.

Proposition 1.1.48. *Let A be a subset of a topological space (X, τ), and $x \in X$. Then the following hold:*

$$x \in A' \iff \textit{there exists a net in } A \setminus \{x\} \textit{ which converges to } x \tag{1.1.41}$$

and

$$x \in \overline{A} \iff \textit{there exists a net in } A \textit{ which converges to } x. \tag{1.1.42}$$

Proof. We start with the direct implication in (1.1.41) and pick $x \in A'$. Then for every $N \in \mathcal{N}(x;\tau)$ we have $N \cap (A \setminus \{x\}) \neq \varnothing$, so we can select $x_N \in N \cap (A \setminus \{x\})$. Since the collection $\mathcal{N}(x;\tau)$ is a directed set when equipped with the binary relation $\supseteq$, we ultimately obtain that $\{x_N\}_{N \in \mathcal{N}(x;\tau)}$ is a net in $A \setminus \{x\}$. We claim that this net converges to x. Indeed, for each $N \in \mathcal{N}(x;\tau)$ there exists $M_0 := N \in \mathcal{N}(x;\tau)$ such that

$$x_M \in M \subseteq N \quad \text{whenever} \quad M \subseteq M_0, \tag{1.1.43}$$

where the membership in (1.1.43) is a consequence of the way in which the point x_M has been selected, while the subsequent inclusion follows from the fact that we have $M \subseteq M_0 = N$. This completes the left-to-right implication in (1.1.41).

Conversely, if $\{x_\alpha\}$ is a net in $A \setminus \{x\}$ such that $\{x_\alpha\}$ converges to $x \in X$, then by definition (1.1.29) holds. In particular, each $N \in \mathcal{N}(x;\tau)$ contains some x_α so

$$N \cap (A \setminus \{x\}) \neq \varnothing, \tag{1.1.44}$$

showing that $x \in A'$. This gives the right-to-left implication in (1.1.41), which is now established.

The left-to-right implication in (1.1.42) follows from Theorem 1.1.38, (1.1.41), and the observation right before the statement of the proposition regarding limits of constant nets. For the right-to-left implication in (1.1.42) let $x \in X$ such that there exists a net $\{x_\alpha\}_{\alpha \in D}$ in A, indexed over a directed set D, which converges to x. By definition this entails that for each $N \in \mathcal{N}(x;\tau)$ there exists $\alpha_0 \in D$ such that $x_\alpha \in N$ for each $\alpha \in D$ such that $\alpha \geq \alpha_0$. In particular, $N \cap A \neq \varnothing$ for each $N \in \mathcal{N}(x;\tau)$, yielding $x \in \overline{A}$, as desired. □

We next discuss the notion of a subnet, which plays a role analogous to that of a subsequence in the theory of sequences. Just as studying subsequences often reveals finer structural properties of sequences (such as convergence behavior and limit points), examining subnets allows us to probe deeper into the convergence patterns of nets, by providing the flexibility needed to capture these subtleties.

Definition 1.1.49. *A* `subnet` *of a net $\{x_\alpha\}_{\alpha \in D}$ is a net $\{y_\beta\}_{\beta \in E}$ together with a mapping from the index set E to the index set D*

$$E \ni \beta \mapsto \alpha(\beta) \in D \ \textit{satisfying} \ \begin{cases} \forall\, \alpha_0 \in D \ \exists \beta_0 \in E \ \textit{with} \ \beta \geq \beta_0 \Longrightarrow \alpha(\beta) \geq \alpha_0, \\ y_\beta = x_{\alpha(\beta)} \quad \forall\, \beta \in E. \end{cases} \tag{1.1.45}$$

It is not difficult to observe that if a net converges to a point, then every subnet converges to the same point.

The first property listed in (1.1.45) is referred to as *cofinality*. This ensures that the subnet "follows" the original net far enough along the directed set so that it captures the same limiting behavior. One may wonder why the definition of subnet seems so convoluted, and why not just define a subnet to be the original function from the directed index set D restricted to an appropriate subset E of D. The answer is that one needs the more complicated definition of subnet in order to obtain an analog for topological spaces of the sequential characterization of compactness for metric spaces. More specifically, it turns out that *every net in a relatively compact set has a convergent subnet*, and this is something we shall use in the proof of Theorem 2.1.20.

Exercise 1.1.50. Prove that every net in a compact topological space (X, τ) has a subnet convergent to a point in X.

Hint: Let $\{x_\alpha\}_{\alpha \in D}$ be a net in X, where D is a directed set. For each $\alpha \in D$, define the closed set

$$F_\alpha := \overline{\{x_\beta : \beta \geq \alpha\}}. \tag{1.1.46}$$

Show that the collection $\{F_\alpha\}_{\alpha \in D}$ satisfies the following finite intersection property:

$$\bigcap_{i=1}^{n} F_{\alpha_i} \neq \varnothing \quad \text{for every finite subset } \{\alpha_1, \ldots, \alpha_n\} \subset D. \tag{1.1.47}$$

To see this, observe that since D is directed, every finite subset $\{\alpha_1, \ldots, \alpha_n\} \subset D$ has an upper bound α^* and check that $x_{\alpha^*} \in \bigcap_{i=1}^{n} F_{\alpha_i}$. Next, since X is compact and $\{F_\alpha\}_{\alpha \in D}$ is a family of closed subsets of X which has the finite intersection property, conclude that

$$\bigcap_{\alpha \in D} F_\alpha \neq \varnothing. \tag{1.1.48}$$

This makes it possible to select a point $x \in \bigcap_{\alpha \in D} F_\alpha$. To build a subnet that converges to x, denote by $\mathcal{N}(x; \tau)$ the family of neighborhoods of x in (X, τ). Define

$$E := \{(\alpha, U) \in D \times \mathcal{N}(x; \tau) : x_\alpha \in U\} \tag{1.1.49}$$

ordered by

$$(\alpha_1, U_1) \leq (\alpha_2, U_2) \text{ if } \alpha_1 \leq \alpha_2 \text{ and } U_2 \subseteq U_1. \tag{1.1.50}$$

Show that this makes E a directed set, and the map $E \ni \beta = (\alpha, U) \mapsto \gamma(\beta) := \alpha \in D$ defines a subnet $\{x_{\gamma(\beta)}\}_{\beta \in E}$ that converges to x.

1.2 Metric Spaces

Here, the definition of metric spaces is given, along with a number of relevant properties such as convergence, Cauchy sequences, completeness, and equivalence of metrics.

Definition 1.2.1. *Given a non-empty set X, a function $d : X \times X \longrightarrow [0, \infty)$ is called a* `distance`*, or* `metric`*, on X provided it satisfies the following properties:*

(1) d is `non-degenerate`*, i.e., $d(x, y) = 0$ if and only if $x = y$.*
(2) d is `symmetric`*, i.e., for each $x, y \in X$ one has $d(x, y) = d(y, x)$.*
(3) d satisfies the `triangle inequality`*, i.e., $d(x, z) \leq d(x, y) + d(y, z)$ for every triplet of points $x, y, z \in X$.*

The pair (X, d), where X is a non-empty set and d is a distance on X, is called a `metric space`*. In this context, if $x \in X$ and $r > 0$ then*

$$B_d(x, r) := \{y \in X : d(x, y) < r\}, \tag{1.2.51}$$

is called the `ball with center` x `and radius` r.

Definition 1.2.2. *Let X be a set and d be a distance on X. Then the distance d is called* `translation invariant` *provided*

$$d(x+z, y+z) = d(x,y), \quad \forall\, x,y,z \in X. \tag{1.2.52}$$

Exercise 1.2.3. Let X be a set. Then the following hold.

(a) If $d : X \times X \longrightarrow [0,\infty)$ is a function which satisfies the triangle inequality (cf. item *(3)* in Definition 1.2.1) then the function

$$\widetilde{d} : X \times X \longrightarrow [0,\infty) \text{ given by } \widetilde{d}(x,y) := \frac{d(x,y)}{1+d(x,y)}, \qquad \forall\, x,y \in X, \tag{1.2.53}$$

satisfies the triangle inequality as well.

(b) If (X,d) is a metric space then $(X,\widetilde{d})$, with $\widetilde{d}$ as in (1.2.53), is also a metric space.

Hint: Show that the function $f(t) := \frac{t}{1+t}$ for $t \in [0,\infty)$ is both non-decreasing and subadditive.

Definition 1.2.4. *Let X be a set. Two distances $d_1, d_2 : X \times X \to [0,\infty)$ on X are called* `equivalent` *provided there exist constants $c, C \in (0,\infty)$ such that*

$$c d_2(x,y) \leq d_1(x,y) \leq C d_2(x,y) \quad \textit{for all} \quad x,y \in X. \tag{1.2.54}$$

Proposition 1.2.5. *Let (X,d) be a metric space and define*

$$\tau_d := \{A \subseteq X : \forall\, a \in A\ \exists\, r_a > 0 \ \textit{ such that } \ B_d(a,r_a) \subseteq A\}, \tag{1.2.55}$$

where $B_d(a,r_a)$ is the ball with center at a and radius r_a as introduced in (1.2.51). *Then*

(1) τ_d is a topology on X, called `the induced topology` *by the distance d.*
(2) For each $x \in X$ and each $r > 0$ the ball $B_d(x,r)$ is open, i.e., $B_d(x,r) \in \tau_d$.

Proof. Clearly, the empty set and X itself belong to τ_d. A moment's reflection also shows that τ_d is stable under arbitrary unions and finite intersections. Thus, τ_d is indeed a topology on X. Finally, *(2)* is justified by observing that

$$B_d(y, r - d(x,y)) \subseteq B_d(x,r) \tag{1.2.56}$$

for all $x \in X$, $r \in (0,\infty)$, and $y \in B_d(x,r)$. □

The following definition will play an important role later on.

Definition 1.2.6. *Given a topological space (X,τ), call the topology τ* `metrizable` *if there exists a distance d on X with the property that $\tau = \tau_d$.*

Significant examples are given in §1.10-§1.11. We continue the presentation in this section by recording another basic definition.

Definition 1.2.7. *Let (X, d) be a metric space.*

(a) A sequence $\{x_n\}_{n\in\mathbb{N}} \subseteq X$ is called `convergent` *provided there exists $x \in X$ such that the following is satisfied. For each $\varepsilon > 0$ there exists $N_\varepsilon \in \mathbb{N}$ such that for each $n \in \mathbb{N}$ such that $n \geq N_\varepsilon$ one has $d(x_n, x) \leq \varepsilon$. The point x is called a* `limit` *of the sequence $\{x_n\}_{n\in\mathbb{N}}$.*
(b) A sequence $\{x_n\}_{n\in\mathbb{N}} \subseteq X$ is called `Cauchy` *if for each $\varepsilon > 0$ there exists $N_\varepsilon \in \mathbb{N}$ such that for each $n, m \geq N_\varepsilon$ one has $d(x_n, x_m) \leq \varepsilon$.*
(c) The metric space (X, d) is called `complete` *provided every Cauchy sequence in X is convergent.*

The reader is invited to check that the definition of convergence in item *(a)* of Definition 1.2.7 is, in light of Proposition 1.2.5, compatible with the notion of convergence introduced earlier in Definition 1.1.41 for generic topological spaces.

Exercise 1.2.8. If (X, d) is a metric space and τ_d is the induced topology on X by the distance d, show that (X, τ_d) is a Hausdorff topological space. Conclude that every convergent sequence of (X, d) has a unique limit in X.

Hint: Verify that $B_d(x, \frac{1}{2}d(x, y)) \cap B_d(y, \frac{1}{2}d(x, y)) = \varnothing$ for any two distinct points $x, y \in X$.

Exercise 1.2.9. Let X be a set and $d_1, d_2 : X \times X \longrightarrow [0, \infty)$ be two equivalent distances on X. Show that

(a) The induced topologies on X by d_1 and d_2 are identical, i.e., $\tau_{d_1} = \tau_{d_2}$.
(b) If one of the spaces (X, d_1), (X, d_2) is complete, then so is the other.

Exercise 1.2.10. For each $i \in \mathbb{N}$ let (X_i, d_i) be a metric space and define X to be the collection of all sequences $x = \{x_k\}_{k\in\mathbb{N}}$ with the property that $x_k \in X_k$ for each $k \in \mathbb{N}$.

(a) Show that $d : X \times X \longrightarrow [0, \infty)$ given by

$$d(x, y) := \sum_{j=1}^{\infty} 2^{-j} \cdot \frac{d_j(x_j, y_j)}{1 + d_j(x_j, y_j)}, \qquad \forall\, x, y \in X, \quad x = \{x_k\}_{k\in\mathbb{N}} \text{ and } y = \{y_k\}_{k\in\mathbb{N}}, \tag{1.2.57}$$

is a distance on X.
(b) If for each $i \in \mathbb{N}$ one has $x^i := \{x^i_k\}_{k\in\mathbb{N}} \in X$ then the sequence $\{x^i\}_{i\in\mathbb{N}} \subseteq X$ converges to $\eta = \{\eta_k\}_{k\in\mathbb{N}} \in X$ if and only if for each $k \in \mathbb{N}$ one has

$$\lim_{i\to\infty} d_k(x^i_k, \eta_k) = 0. \tag{1.2.58}$$

(c) If (X_k, d_k) is a compact metric space for each $k \in \mathbb{N}$ then (X, d) is a compact metric space as well.

Hint: For part *(a)* use Exercise 1.2.3.

We continue by presenting a sequential characterization of compactness and relative compactness in metric spaces. Although Theorem 1.1.40 could be invoked for this purpose, we prefer to pursue a direct, self-contained approach, which proves to be both clearer and more illuminating.

Theorem 1.2.11. *Let (X, d) be a metric space. Then the topological space (X, τ_d) is compact if and only if every sequence $\{x_j\}_{j\in\mathbb{N}}$ in X has a subsequence which is convergent in τ_d to a point in X.*

Proof. To prove the left-to-right implication, assume that (X, τ_d) is compact, and fix a sequence $\{x_j\}_{j\in\mathbb{N}} \subseteq X$. If this contains a stationary subsequence then this is obviously convergent, so we are done. If that's not the case, it follows that each term in $\{x_j\}_{j\in\mathbb{N}}$ has finite multiplicity. Weeding out repetitions, there is therefore no loss of generality in assuming that the terms of $\{x_j\}_{j\in\mathbb{N}} \subseteq X$ are mutually distinct. In particular, $A := \{x_j : j \in \mathbb{N}\}$ is a well-defined subset of X of infinite cardinality. Suppose $A' = \varnothing$. From (1.1.23) we then conclude that

$$\begin{gathered}\text{for each } y \in X \text{ there exists } r_y > 0 \text{ with the property that}\\ B_d(y, r_y) \cap A \subseteq \{y\}.\end{gathered} \tag{1.2.59}$$

Since all d-balls are open in τ_d (cf. item *(2)* in Proposition 1.2.5), it follows that $\{B_d(y, r_y)\}_{y\in X}$ is an open cover of the compact set X. Consequently, there exists a finite family $\{y_i\}_{1\leq i\leq N}$ of points in X such that

$$A \subseteq X \subseteq \bigcup_{i=1}^{N} B_d(y_i, r_{y_i}). \tag{1.2.60}$$

Hence,

$$A = \bigcup_{i=1}^{N} \big(B_d(y_i, r_{y_i}) \cap A\big) \subseteq \bigcup_{i=1}^{N} \{y_i\} \tag{1.2.61}$$

with the final inclusion coming from (1.2.59). However, this forces A to be finite, contradicting the fact that $A = \{x_j : j \in \mathbb{N}\}$ has infinite cardinality. The above analysis shows that we actually have $A' \neq \varnothing$. Pick a point $x_* \in A'$ and, for each integer $k \in \mathbb{N}$ select some $x_{j_k} \in A \cap B_d(x_*, 1/k)$. Matters may be arranged to ensure that $1 \leq j_1 < j_2 < \cdots < j_k < j_{k+1} < \cdots$, so $\{x_{j_k}\}_{k\in\mathbb{N}}$ is a subsequence of the original sequence which converges in τ_d to $x_* \in X$. This concludes the proof of the left-to-right implication.

Consider next the right-to-left implication. Work under the assumption that

$$\begin{gathered}\text{every sequence } \{x_j\}_{j\in\mathbb{N}} \text{ in } X \text{ has a subsequence}\\ \text{which is convergent in } \tau_d \text{ to a point in } X.\end{gathered} \tag{1.2.62}$$

The first claim we make is that

$$\text{for each } \varepsilon > 0 \text{ we can cover } X \text{ by finitely many } d\text{-balls of radius } \varepsilon. \tag{1.2.63}$$

To justify this claim, fix an arbitrary $\varepsilon > 0$. If X cannot be covered by finitely many d-balls of radius ε then we can inductively construct a sequence $x_1, x_2, \ldots$ in X such that $x_{j+1} \in X \setminus \left(\bigcup_{1 \le i \le j} B_d(x_i, \varepsilon) \right)$. Then $\{x_j\}_{j \in \mathbb{N}} \subseteq X$ is a sequence with the property that

$$d(x_j, x_k) \ge \varepsilon \quad \text{for all } j, k \in \mathbb{N} \text{ with } j \ne k. \tag{1.2.64}$$

This property precludes $\{x_j\}_{j \in \mathbb{N}}$ from possessing a subsequence which converges in (X, τ_d), contradicting (1.2.62). Thus, (1.2.63) is established.

We next use this to construct a countable dense subset of (X, τ_d). Specifically, for each $n \in \mathbb{N}$ denote by $\mathcal{B}_n$ a finite collection of d-balls of radius $1/n$ covering X (which exists by (1.2.63)). Then, if D is the collection of the centers of all balls in $\bigcup_{n \in \mathbb{N}} \mathcal{B}_n$, it follows that D is indeed a countable dense subset of (X, τ_d). For further use, write this explicitly as $D = \{z_j\}_{j \in \mathbb{N}}$.

Suppose now $\mathcal{O} = (O_\alpha)_{\alpha \in I}$ is an arbitrary open cover of (X, τ_d), and fix a reference index $\alpha_* \in I$. For each $j \in \mathbb{N}$ and each $r \in \mathbb{Q}_+$, let $I_{j,r} := \{\alpha \in I : B_d(z_j, r) \subseteq O_\alpha\}$ and use this to define $O_{j,r} \in \mathcal{O}$ as follows: if $I_{j,r} \ne \varnothing$ set $O_{j,r} := O_\alpha$ for some $\alpha \in I_{j,r}$, and if $I_{j,r} = \varnothing$, set $O_{j,r} := O_{\alpha_*}$. Clearly,

$$\widetilde{\mathcal{O}} := \{O_{j,r} : j \in \mathbb{N} \text{ and } r \in \mathbb{Q}_+\} \tag{1.2.65}$$

is a countable subset of $\mathcal{O}$, and we claim that it covers X. To show this, pick an arbitrary $x \in X$. Since $(O_\alpha)_{\alpha \in I}$ covers X, there exists $\alpha_x \in I$ such that $x \in O_{\alpha_x}$. Next, since by design O_{α_x} is open in τ_d, we can find some $\rho_x > 0$ with the property that $B_d(x, \rho_x) \subseteq O_{\alpha_x}$. Given that $D = \{z_j\}_{j \in \mathbb{N}}$ is dense in (X, τ_d), there exists some $j_x \in \mathbb{N}$ for which $d(z_{j_x}, x) < \rho_x/4$. Pick a rational number $r_x \in (\rho_x/4, \rho_x/2)$. Such a choice ensures that

$$x \in B_d(z_{j_x}, r_x) \subseteq B_d(x, \rho_x) \subseteq O_{\alpha_x}. \tag{1.2.66}$$

In particular, $\alpha_x \in I_{j_x,r_x}$ so $I_{j_x,r_x} \ne \varnothing$. The latter guarantees that $B_d(z_{j_x}, r_x) \subseteq O_{j_x,r_x}$ hence, ultimately, $x \in O_{j_x,r_x}$. This finishes the proof of the fact that (1.2.65) is a countable subset of $\mathcal{O} = (O_\alpha)_{\alpha \in I}$ which covers X.

Let us summarize our progress. The argument so far, shows that if (1.2.62) holds then every open cover of (X, τ_d) may be refined to a countable subcover. The end-game in the proof of the right-to-left implication in the statement of the theorem is as follows. Let $\mathcal{U}$ be an arbitrary open cover of (X, τ_d). Given what we have just established, there is no loss of generality in assuming that this is countable, say $\mathcal{U} = \{U_j\}_{j \in \mathbb{N}}$. Our goal is to extract from this a finite subcover of X. Reasoning by contradiction, assume this is not possible. Then, inductively, we can choose $x_1, x_2, \dots$ in X such that $x_{j+1} \in X \setminus \left(\bigcup_{1 \le i \le j} U_i \right)$. By our working assumption in (1.2.62), we can find a subsequence $\{x_{j_k}\}_{k \in \mathbb{N}}$ converging to some point in $x_* \in X$. Since $\{U_j\}_{j \in \mathbb{N}}$ covers X, there exists $j_* \in \mathbb{N}$ such that $x_* \in U_{j_*}$. In turn, since $U_{j_*} \in \tau_d$, this implies that there exists $k_* \in \mathbb{N}$ for which

$$x_{j_k} \in U_{j_*} \text{ whenever } k \ge k_*. \tag{1.2.67}$$

On the other hand, by design, $x_j \notin U_{j_*}$ if $j > j_*$. This, however, contradicts (1.2.67) if we choose $j := j_k$ with $k > \max\{j_*, k_*\}$, since then $j = j_k \ge k > j_*$. The contradiction we reached shows that we can extract from $\mathcal{U} = \{U_j\}_{j \in \mathbb{N}}$ a finite subcover of X, and this finishes the proof of the right-to-left implication in the statement of the theorem. □

In turn, Theorem 1.2.11 readily adapts to produce a sequential characterization of relative compactness for subsets of metric spaces.

Corollary 1.2.12. *Let (X, d) be a metric space, and consider a set $A \subseteq X$. Then A is relatively compact in (X, τ_d) if and only if every sequence $\{x_j\}_{j\in\mathbb{N}} \subseteq A$ has a subsequence which is convergent in τ_d to a point in X.*

Proof. First, suppose A is a relatively compact subset of (X, τ_d), and consider a sequence $\{x_j\}_{j\in\mathbb{N}} \subseteq A$. Then $\overline{A}$ is compact in (X, τ_d), so $\overline{A}$ with the distance induced by d becomes a compact metric space. Given that $\{x_j\}_{j\in\mathbb{N}} \subseteq \overline{A}$, Theorem 1.2.11 applies and shows that there is a subsequence which is convergent in the topology induced by d to a point in $\overline{A}$.

Conversely, suppose every sequence in A has a subsequence which is convergent in τ_d to a point in X. Pick a sequence $\{x_j\}_{j\in\mathbb{N}} \subseteq \overline{A}$. Then for each $j \in \mathbb{N}$ there exists $y_j \in A$ with $d(x_j, y_j) < 1/j$. This gives rise to a sequence $\{y_j\}_{j\in\mathbb{N}} \subseteq A$ which, thanks to the current working hypothesis, has a subsequence $\{y_{j_k}\}_{k\in\mathbb{N}}$ which is convergent in τ_d to a point $z \in X$. Since for each $k \in \mathbb{N}$ we have

$$d(z, x_{j_k}) \leq d(z, y_{j_k}) + d(y_{j_k}, x_{j_k}) < d(z, y_{j_k}) + 1/j_k \leq d(z, y_{j_k}) + 1/k, \tag{1.2.68}$$

it follows that $\{x_{j_k}\}_{k\in\mathbb{N}}$ is also convergent in τ_d to $z \in X$. Note that the latter property places z in $\overline{A}$. In summary, we have shown that every sequence in $\overline{A}$ has a subsequence which is convergent in τ_d to a point in $\overline{A}$. Theorem 1.2.11 then applies and shows that $\overline{A}$ is compact in τ_d, ultimately making A a relatively compact subset of (X, τ_d). □

Remark 1.2.13. Given that the conclusion in Corollary 1.2.12 is of a purely topological nature, this result is actually true for any topological space (X, τ) whose topology is metrizable (cf. Definition 1.2.6).

To state the last result in this section requires first making the following definition (formalizing a phenomenon we have already encountered in (1.2.63)).

Definition 1.2.14. *Let (X, d) be a metric space. A subset $A \subseteq X$ is said to be* `totally bounded` *if for every $\varepsilon > 0$ there exist finitely many points $x_1, \dots, x_N$ in X such that*

$$A \subseteq \bigcup_{i=1}^{N} B_d(x_i, \varepsilon), \tag{1.2.69}$$

where $B_d(x, \varepsilon) := \{y \in X : d(y, x) < \varepsilon\}$ denotes the open d-ball of radius ε centered at x.

Call the family of d-balls $\{B_d(x_i, \varepsilon)\}_{1\leq i\leq N}$ an `ε-cover` *of the set A.*

Obviously, every compact subset of a metric space is totally bounded. The extent to which the converse implication is true is studied next.

Theorem 1.2.15. *Let (X, d) be a complete metric space, and let $A \subseteq X$ be totally bounded. Then A is relatively compact in (X, τ_d).*

Proof. By Corollary 1.2.12, it suffices to show that every sequence in A has a convergent subsequence in X. With this goal in mind, fix some sequence $\{a_m\}_{m\in\mathbb{N}} \subseteq A$. As in the proof of Theorem 1.2.11, there is no loss of generality in assuming that the terms of the sequence $\{a_m\}_{m\in\mathbb{N}}$ are mutually distinct. Since A is totally bounded, Definition 1.2.14 tells us that for each $k \in \mathbb{N}$ there exist points $x_1^{(k)}, \dots, x_{N_k}^{(k)} \in X$ such that

$$A \subseteq \bigcup_{i=1}^{N_k} B_d\Big(x_i^{(k)}, \frac{1}{2^k}\Big), \tag{1.2.70}$$

where, as usual, $B_d(x, r) := \{y \in X : d(y, x) < r\}$ for each $x \in X$ and $r > 0$.

We now proceed to construct inductively a nested sequence of subsets of the set $\{a_m;\ m \in \mathbb{N}\}$ in the following fashion. In the first step, cover A as in (1.2.70) corresponding to $k := 1$ and choose an index $i_1 \in \{1, \dots, N_1\}$ such that infinitely many terms of $\{a_m\}_{m\in\mathbb{N}}$ lie in the ball $B_d(x_{i_1}^{(1)}, 2^{-1})$. Let $A_1 := \{a_m : a_m \in B_d(x_{i_1}^{(1)}, 2^{-1})\}$ be this infinite subsequence. In the second step, cover A as in (1.2.70) with $k := 2$ and choose $i_2 \in \{1, \dots, N_2\}$ such that infinitely many terms of A_1 lie in $B_d(x_{i_2}^{(2)}, 2^{-2})$; denote that infinite subsequence by A_2. Proceed inductively: given the infinite subsequence A_{k-1} for some $k \geq 2$, cover A as in (1.2.70) and choose $i_k \in \{1, \dots, N_k\}$ so that infinitely many terms of A_{k-1} lie in $B_d(x_{i_k}^{(k)}, 2^{-k})$; let A_k be that infinite subsequence.

Once the sets $A_1, A_2, \dots$ have been constructed, list their elements as sequences, preserving the order in which their respective terms appear in the original sequence $\{a_m\}_{m\in\mathbb{N}}$. This gives rise to an infinite matrix (with the elements of A_k listed as the k-th row, for each k), and we select the diagonal in this matrix, which is a genuine subsequence $\{a_{m_k}\}_{k\in\mathbb{N}}$ of the original sequence. By design, this has the property that $a_{m_k} \in A_k$ for each $k \in \mathbb{N}$ and, hence,

$$a_{m_k} \in B_d\Big(x_{i_k}^{(k)}, \frac{1}{2^k}\Big) \quad \text{for all } \ k \in \mathbb{N}. \tag{1.2.71}$$

In addition, for $j > k$ both a_{m_k} and a_{m_j} lie in $A_j \subseteq A_k$, hence in the same ball $B_d(x_{i_k}^{(k)}, 2^{-k})$. Therefore,

$$d(a_{m_j}, a_{m_k}) \leq d(a_{m_j}, x_{i_k}^{(k)}) + d(a_{m_k}, x_{i_k}^{(k)}) < \frac{1}{2^k} + \frac{1}{2^k} = \frac{1}{2^{k-1}}. \tag{1.2.72}$$

Given an arbitrary $\varepsilon > 0$, choose $k \in \mathbb{N}$ large enough so that $2^{-(k-2)} < \varepsilon$. Then for all $j, \ell > k$ we have

$$d(a_{m_j}, a_{m_\ell}) < d(a_{m_j}, a_{m_k}) + d(a_{m_\ell}, a_{m_k}) < \frac{1}{2^{k-1}} + \frac{1}{2^{k-1}} = \frac{1}{2^{k-2}} < \varepsilon, \tag{1.2.73}$$

which shows that $\{a_{m_k}\}_{k\in\mathbb{N}}$ is Cauchy in X. Since X is complete, there exists some $x \in X$ such that $a_{m_k} \to x$ as $k \to \infty$. Parenthetically, note that because $a_{m_k} \in A$

for all $k \in \mathbb{N}$, we have $x \in \overline{A}$. We have shown that every sequence in A has a subsequence convergent to a point in $\overline{A}$. Thanks to Corollary 1.2.12, we conclude that A is relatively compact. □

1.3 Vector Spaces

This section introduces the notion of vector space and explores a number of fundamental notions, such as the concept of basis, linear independence, span, and dimension.

Throughout this monograph, we consider vector spaces over $\mathbb{K}$, with the understanding that $\mathbb{K}$ represents the field of real numbers, or the field of complex numbers. Elements in $\mathbb{K}$ are called `scalars`. We shall employ the terminologies "vector space" and "linear space" interchangeably. Further, the reader is alerted that anytime we are considering a subspace of a vector space, we actually demand it to be a *linear* subspace.

Definition 1.3.1. *A* `vector space`, *or* `linear space`, *over* $\mathbb{K}$ $(:= \mathbb{R}$ *or* $\mathbb{C})$ *is a set X, whose elements are called* `vectors`, *together with an operation + mapping* $X \times X$ *into X, called* `vector addition`, *and an operation · mapping from* $\mathbb{K} \times X$ *into X, called* `scalar multiplication`, *satisfying the following axioms:*

(1) The vector addition + is commutative and associative, i.e.

$$x + y = y + x \text{ and } x + (y + z) = (x + y) + z, \qquad \forall\, x, y, z \in X. \tag{1.3.74}$$

(2) There exists a `zero vector` *in X, denoted by* 0*, with the property that*

$$x + 0 = x, \qquad \forall\, x \in X. \tag{1.3.75}$$

(3) For each vector $x \in X$ *there exists a vector* $y \in X$ *such that* $x + y = 0$*, called an* `additive inverse` *of the vector x (this will be shown to be unique, and will be denoted by* $-x$*).*

(4) For each $\alpha, \beta \in \mathbb{K}$ *and each* $x, y \in X$*, one has*

$$\alpha \cdot (x + y) = \alpha \cdot x + \alpha \cdot y, \text{ and } (\alpha + \beta) \cdot x = \alpha \cdot x + \beta \cdot x, \tag{1.3.76}$$

and

$$\alpha \cdot (\beta \cdot x) = (\alpha\beta) \cdot x. \tag{1.3.77}$$

(5) For every vector $x \in X$*, one has* $1 \cdot x = x$*.*

From Definition 1.3.1, one may immediately show that the zero vector of a vector space X and the additive inverse of any vector $x \in X$ are unique, and that $-x = (-1) \cdot x$ for each $x \in X$. The difference $x - y$ shall denote in the sequel the vector $x + (-y)$.

Definition 1.3.2. *A subset Y of a vector space X is called a* `subspace` *of X if Y is a vector space (over the same field) with the operations of vector addition and scalar multiplication defined on X.*

It is automatic that every vector space X has at least two subspaces; namely $\{0\}$ and X. The former is called the `trivial subspace` of X.

Theorem 1.3.3. *Let X be a vector space over $\mathbb{K}$ and Y be a subset of X. Then Y is a subspace of X if and only if the following properties hold (for the operations defined in X):*

(1) 0*, the zero vector in X, belongs to Y;*
(2) $y_1 + y_2 \in Y$*, whenever* $y_1, y_2 \in Y$*;*
(3) $\alpha \cdot y \in Y$*, whenever* $\alpha \in \mathbb{K}$ *and* $y \in Y$*.*

Proof. Suppose that Y is a subspace of the vector space X. Thus *(2)-(3)* hold as a consequence of the fact that Y is a vector space. Also, there exists a vector $o \in Y$ such that $y + o = y$ for each $y \in Y$. However, $y + 0 = y$ for every $y \in Y$, and thus, by uniqueness of the zero vector of X, we obtain that 0 must be o, and as such $0 \in Y$. This establishes *(1)*.

Conversely, assume properties *(1)-(3)* hold. To prove that Y is a vector space, it suffices to show that each vector in Y has an additive inverse in Y. However, if $y \in Y$, then $-y := (-1) \cdot y \in Y$ by property *(3)*, and this is an additive inverse of y. □

Henceforth, if X is a vector space over $\mathbb{K}$, given any $x \in X$ and $\alpha \in \mathbb{K}$ we shall denote by αx the vector $\alpha \cdot x$. If $A \subseteq X$, $\alpha \in \mathbb{K}$, and $x \in X$, we set

$$\alpha A := \{\alpha y : y \in A\} \quad \text{and} \quad x + A := \{x + y : y \in A\}. \tag{1.3.78}$$

Also, if $A, B \subseteq X$ we denote by $A + B$ the algebraic sum of A and B given by

$$A + B := \Big\{x + y : x \in A \quad \text{and} \quad y \in B\Big\}. \tag{1.3.79}$$

Definition 1.3.4. *Let X be a vector space over the field $\mathbb{K}$ and S a nonempty subset of X. A vector $x \in X$ is called a* `linear combination` *of vectors of S provided there exist a finite number of vectors $x_1, x_2, \dots, x_n \in S$ and $\alpha_1, \alpha_2, \dots, \alpha_n \in \mathbb{K}$ with the property that*

$$x = \sum_{i=1}^{n} \alpha_i x_i. \tag{1.3.80}$$

When (1.3.80) *holds, x is called a linear combination of the vectors $x_1, x_2, \dots, x_n$ and call $\alpha_1, \alpha_2, \dots, \alpha_n$ the* `coefficients` *of the linear combination.*

Since in every vector space X we have that $0x = 0$, for each vector $x \in X$, the zero vector is a linear combination of any nonempty subset of X.

Given vector space X, a set $V \subseteq X$ is said to be `convex` provided

$$\alpha V + (1 - \alpha)V \subseteq V \text{ for each } \alpha \in [0, 1]. \tag{1.3.81}$$

Definition 1.3.5. *Let X be a vector space over the field $\mathbb{K}$. Then*

(a) A set $S \subseteq X$ is called `linearly dependent` *if there exist a finite number of distinct vectors $x_1, x_2, \ldots, x_n \in S$ and scalars $\alpha_1, \alpha_2, \ldots, \alpha_n \in \mathbb{K}$, not all zero, such that $\sum_{i=1}^{n} \alpha_i x_i = 0$.*

(b) If $S \subseteq X$ is not linearly dependent, then S is called `linearly independent.`

(c) If $S \subseteq X$ is nonempty then $\operatorname{span}_{\mathbb{K}}(S)$ *consists of all finite linear combinations of elements of S, i.e.,*

$$y \in \operatorname{span}_{\mathbb{K}}(S) \iff \begin{cases} \exists\, m \in \mathbb{N} \ \textit{ and } \ x_1, \ldots, x_m \in S \ \textit{ and } \ \alpha_1, \ldots, \alpha_m \in \mathbb{K} \\ \textit{such that } \ y = \sum_{j=1}^{m} \alpha_j x_j. \end{cases} \tag{1.3.82}$$

If S is empty then set $\operatorname{span}_{\mathbb{K}}(\varnothing) := \{0\}$. *Occasionally, when no confusion may arise, we shall drop the subscript $\mathbb{K}$ and simply write* $\operatorname{span}(S)$ *for the span of the set S.*

(d) A set $S \subseteq X$ which is linearly independent and $\operatorname{span}_{\mathbb{K}}(S) = X$ *is called a* `basis` *of X.*

Observe that, in every vector space, the empty set is linearly independent, and a singleton is linearly dependent precisely when this coincides with the trivial subspace. Also, it is immediate that

$$\text{if } Y \text{ is a subspace of } X \text{ then } \operatorname{span}(Y) = Y. \tag{1.3.83}$$

Proposition 1.3.6. *Let X be a vector space over the field $\mathbb{K}$, and suppose that S_1, S_2 are subsets of X such that $S_1 \subseteq S_2$. Then the following hold.*

(1) If S_1 is linearly dependent then S_2 is linearly dependent.

(2) If S_2 is linearly independent then S_1 is linearly independent.

Proof. It is clear that *(1)* and *(2)* are equivalent to one another. We shall proceed to prove *(2)*. To this end, suppose S_2 is a linearly independent subset of X, and let $x_1, x_2, \ldots, x_n$ be vectors in S_1, for some $n \in \mathbb{N}$, and $\alpha_1, \alpha_2, \ldots, \alpha_n \in \mathbb{K}$ be such that $\sum_{i=1}^{n} \alpha_i x_i = 0$. Since $S_1 \subseteq S_2$, the vectors $x_1, x_2, \ldots, x_n$ are vectors in S_2. Linear independency of S_2 then entails $\alpha_j = 0$ for each $j \in \{1, \ldots, n\}$, as desired. □

Proposition 1.3.7. *Let S be a linearly independent subset of a vector space X over the field $\mathbb{K}$, and let $x \in X \setminus S$. Then*

$$S \cup \{x\} \textit{ is linearly dependent} \iff x \in \operatorname{span}_{\mathbb{K}}(S). \tag{1.3.84}$$

Proof. In the case when $S \cup \{x\}$ is linearly dependent, we can select a finite family of distinct vectors $x_1, x_2, \ldots, x_n \in S \cup \{x\}$ with the property that

$$\sum_{i=1}^{n} \alpha_i x_i = 0, \tag{1.3.85}$$

for some scalars $\alpha_1, \alpha_2, \ldots, \alpha_n \in \mathbb{K}$, not all zero. The linear independency of the set S guarantees that one of the x_i's, say x_1, must be the vector x and must have a non-zero coefficient α_1. As a result,

$$\alpha_1 x + \sum_{i=2}^{n} \alpha_i x_i = 0, \tag{1.3.86}$$

or equivalently,

$$x = -\alpha_1{}^{-1} \sum_{i=2}^{n} \alpha_i x_i = \sum_{i=2}^{n} (-\alpha_1{}^{-1}\alpha_i)x_i. \tag{1.3.87}$$

From (1.3.87) and the fact that $x_2, \ldots, x_n \in S$ we obtain $x \in \text{span}\,(S)$. This establishes the left-to-right implication in (1.3.84).

Conversely, assume that $x \in \text{span}_{\mathbb{K}}(S)$. Then there exist vectors $x_1, \ldots, x_n \in S$ and scalars $\alpha_1, \ldots, \alpha_n \in \mathbb{K}$ such that x is the linear combination of the x_i's with the coefficients α_i, $i \in \{1, \ldots, n\}$. In particular,

$$(-1)x + \sum_{i=1}^{n} \alpha_i x_i = 0. \tag{1.3.88}$$

Since $x \notin S$ it follows that $x \notin \{x_1, x_2, \ldots, x_n\}$. This combined with the fact that the coefficient of x in (1.3.88) is nonzero implies that the set $\{x, x_1, x_2, \ldots, x_n\}$ is linearly dependent. By virtue of the fact that $\{x, x_1, x_2, \ldots, x_n\} \subseteq S \cup \{x\}$, it follows from Theorem 1.3.6 that $S \cup \{x\}$ is linearly dependent. This establishes the equivalence claimed in (1.3.84) and thus completes the proof of the proposition. □

Proposition 1.3.8. *Assume that X is a vector space over the field $\mathbb{K}$. The following hold.*

(1) If A is a nonempty subset of X, then A is linearly independent if and only if every finite subset of A is linearly independent.
(2) If $A \subseteq X$ then $\text{span}_{\mathbb{K}}(A)$ *is a subspace of X.*
(3) If $A \subseteq Y$, where Y is a subspace of X then $\text{span}_{\mathbb{K}}(A) \subseteq Y$.

Proof. The left-to-right implication in *(1)* is immediate by Theorem 1.3.6 while the right-to-left implication is a direct consequence of parts *(a)-(b)* in Definition 1.3.5.

Moving on to *(2)*, the plan is to prove that $\text{span}_{\mathbb{K}}(A)$ is a subspace of X by showing that it satisfies properties *(i)-(iii)* of Theorem 1.3.3. First note that those axioms trivially hold true when $A = \varnothing$, since $\text{span}_{\mathbb{K}}(\varnothing) = \{0\}$ (see part *(c)* in Definition 1.3.5). Assume next that A is a nonempty subset of X. Since for every $x \in A$ we have $0 = 0x$ it follows that $0 \in \text{span}_{\mathbb{K}}(A)$, i.e., *(i)* of Theorem 1.3.3 is valid. Fix now $z, w \in \text{span}_{\mathbb{K}}(A)$ and let vectors $x_1, \ldots, x_m \in A$ and $y_1, \ldots, y_n \in A$ along with scalars $\alpha_1, , \ldots, \alpha_m \in \mathbb{K}$ and $\beta_1, \ldots, \beta_n \in \mathbb{K}$ such that

$$x = \sum_{i=1}^{m} \alpha_i x_i \quad \text{and} \quad y = \sum_{j=1}^{n} \beta_j y_j. \tag{1.3.89}$$

Then

$$x+y=\sum_{i=1}^{m}\alpha_i x_i+\sum_{j=1}^{n}\beta_j y_j\in \operatorname{span}_{\mathbb{K}}(A). \tag{1.3.90}$$

This shows that property *(ii)* of Theorem 1.3.3 holds. Going further, select some $\alpha \in \mathbb{K}$ and pick an arbitrary vector $x \in \operatorname{span}_{\mathbb{K}}(A)$. Then $x = \sum_{i=1}^{m}\beta_i x_i$ for some vectors $x_1, x_2, \ldots, x_m \in A$ and some scalars $\beta_1, \beta_2, \ldots, \beta_m \in \mathbb{K}$. Thus

$$\alpha x=\alpha\sum_{i=1}^{n}\beta_i x_i=\sum_{i=1}^{n}(\alpha\beta_i)x_i\in \operatorname{span}_{\mathbb{K}}(A), \tag{1.3.91}$$

and hence axiom *(iii)* of Theorem 1.3.3 holds. Consequently, $\operatorname{span}_{\mathbb{K}}(A)$ is a subspace of X. This finishes the proof of part *(2)*.

Finally let Y be a subspace of X containing the set A. Then clearly

$$\operatorname{span}_{\mathbb{K}}(A)\subseteq \operatorname{span}_{\mathbb{K}}(Y)=Y, \tag{1.3.92}$$

where the equality follows from the fact that Y is a subspace of X. This completes the proof of the theorem. □

Theorem 1.3.9. *If A is a linearly independent subset of a vector space X over the field $\mathbb{K}$, then there exists a basis B for X over $\mathbb{K}$ with the property that $B \supseteq A$.*

Proof. Suppose that A is a linearly independent subset of a vector space X. Associated with it, introduce the collection of all linearly independent subsets of X containing A,

$$C_A := \{B \subseteq X : B \text{ is linearly independent and } B \supseteq A\}. \tag{1.3.93}$$

Note that C_A is a nonempty, since $A \in C_A$, and that $(C_A, \subseteq)$ is a partially ordered set.

Step I: $(C_A, \subseteq)$ has a maximal element.

This is immediately seen in the degenerate case $X = \{0\}$. The only linearly independent subset of X is then $A = \varnothing$ and $C_\varnothing = \{\varnothing\}$. In this case $\varnothing$ is the maximal element of $C_\varnothing$.

Suppose next $X \neq \{0\}$, in which case the collection C_A contains a nonempty set. We claim next that every chain $\mathcal{B} \subseteq C_A$, i.e., every ordered subset $\mathcal{B}$ of $(C_A, \subseteq)$, has an upper bound in C_A. Zorn's Lemma (cf., e.g., [14, p. 5]) guarantees then that the partially ordered set $(C_A, \subseteq)$ has a maximal element. The case when $\mathcal{B}$ only contains the empty set is trivial, so assume further that $\mathcal{B}$ contains at least one nonempty set and choose a finite set $S \neq \varnothing$ in X with the property that

$$S\subseteq \bigcup_{B\in\mathcal{B}} B. \tag{1.3.94}$$

Denote by n the cardinality of S and note that, since the collection $\mathcal{B}$ is ordered and (1.3.94) is satisfied, we may assume without loss of generality that

$$S = \{x_i : i \in \{1,\dots,n\}\} \text{ with } x_i \in B_i \in \mathcal{B} \text{ for each } i \in \{1,\dots,n\}, \quad \text{and } B_1 \subseteq B_2 \subseteq \cdots \subseteq B_n. \tag{1.3.95}$$

In particular $S \subseteq B_n$ and since $B_n \in C_A$ is linearly independent, so is S, by Theorem 1.3.6. This shows that every finite subset of $\bigcup_{B\in\mathcal{B}} B$ is linearly independent, and hence

$$\bigcup_{B\in\mathcal{B}} B \text{ is a linearly independent set in } X, \tag{1.3.96}$$

by Proposition 1.3.8. Furthermore, since $\mathcal{B} \subseteq C_A$, it follows that each $B \in \mathcal{B}$ contains the set A and as such so does $\bigcup_{B\in\mathcal{B}} B$. Combining this with (1.3.96) we may conclude that

$$\bigcup_{B\in\mathcal{B}} B \in C_A. \tag{1.3.97}$$

Since $\bigcup_{B\in\mathcal{B}} B$ is clearly an upper bound for $\mathcal{B}$, using (1.3.97) we conclude that $\mathcal{B}$ has an upper bound in C_A, as desired. This completes *Step I.*

Step II: The set $B \in C_A$ is maximal in C_A if and only if B is a basis for X.

We start again with the degenerate case $X = \{0\}$, in which scenario $A = \varnothing$ is the only linear independent subset of X. In this case $C_A = \{\varnothing\}$ and $B = \varnothing$ is the only maximal element in C_A and

$$\operatorname{span}_{\mathbb{K}}(B) = \operatorname{span}_{\mathbb{K}}(\varnothing) = \{0\} = X, \tag{1.3.98}$$

where the second equality in (1.3.98) is a consequence of part *(c)* in Definition 1.3.5. Since $\varnothing$ is linearly independent and spans X, this is a basis for X as desired. The converse implication is immediate as the only element in C_A is the empty set, which is maximal in C_A.

Consider next the case when $X \neq \{0\}$ and choose $B \in C_A$ maximal. Then B is linearly independent and contains A. If $\operatorname{span}_{\mathbb{K}}(B) \neq X$, pick $x \in X \setminus \operatorname{span}_{\mathbb{K}}(B)$. Invoking Theorem 1.3.7 it follows that $B \cup \{x\}$ is a linearly independent set containing A, and thus $B \cup \{x\} \in C_A$ and $B \subset B \cup \{x\}$. Therefore B is not maximal in C_A, leading to a contradiction. This shows that $\operatorname{span}_{\mathbb{K}}(B) = X$, completing the proof that B is a basis of X, and establishing the left-to-right implication.

For the converse, assume that $B \in C_A$ is a basis for X. Let $B_o \in C_A$ be a maximal element (its existence is ensured by *Step I*) and assume by contradiction that $B_o \neq B$. Then $B \subset B_o$ due to the maximality of B_o. Since $\operatorname{span}_{\mathbb{K}}(B) = X$ then $B \cup \{x\}$ is a linearly dependent set for every $x \in B_o \setminus B$ by Theorem 1.3.7. However we have $B \cup \{x\} \subseteq B_o$ and since B_o is linearly independent by part *(1)* in Proposition 1.3.6 leads to a contradiction. This completes the proof of the fact that B is maximal in $\mathbb{C}$, and finishes the right-to-left implication in *Step II.*

Having disposed of *Step I* and *Step II* we finally notice that these combined ensure that, for every linearly independent subset A of X, there exists a basis B for X with the property that $B \supseteq A$, finishing the proof of the theorem. □

Since the empty set is linearly independent, it follows that Theorem 1.3.9 holds for $A = \varnothing$. As a consequence, we have the following result:

Corollary 1.3.10. *Every vector space over the field* $\mathbb{K}$ *has a basis.*

Theorem 1.3.11. *Assume that* X *is a vector space over the field* $\mathbb{K}$. *Then the following implication holds*

$$A \subseteq X \text{ and } \operatorname{span}_{\mathbb{K}}(A) = X \Longrightarrow \exists\, B \subseteq X \text{ basis for } X \text{ such that } B \subseteq A. \tag{1.3.99}$$

Proof. Fix $A \subseteq X$ such that $\operatorname{span}_{\mathbb{K}}(A) = X$. Notice that if A is linearly independent we are done as A is then a basis for X containing itself. Assume next that A is not linearly independent and consider the collection of sets

$$\mathfrak{C}_A := \{B \subseteq X : B \text{ is linearly independent and } B \subseteq A\}, \tag{1.3.100}$$

We claim that this collection, when equipped with the partial order $\subseteq$, has a maximal element. This is easily seen when $X = \{0\}$. Then either $A = \varnothing$ or $A = \{0\}$ and in both scenarios $\mathfrak{C}_A = \{\varnothing\}$. Then trivially $\varnothing$ is a maximal element of $\mathfrak{C}_A$.

Consider now the case when $X \neq \{0\}$. Consequently A contains a nonzero vector since $\operatorname{span}_{\mathbb{K}}(A) = X$. Also, it is immediate that

$$\forall\, x \in A \text{ such that } x \neq 0 \text{ one has } \{x\} \in \mathfrak{C}_A. \tag{1.3.101}$$

The argument proceeds now in a very similar fashion to that of the proof of *Step I* in Theorem 1.3.9 and we include it here for the sake of completeness. As before, we shall show that every chain in $\mathfrak{C}_A$ has an upper bound in $\mathfrak{C}_A$. To this end, let $\mathcal{B} \subseteq \mathfrak{C}_A$ be a chain which contains at least one nonempty set and let $S \neq \varnothing$ be finite and such that (1.3.94)-(1.3.95) are satisfied. As before, $S \subseteq B_n$ and since $B_n \in \mathfrak{C}_A$ is linearly independent, so is S. Appealing again to Theorem 1.3.6 we obtain that $\bigcup_{B \in \mathcal{B}} B$ is linearly independent and, since each $B \in \mathcal{B}$ is contained in A, so is $\bigcup_{B \in \mathcal{B}} B$. This ultimately implies $\bigcup_{B \in \mathcal{B}} B \in \mathfrak{C}_A$, and this is clearly an upper bound for the chain $\mathcal{B}$. Consequently, by Zorn's Lemma $\mathfrak{C}_A$ has a maximal element.

Denote by A_0 the maximal element of $\mathfrak{C}_A$. Then $A_0 \subseteq A$ and since A is not linearly independent, A_0 is in fact a proper subset of A. Choose an arbitrary $a \in A \setminus A_0$ and consider the set $A_0 \cup \{a\} \subseteq A$, which is not linearly independent due to the fact that A_0 is maximal in C'_A. Since A_0 is linearly independent, it follows from Theorem 1.3.7 that $a \in \operatorname{span}_{\mathbb{K}}(A_0)$. As a result, since a was arbitrary in $A \setminus A_0$, we have $A \setminus A_0 \subseteq \operatorname{span}_{\mathbb{K}}(A_0)$, and thus

$$A = A_0 \cup (A \setminus A_0) \subseteq \operatorname{span}_{\mathbb{K}}(A_0). \tag{1.3.102}$$

Therefore

$$\operatorname{span}_{\mathbb{K}}(A) \subseteq \operatorname{span}(\operatorname{span}_{\mathbb{K}}(A_0)) = \operatorname{span}_{\mathbb{K}}(A_0) \subseteq \operatorname{span}_{\mathbb{K}}(A), \tag{1.3.103}$$

which implies that $\operatorname{span}_{\mathbb{K}}(A_0) = \operatorname{span}_{\mathbb{K}}(A) = X$. In conclusion, since A_0 is also linearly independent, A_0 is a basis for X, finishing the proof of the theorem. □

Theorem 1.3.12. *Let X be a vector space over the field $\mathbb{K}$. Suppose that $S, U \subseteq X$ satisfy*

$$\#S = n, \text{ for some } n \in \mathbb{N}, \text{ and } \operatorname{span}_{\mathbb{K}}(S) = X, \tag{1.3.104}$$

and

$$\#U = m, \text{ for some } m \in \mathbb{N}, \text{ and } U \text{ is linearly independent in } X, \tag{1.3.105}$$

where by # we have denoted cardinality.

Then $m \leq n$ and there exists a set $V \subseteq S$ containing exactly $n - m$ vectors having the property that $\operatorname{span}_{\mathbb{K}}(U \cup V) = X$.

Proof. The proof is by mathematical induction on m, the cardinality of the set U. For $m = 0$ we have $U = \varnothing$ and hence $0 \leq n$. In this case, pick $V := S$ to obtain that

$$\operatorname{span}_{\mathbb{K}}(U \cup V) = \operatorname{span}_{\mathbb{K}}(S) = X, \tag{1.3.106}$$

by the second part of (1.3.104).

Suppose that the theorem holds true for some $m \in \mathbb{N}$. We shall show that the theorem still holds for $m + 1$. For this reason, assume that $U := \{x_1, x_2, \dots, x_{m+1}\}$ is a linearly independent subset of X. Theorem 1.3.6 yields that $\{x_1, x_2, \dots, x_m\}$ is also a linearly independent set. Hence, by the induction hypothesis, $m \leq n$ and there exists a subset $Y : \{y_1, y_2, \dots, y_{n-m}\}$ of S with the property that

$$\operatorname{span}_{\mathbb{K}}(\{x_1, x_2, \dots, x_m\} \cup \{y_1, y_2, \dots, y_{n-m}\}) = X. \tag{1.3.107}$$

If $m = n$ then $Y = \varnothing$ and using (1.3.107), we may conclude that

$$x_{m+1} \in \operatorname{span}_{\mathbb{K}}(\{x_1, x_2, \dots, x_m\}). \tag{1.3.108}$$

Using Theorem 1.3.7, (1.3.108) contradicts that U is linearly independent. Hence $m < n$ so $m + 1 \leq n$ as desired.

Going back to (1.3.107), there exist two families of scalars, $\alpha_1, \alpha_2, \dots, \alpha_m \in \mathbb{K}$ and $\beta_1, \beta_2, \dots, \beta_{n-m} \in \mathbb{K}$, such that

$$\sum_{i=1}^{m} \alpha_i x_i + \sum_{i=1}^{n-m} \beta_i y_i = x_{m+1}. \tag{1.3.109}$$

Appealing again to Theorem 1.3.7 and the linear independence of U, (1.3.109) ensures that at least one of the β's is nonzero. Without loss of generality, assume that this scalar is β_1. Thus (1.3.109) may be written as

$$y_1 = \sum_{i=1}^{m+1} (-\beta_1^{-1} \alpha_i) x_i + \sum_{i=2}^{n-m} (-\beta_1^{-1} \beta_i) y_i, \tag{1.3.110}$$

where $\alpha_{m+1} := 1$. Introduce the set $V := \{y_2, y_3, \dots, y_{n-m}\}$, and observe that, by (1.3.110), y_1 belongs to $\operatorname{span}_{\mathbb{K}}(U \cup V)$, and thus

$$\{x_1, x_2, \ldots, x_m\} \cup \{y_1, y_2, \ldots, y_{n-m}\} \subseteq \text{span}_{\mathbb{K}}(U \cup V). \tag{1.3.111}$$

In concert, (1.3.107) and (1.3.111) imply that $\text{span}_{\mathbb{K}}(U \cup V) = X$. Since V is a subset of S containing $n - (m + 1)$ vectors, the theorem is true for $m + 1$. This completes the mathematical induction and subsequently the proof of the theorem. □

Although a vector space may have distinct (finite) bases, the next result asserts that the number of vectors in a basis is an intrinsic characteristic of the vector space.

Corollary 1.3.13. *Let X be a vector space over the field $\mathbb{K}$ having a finite basis. Then every basis for X contains the same number of vectors.*

Proof. Suppose that B is a finite basis for X such that $\#B = n$ for some $n \in \mathbb{N}$, and let $\widetilde{B}$ be another basis for X. Assume first that $\#\widetilde{B} > n$ and pick a subset L of $\widetilde{B}$ of cardinality $n + 1$. Using Theorem 1.3.6, since $\widetilde{B}$ is a linearly independent set, L is too linearly independent. Because $\text{span}_{\mathbb{K}}(B) = X$, it follows from Theorem 1.3.12 (applied to the sets B and L) that $n + 1 \leq n$, a contradiction. This shows that

$$\#\widetilde{B} \leq \#B. \tag{1.3.112}$$

Reversing the roles of the bases B and $\widetilde{B}$ and proceeding as above, we infer that $\#B \leq \#\widetilde{B}$. All in all, the two bases have the same number of vectors, as wanted. □

By virtue of Corollary 1.3.13, the next definition is meaningful.

Definition 1.3.14. *Let X be a vector space over the field $\mathbb{K}$. Then*

(1) X is called `finite dimensional` *if it has a finite basis. The unique number of vectors in every basis for X is called the* `dimension` *of X and is denoted by* $\dim_{\mathbb{K}} X$.
(2) If X fails to be finite dimensional, it is called `infinite dimensional`. *In such a scenario, one defines* $\dim_{\mathbb{K}} X := \infty$.

Hence, $\dim_{\mathbb{K}} X \in \mathbb{N} \cup \{0, \infty\}$ is always well defined, for each vector space X over the field $\mathbb{K}$.

Exercise 1.3.15. Let X be an infinite-dimensional vector space over the field $\mathbb{K}$. Then for each $n \in \mathbb{N}$ there exists a linearly independent subset S_n of X with $\#S_n = n$.

Hint: Proceed by induction on n.

Definition 1.3.16. *Assume that X and Y are vector spaces over the field $\mathbb{K}$. Then*

(a) A function $T : X \longrightarrow Y$ is called a `linear transformation` *from X to Y provided*[3]

$$T(\alpha x + \beta y) = \alpha T(x) + \beta T(y) \qquad \forall\, x, y \in X \text{ and } \forall\, \alpha, \beta \in \mathbb{K}. \tag{1.3.113}$$

---

[3] we shall denote by $T(x)$, or simply Tx, the output under the mapping T corresponding to the input x

(b) If T is a linear transformation from X to Y then the `kernel` *of T is defined by setting*

$$\ker T := \{x \in X : Tx = 0\}. \tag{1.3.114}$$

(c) If T is a linear transformation from X to Y then the `range` *of T is defined by setting*

$$\operatorname{ran} T := \{y \in Y : \exists\, x \in X \ \textit{ such that } \ Tx = y\}. \tag{1.3.115}$$

(d) A linear transformation from X into itself is called a `linear operator` *on X. Throughout, the symbol I (or I_X if more specificity is required) denotes the identity operator on the vector space X.*

Proposition 1.3.17. *Assume that X, Y are vector spaces over the field $\mathbb{K}$ and suppose $T : X \to Y$ is a linear transformation from X into Y. Then T maps the zero vector of X to the zero vector of Y.*

Proof. Let 0_X denote the zero vector of X and note that, since $0_X = 0_X + 0_X$, part *(a)* of Definition 1.3.16 yields $T(0_X) = T(0_X) + T(0_X)$. The desired result is then obtained by adding $-T(0_X)$, which is the additive inverse in Y of $T(0_X)$, to both sides of the latter equality. □

Proposition 1.3.18. *Let X and Y be vector spaces over the field $\mathbb{K}$ and $T : X \longrightarrow Y$ be a linear transformation from X into Y. Then* $\ker T$ *and* $\operatorname{ran} T$ *are subspaces of X and Y, respectively.*

Proof. Let 0_X and 0_Y denote the zero vectors of X and Y, respectively. Since we know that $T(0_X) = 0_Y$ (by Proposition 1.3.17) it follows that $0_X \in \ker T$. Next, let $x, y \in \ker T$ and $\alpha, \beta \in \mathbb{K}$ and write

$$T(\alpha x + \beta y) = \alpha T(x) + \beta T(y) = \alpha \cdot 0_Y + \beta \cdot 0_Y = 0_Y. \tag{1.3.116}$$

Hence $\alpha x + \beta y \in \ker T$. Consequently $\ker T$ is a subspace of X, by Theorem 1.3.3.

Our next goal is to show that $\operatorname{ran} T$ is a subspace of the vector space Y. By virtue of Proposition 1.3.17, we have $0_Y \in \operatorname{ran} T$. If $x, y \in \operatorname{ran} T$ and $\alpha, \beta \in \mathbb{K}$ then, according to (1.3.115), there exist $z, w \in X$ such that $T(z) = x$ and $T(w) = y$. Thus

$$T(\alpha z + \beta w) = \alpha T(z) + \beta T(w) = \alpha x + \beta y, \tag{1.3.117}$$

As a result, $\alpha x + \beta y \in \operatorname{ran} T$, and hence $\operatorname{ran} T$ is a subspace of Y (again by Theorem 1.3.3). □

Our next result asserts that a linear transformation between vector spaces is injective if and only if it has trivial null space.

Theorem 1.3.19. *Assume X, Y are vector spaces over the field $\mathbb{K}$ and $T : X \to Y$ is a linear transformation from X into Y. Then*

$$T \ \textit{is injective} \iff \ker T = \{0\}. \tag{1.3.118}$$

Proof. By Proposition 1.3.17,

$$\{0\} \subseteq \ker T, \tag{1.3.119}$$

and clearly, if T is in addition injective, then $\ker T = \{0\}$. Conversely, assume that $\ker T = \{0\}$ and let $x_1, x_2 \in X$ and use the linearity of the operator T to write $T(x_1) - T(x_2) = T(x_1 - x_2)$. Thus, if $T(x_1) = T(x_2)$ then $T(x_1 - x_2) = 0$ and hence $x_1 - x_2 \in \ker T = \{0\}$. This implies $x_1 = x_2$, showing that T is injective. □

Theorem 1.3.20. *Let X, Y be vector spaces over the field $\mathbb{K}$ and assume $T : X \to Y$ is a linear transformation from X into Y. Then the following hold.*

(1) If $B \subseteq X$ is such that $X = \mathrm{span}_{\mathbb{K}}(B)$, then $\mathrm{ran}\, T = \mathrm{span}_{\mathbb{K}}(T(B))$.
(2) If T is injective and B is a linearly independent subset of X, then $T(B)$ is a linearly independent subset of Y.

Proof. In order to prove the statement made in part *(1)* we start with the observation that clearly $T(B) \subseteq \mathrm{ran}\, T$. This, and the fact that $\mathrm{ran}\, T$ is a subspace of Y further imply that

$$\mathrm{span}_{\mathbb{K}}\,(T(B)) \subseteq \mathrm{span}_{\mathbb{K}}(\mathrm{ran}\, T) = \mathrm{ran}\, T. \tag{1.3.120}$$

For the converse inclusion, suppose that $z \in \mathrm{ran}\, T$. Then $z = T(x)$ for some $x \in X$. Since B spans X, there exist a number $n \in \mathbb{N}$, vectors $x_1, x_2, \dots, x_n \in B$ and scalars $\alpha_1, \alpha_2, \dots, \alpha_n \in \mathbb{K}$ with the property that $x = \sum_{i=1}^{n} \alpha_i x_i$. Since T is linear transformation, this gives that

$$T(x) = \sum_{i=1}^{n} \alpha_i T(x_i). \tag{1.3.121}$$

However $T(x_i)$ belongs to $T(B)$ for each $i \in \{1, \dots, n\}$, and combined with (1.3.121) this shows that $z = T(x) \in \mathrm{span}_{\mathbb{K}}(T(B))$. This shows that $\mathrm{ran}\, T$ is contained in $\mathrm{span}_{\mathbb{K}}(T(B))$, finishing the proof of *(1)*.

Turning our attention to part *(2)*, assume that T is injective and that $B \subseteq X$ is a linearly independent set. Assume by contradiction that $T(B)$ is not linearly independent. Then there exist a number $n \in \mathbb{N}$, vectors $x_1, \dots x_n \in B$, and scalars $\alpha_1, \dots, \alpha_n \in \mathbb{K}$, with the property that one can find $j \in \{1, \dots, n\}$ for which $\alpha_j \neq 0$ and $\sum_{i=1}^{n} \alpha_i T(x_i) = 0$. Since T is linear and injective, this equality may be equivalently written as

$$\sum_{i=1}^{n} \alpha_i x_i = 0, \tag{1.3.122}$$

contradicting the fact that B is a linearly independent set. This finishes the proof of the statement in *(2)*. □

Exercise 1.3.21. Let X be a vector space over the field $\mathbb{K}$. Then for each linearly independent subset $S \subseteq X$ one has $\#S \leq \dim_{\mathbb{K}} X$.

Hint: Use Theorem 1.3.12 and Definition 1.3.14.

Exercise 1.3.22. Let X and Y be vector spaces over the field $\mathbb{K}$ and let $T : X \longrightarrow Y$ be an injective linear mapping. Then $\dim_{\mathbb{K}} Y \geq \dim_{\mathbb{K}} X$.

Hint: Use item *(2)* in Theorem 1.3.20 and Exercise 1.3.21.

Two environments are indistinguishable from the point of view of functional analysis when they happen to be isomorphic, in the sense described below.

Definition 1.3.23. *Let X and Y be vector spaces over the field $\mathbb{K}$. Then*

(a) A function $T : X \longrightarrow Y$ is called a `vector space isomorphism`, *or simply an* `isomorphism`, *between X and Y whenever T is a linear transformation which is both injective and surjective.*
(b) The vector spaces X and Y are said to be `isomorphic` *if there exists an isomorphism of X onto Y.*

Exercise 1.3.24. Let X and Y be vector spaces over the field $\mathbb{K}$ and let $T : X \longrightarrow Y$ be an bijective linear mapping. Then $T^{-1} : Y \longrightarrow X$ is also a (bijective) linear mapping.

Corollary 1.3.25. *Every two isomorphic vector spaces have the same dimension.*

Proof. This is immediate by Theorem 1.3.20. □

Proposition 1.3.26. *Assume that X is a vector space over the field $\mathbb{K}$ and that M and N are subspaces of X. If B_M and B_N denote bases for M and N, respectively, then the following hold.*

(i) $M \cap N = \{0\}$ if and only if $B_M \cap B_N = \varnothing$ and $B_M \cup B_N$ is linearly independent.
(ii) $M + N = X$ and $B_M \cup B_N$ is linearly independent if and only if $B_M \cup B_N$ is a basis for X.

Proof. To prove *(i)* we start with the observation that if M and N are subspaces of X with bases B_M and respectively B_N, then

$$\{0\} \subseteq \operatorname{span}_{\mathbb{K}}(B_M \cap B_N) \subseteq \operatorname{span}_{\mathbb{K}}(M \cap N) = M \cap N. \tag{1.3.123}$$

Hence $M \cap N = \{0\}$ implies $\operatorname{span}_{\mathbb{K}}(M \cap N) = \{0\}$, which, in turn, yields $B_M \cap B_N = \varnothing$.

Furthermore, if $M \cap N = \{0\}$, then it is not difficult to see that the union of the linearly independent sets B_M and B_N is a linearly independent set in X. Indeed, assume that $\{x_1, \dots, x_m\} \subseteq B_M$ and $\{y_1, \dots, y_n\} \subseteq B_N$ and $\alpha_1, \dots, \alpha_m \in \mathbb{K}$ and $\beta_1, \dots, \beta_n \in \mathbb{K}$ are such that

$$\sum_{i=1}^{m} \alpha_i x_i + \sum_{j=1}^{n} \beta_j y_j = 0. \tag{1.3.124}$$

Since $\sum_{i=1}^{m} \alpha_i x_i \in M$ and $\sum_{j=1}^{n} \beta_j y_j \in N$, identity (1.3.124) combined with the fact that $M \cap N = \{0\}$ gives that $\sum_{i=1}^{m} \alpha_i x_i = 0$ and $\sum_{j=1}^{n} \beta_j y_j = 0$. The linear independence of the sets B_M and B_N implies that all coefficients α_i and β_j in (1.3.124) are zero, giving the linear independence of the set $B_M \cup B_N$. This finishes the proof of the direct implication in *(1)*.

Next, we claim that if $B_M \cup B_N$ is a linearly independent set in X then

$$\mathrm{span}_{\mathbb{K}}(B_M) \cap \mathrm{span}_{\mathbb{K}}(B_N) = \mathrm{span}_{\mathbb{K}}(B_M \cap B_N). \tag{1.3.125}$$

Notice that the right-to-left inclusion in (1.3.125) is clear. To establish the left-to-right inclusion in (1.3.125) start by selecting $z \in \mathrm{span}_{\mathbb{K}}(B_M) \cap \mathrm{span}_{\mathbb{K}}(B_N)$. Then there exist numbers $m, n \in \mathbb{N}$, vectors $x_1, \dots, x_m \in B_M$ and $y_1, \dots, y_n \in B_N$, along with two families of scalars $\alpha_1, \dots, \alpha_m \in \mathbb{K}$ and $\beta_1, \dots, \beta_n \in \mathbb{K}$ such that

$$z = \sum_{i=1}^{m} \alpha_i x_i = \sum_{j=1}^{n} \beta_j y_j. \tag{1.3.126}$$

Without loss of generality assume that $x_1, \dots, x_m \in B_M$ and $y_1, \dots, y_n \in B_N$ are such that

$$\{x_1, \dots, x_{m_o}\} \subseteq B_M \cap B_N \quad \text{and} \quad \{x_{m_o+1}, \dots, x_m\} \subseteq B_M \setminus B_N, \tag{1.3.127}$$

for some $m_o \in \{0, 1, \dots, m\}$, and

$$\{y_1, \dots, y_{n_o}\} \subseteq B_M \cap B_N \quad \text{and} \quad \{y_{n_o+1}, \dots, y_n\} \in B_N \setminus B_M, \tag{1.3.128}$$

for some $n_o \in \{0, 1, \dots, n\}$. In light of (1.3.126)-(1.3.128), and making use of the fact that $B_M \cup B_N$ is linearly independent, we obtain that

$$\alpha_i = 0 \;\text{ when }\; i \in \{m_o+1, \dots, m\} \;\text{ and }\; \beta_j = 0 \;\text{ when }\; j \in \{n_o+1, \dots, n\}. \tag{1.3.129}$$

In turn (1.3.129) and (1.3.126) give that $z \in \mathrm{span}_{\mathbb{K}}(\{x_1, \dots, x_{m_o}\})$ which, by (1.3.127), is a subset of $\mathrm{span}_{\mathbb{K}}(B_M \cap B_N)$. Thus $z \in \mathrm{span}_{\mathbb{K}}(B_M \cap B_N)$, completing the proof of the left-to-right inclusion in (1.3.125). The proof of (1.3.125) is therefore finished.

Assume next that $B_M \cap B_N = \varnothing$ and $B_M \cup B_N$ is a linearly independent set in X. First, we trivially have

$$\{0\} \subseteq M \cap N = \mathrm{span}_{\mathbb{K}}(B_M) \cap \mathrm{span}_{\mathbb{K}}(B_N). \tag{1.3.130}$$

Second, keeping in mind that we are presently assuming $B_M \cap B_N = \varnothing$, invoking (1.3.125) we obtain

$$\mathrm{span}_{\mathbb{K}}(B_M) \cap \mathrm{span}_{\mathbb{K}}(B_N) = \mathrm{span}_{\mathbb{K}}(\varnothing) = \{0\}. \tag{1.3.131}$$

In concert, (1.3.130) and (1.3.131) show that $M \cap N = \{0\}$. This completes the proof of the reverse implication in *(1)*.

Turning our attention to proving the equivalence in *(2)*, we start by observing that, since B_M and B_N are bases of M and N, respectively, we have

$$\mathrm{span}_{\mathbb{K}}(B_M \cup B_N) = \mathrm{span}_{\mathbb{K}}(M \cup N) = M + N \subseteq X. \tag{1.3.132}$$

Thus, if we assume that $B_M \cup B_N$ is a basis for X, then $B_M \cup B_N$ is linearly independent. In this case we also have $X = \mathrm{span}_{\mathbb{K}}(B_M \cup B_N)$ and in concert with (1.3.132) we obtain that $M + N = X$. This establishes the right-to-left implication in *(2)*.

Looking at the reverse implication in *(2)*, assume that $M+N = X$ and that $B_M \cup B_N$ is a linearly independent set in X. On the one hand, since $\mathrm{span}_{\mathbb{K}}(B_M \cup B_N) = X$ we may invoke Theorem 1.3.11 to conclude that there exists a basis B' for X with the property that $B' \subseteq B_M \cup B_N$. On the other hand, since $B_M \cup B_N$ is linearly independent, Theorem 1.3.9 guarantees the existence of a basis B for X such that $B_M \cup B_N \subseteq B$. In particular $B' \subseteq B$. However, a basis is maximal with respect to inclusion (see Step II in the proof of Theorem 1.3.9) so $B' = B$. As a result, $B_M \cup B_N = B$ is a basis for X, establishing the left-to-right implication in *(2)*. The proof of Proposition 1.3.26 is now complete. □

We continue by formalizing the notions of algebraic direct sum and algebraic complement.

Definition 1.3.27. *Assume X is a vector space. Then for any two subspaces M, N of X satisfying $M \cap N = \{0\}$ use the symbol $M \boxplus N$, and refer to it as the* `algebraic direct sum` *of M and N, for the linear subspace $M + N$ of X (cf.* (1.3.79)*), as a means of attracting attention to the fact that M and N are linearly independent* (*in the sense that they only have in common the zero vector*)*. Succinctly put,*

$$M \boxplus N := \{x + y : x \in M, \ y \in N\} \ \textit{whenever} \quad M, N \textit{ are two subspaces of } X \textit{ with } M \cap N = \{0\}. \tag{1.3.133}$$

Two subspaces M and N of the vector space X are said to form a pair of `algebraic complements` *in X, or said to be* `algebraic complements` *of each other in X, or N is called an* `algebraic complement` *of M, or M is called an* `algebraic complement` *of N, if*

$$X = M + N \quad \textit{and} \quad M \cap N = \{0\}. \tag{1.3.134}$$

Thus, in the context of (1.3.134)*, the space X is the algebraic direct sum of M and N:*

$$X = M \boxplus N. \tag{1.3.135}$$

Exercise 1.3.28. If X is a vector space over the field $\mathbb{K}$ and M, N are two subspaces of X with the property that $X = M \boxplus N$ then $\dim_{\mathbb{K}} X = \dim_{\mathbb{K}} M + \dim_{\mathbb{K}} N$.

Hint: If B_M is a basis for M over $\mathbb{K}$, and B_N is a basis for N over $\mathbb{K}$, show that $B_M \cup B_N$ is a basis for X over $\mathbb{K}$.

Theorem 1.3.29. *Every subspace of a vector space is algebraically complemented* (*i.e., has an algebraic complement*)*.*

Proof. Let X be a vector space over the field $\mathbb{K}$ and assume that M is a subspace of X with basis B_M. Upon appealing to Theorem 1.3.9, choose a basis B for X such that $B_M \subseteq B$, and introduce the set

$$B_N := B \setminus B_M. \tag{1.3.136}$$

By Theorem 1.3.6, since B is linearly independent so is B_N, and by Proposition 1.3.8 part *(2)*, $N := \mathrm{span}_{\mathbb{K}}(B_N)$ is a subspace of X. Therefore B_N is a basis for the subspace N. Since

$$B_M \cap B_N = \varnothing \quad \text{and} \quad B_M \cup B_N = B, \tag{1.3.137}$$

it follows from Proposition 1.3.26 that N is an algebraic complement of M, establishing the proof of the theorem. □

Linear subspaces possessing a finite-dimensional algebraic complement are said to have finite codimension.

Definition 1.3.30. *Let X be a vector space and Y be a linear subspace of X. Then Y is said to be* `of finite codimension` *provided there exists a finite-dimensional linear subspace Z of X with the property that $X = Y \boxplus Z$.*

Proposition 1.3.31. *Let X be a vector space over the field $\mathbb{K}$ and Y, Z be subspaces of X such that*

$$X = Y + Z \ \ \textit{and} \ \ Z \ \ \textit{has finite dimension}. \tag{1.3.138}$$

Then Y has finite codimension in X.

Proof. Let $\{v_1, \dots, v_N\}$ be a basis in Z, where $N := \dim_{\mathbb{K}} Z \in \mathbb{N}$. Consider the sequence of subsets $\{J_j\}_{0 \le j \le N}$ of $\{0, 1, \dots, N\}$ and the sequence of subspaces $\{Y_j\}_{0 \le j \le N}$ of X defined iteratively as follows. Start with $J_0 := \varnothing$ and $Y_0 := Y$ then, assuming that the sets $\{J_j\}_{0 \le j \le k}$ and the subspaces $\{Y_j\}_{0 \le j \le k}$ have been already constructed for some $k \in \{0, 1, \dots, N-1\}$, set

$$J_{k+1} := \begin{cases} J_k \ \text{ if } \ v_{k+1} \in Y_k, \\ J_k \cup \{k+1\} \ \text{ if } \ v_{k+1} \notin Y_k, \end{cases} \tag{1.3.139}$$

and

$$Y_{k+1} := \mathrm{span}(Y_k \cup \{v_{k+1}\}). \tag{1.3.140}$$

As an immediate consequence of the manner in which the sets J_k and subspaces Y_k have been defined for $0 \le k \le N$, we have the following properties:

$$\varnothing = J_0 \subseteq J_1 \subseteq \cdots \subseteq J_k \subseteq J_{k+1} \subseteq \cdots \subseteq J_N \subseteq \{1, \dots, N\}, \tag{1.3.141}$$

$$Y = Y_0 \subseteq Y_1 \subseteq \cdots \subseteq Y_k \subseteq Y_{k+1} \subseteq \cdots \subseteq Y_N \subseteq X, \tag{1.3.142}$$

$$\begin{gathered} \text{for each } j \in \{1, \dots, N\} \text{ one has} \\ j \in J_j \iff v_j \notin Y_{j-1}, \end{gathered} \tag{1.3.143}$$

$$\begin{gathered} \text{whenever } 1 \le j \le k \le N \text{ one has} \\ j \in J_k \iff j \in J_j. \end{gathered} \tag{1.3.144}$$

and

$$Y_k = \operatorname{span}(Y \cup \{v_j\}_{1\leq j\leq k}) \quad \text{for each} \quad k \in \{1, \dots, N\}. \tag{1.3.145}$$

Regarding (1.3.145), we claim that the following related property holds:

$$Y_k = \operatorname{span}(Y \cup \{v_j\}_{j\in J_k}) \quad \text{for each} \quad k \in \{1, \dots, N\}. \tag{1.3.146}$$

The proof of (1.3.146) proceeds by induction on k. Corresponding to $k = 1$, the fact that

$$Y_1 = \operatorname{span}(Y \cup \{v_j\}_{j\in J_1}) \tag{1.3.147}$$

is a direct consequence of (1.3.139)-(1.3.140) with $k := 0$ (bearing in mind that $J_0 := \varnothing$ and $Y_0 := Y$). Assume next that $Y_k = \operatorname{span}(Y \cup \{v_j\}_{j\in J_k})$ for some integer $k \in \{1, \dots, N-1\}$. Then

$$\begin{aligned} \operatorname{span}(Y \cup \{v_j\}_{j\in J_{k+1}}) &= \operatorname{span}\Big((Y \cup \{v_j\}_{j\in J_k}) \cup \{v_j\}_{j\in J_{k+1}\setminus J_k}\Big) \\ &= \operatorname{span}\Big(Y_k \cup \{v_j\}_{j\in J_{k+1}\setminus J_k}\Big). \end{aligned} \tag{1.3.148}$$

On the one hand, if $J_{k+1} \setminus J_k \neq \varnothing$ then $J_{k+1} \setminus J_k = \{k+1\}$, a scenario in which we have

$$\operatorname{span}\Big(Y_k \cup \{v_j\}_{j\in J_{k+1}\setminus J_k}\Big) = \operatorname{span}\Big(Y_k \cup \{v_{k+1}\}\Big) = Y_{k+1}, \tag{1.3.149}$$

by (1.3.140). On the other hand, if $J_{k+1} \setminus J_k = \varnothing$ then $v_{k+1} \in Y_k$ and

$$\operatorname{span}\Big(Y_k \cup \{v_j\}_{j\in J_{k+1}\setminus J_k}\Big) = Y_k = Y_{k+1}, \tag{1.3.150}$$

again by (1.3.140). Thus, in all cases, $\operatorname{span}(Y \cup \{v_j\}_{j\in J_{k+1}}) = Y_{k+1}$, finishing the proof of (1.3.146).

We next claim that the finite-dimensional subspace $\operatorname{span}(\{v_j\}_{j\in J_N})$ is an algebraic complement of the subspace Y in X, i.e.,

$$X = Y \boxplus \operatorname{span}(\{v_j\}_{j\in J_N}). \tag{1.3.151}$$

Let us assume for the moment that this is the case. Then (1.3.151), the fact that the cardinality of the set J_N is finite, and Definition 1.3.30 guarantee that the subspace Y has finite codimension in X, as desired.

Finally, turning our attention to the claim made in (1.3.151), observe that (1.3.145) and (1.3.146) imply $\{v_1, \dots, v_N\} \subseteq Y_N = \operatorname{span}(Y \cup \{v_j\}_{j\in J_N})$, from which we further conclude that

$$\begin{aligned} X &= Y + Z = Y + \operatorname{span}(\{v_j\}_{1\leq j\leq N}) \\ &= \operatorname{span}(Y \cup \{v_j\}_{j\in J_N}) = Y + \operatorname{span}(\{v_j\}_{j\in J_N}). \end{aligned} \tag{1.3.152}$$

Assume next that there exist scalars $\{\lambda_j\}_{j\in J_N} \subseteq \mathbb{K}$ such that the linear combination $\sum\limits_{j\in J_N} \lambda_j v_j$ belongs to Y. Assume by contradiction that

$$\{j \in J_N : \lambda_j \neq 0\} \neq \varnothing, \tag{1.3.153}$$

and set

$$j_* := \max\{j \in J_N : \lambda_j \neq 0\}, \tag{1.3.154}$$

the maximum element in J_N with the property that its associated scalar is non-zero. In particular, since $1 \leq j_* \leq N$ and $j_* \in J_N$, we conclude from (1.3.144) (with $k := N$) that $j_* \in J_{j_*}$. In view of (1.3.143) this forces $v_{j_*} \notin Y_{j_*-1}$. On the other hand, since $\sum_{j \in J_N} \lambda_j v_j \in Y$ and $\lambda_{j_*} \neq 0$, it follows that $v_{j_*} \in \operatorname{span}(Y \cup \{v_j\}_{1 \leq j \leq j_*-1}) \subseteq Y_{j_*-1}$ (cf. (1.3.145)), reaching a contradiction. This shows that $\{j \in J_N : \lambda_j \neq 0\} = \varnothing$, hence $\lambda_j = 0$ for all $j \in J_N$. Ultimately, this proves that

$$Y \cap \operatorname{span}(\{v_j\}_{j \in J_N}) = \{0\} \tag{1.3.155}$$

which, in concert with (1.3.152), finishes the proof of (1.3.151). This completes the proof of Proposition 1.3.31. □

Exercise 1.3.32. Let Y and Z be linear subspaces of a vector space X such that Y is of finite codimension in X and $Z \supseteq Y$. Then Z is of finite codimension in X.

Exercise 1.3.33. Show that for any two vector spaces X, Y over the field $\mathbb{K}$, it follows that $X \times Y$ is a vector space with

$$\dim_{\mathbb{K}}(X \times Y) = \dim_{\mathbb{K}} X + \dim_{\mathbb{K}} Y. \tag{1.3.156}$$

Hint: Observe that $X \times Y = (X \times \{0\}) \boxplus (\{0\} \times Y)$ and use Exercise 1.3.28.

Moving on, if X is a vector space over the field $\mathbb{K}$ and M is a subspace of X the following relation on X can be easily checked to be an equivalence relation:

$$x, y \in X \ \text{ satisfy } \ x \sim^M y \iff y \in x + M. \tag{1.3.157}$$

In the sequel we shall denote by X/M the quotient space of all equivalence classes of the relation $\sim^M$. It is easy to check that if $x \in X$ then its equivalence class is precisely $x + M$. Henceforth we agree to abbreviate

$$\operatorname{codim}_{\mathbb{K}} M := \dim_{\mathbb{K}}(X/M). \tag{1.3.158}$$

The projection of X onto X/M is defined by setting

$$\pi : X \to X/M, \qquad \pi(x) := x + M \in X/M, \quad \forall\, x \in X. \tag{1.3.159}$$

Also, we define an addition operation between equivalence classes in X/M by setting

$$(x + M) + (y + M) := (x + y) + M, \quad \forall\, x, y \in X, \tag{1.3.160}$$

along with a scalar multiplication with elements from the field $\mathbb{K}$ given by

$$\lambda \cdot (x + M) := \lambda x + M, \quad \forall\, \lambda \in \mathbb{K} \ \text{ and } \ \forall\, x \in X. \tag{1.3.161}$$

The reader is invited to check that (1.3.160)-(1.3.161) are unambiguously defined using the fact that M is a subspace of X.

Lemma 1.3.34. *Let X be a vector space over the field $\mathbb{K}$. Consider a subspace M of X, and denote the projection of X onto the quotient space X/M (as introduced in* (1.3.159)) *by π. Then*

(1) The quotient space X/M equipped with operations (1.3.160)-(1.3.161) *is a vector space over the field $\mathbb{K}$.*
(2) The projection π is a linear transformation between the vector spaces X and X/M.

Proof. It is straightforward to see that the addition operation introduced in (1.3.160) is commutative and transitive. This is a direct consequence of these properties for the addition in the vector space X and the fact that M is a subspace of X. The zero vector in X/M is the equivalence class of every element $x \in M$, i.e., the subspace M itself. Also, for each $x \in X$, the additive inverse of $x + M$ in X/M is $-x + M$. Fix next $\alpha, \beta \in \mathbb{K}$ and $x, y \in X$. Then, since M is a subspace we may clearly write

$$\begin{aligned}\alpha \cdot \Big\{(x + M) + (y + M)\Big\} &= \alpha \cdot (x + y + M) = \alpha \cdot (x + y) + M \\ &= \alpha \cdot x + \alpha \cdot y + M = \alpha \cdot (x + M) + \alpha \cdot (y + M),\end{aligned} \tag{1.3.162}$$

and

$$\begin{aligned}(\alpha + \beta) \cdot (x + M) &= (\alpha + \beta) \cdot x + M = \alpha \cdot x + \beta \cdot x + M \\ &= \alpha \cdot (x + M) + \beta \cdot (x + M).\end{aligned} \tag{1.3.163}$$

Also,

$$\alpha \cdot \Big\{\beta \cdot (x + M)\Big\} = \alpha\beta x + M = \alpha\beta \cdot (x + M). \tag{1.3.164}$$

These, collectively, show that X/M together with the addition and scalar multiplication introduced in (1.3.160) and (1.3.161) satisfies properties *(1)-(5)* from Definition 1.3.1, and thus is a vector space over the field $\mathbb{K}$, establishing part *(1)* of the lemma.

Turning attention to part *(2)*, fix $\alpha, \beta \in \mathbb{K}$ and $x, y \in X$. Then, using the definition of π from (1.3.159), along with (1.3.160) and (1.3.161) and the fact that M is a subspace of X allows us to conclude

$$\begin{aligned}\pi(\alpha x + \beta y) &= \alpha x + \beta y + M = (\alpha x + M) + (\beta y + M) \\ &= \alpha(x + M) + \beta(y + M) = \alpha \cdot \pi(x) + \beta \cdot \pi(y),\end{aligned} \tag{1.3.165}$$

showing that π is linear. This gives *(2)* and finishes the proof of the lemma. □

Lemma 1.3.35. *Let M be a subspace of a vector space X over the field $\mathbb{K}$. Then every algebraic complement of M is isomorphic to the quotient space X/M. As a corollary,*

$$\text{the dimension over the field } \mathbb{K} \text{ of every algebraic complement of } M \text{ is equal to } \operatorname{codim}_{\mathbb{K}} M \tag{1.3.166}$$

and

$$M \text{ has finite codimension in } X \text{ (in the sense of Definition 1.3.30) if and only if } \operatorname{codim}_{\mathbb{K}} M < \infty. \tag{1.3.167}$$

Proof. Let Y be a subspace of the vector space X. Reasoning as in (1.3.165) it is easy to see that the restriction of π to Y $\pi\big|_Y : Y \to X/M$ is again a linear transformation. We claim that

$$Y \text{ is an algebraic complement of } M \implies \pi\big|_Y \text{ is an isomorphism.} \tag{1.3.168}$$

To prove this claim, fix some $x \in X$. Since $X = M + Y$, it follows that $x = u + y$ for some $u \in M$ and some $y \in Y$. Thus, since M is a subspace of X, we have $x + M = y + M$. In particular, $x + M = \pi\big|_Y(y)$, showing that

$$\pi\big|_Y : Y \to X/M \text{ is surjective.} \tag{1.3.169}$$

Going further, if $\pi\big|_Y(y) = 0 + M$, for some $y \in Y$, then $y + M = M$ and hence y actually belongs to M. Since $M \cap Y = \{0\}$, it follows that $y = 0$. Therefore, $\ker(\pi\big|_Y) = \{0\}$, which by Theorem 1.3.19 implies that $\pi\big|_Y$ is injective. This, combined with (1.3.169), gives the claim made in (1.3.168). Consequently, if Y is an algebraic complement of M then $\pi\big|_Y$ is an isomorphism of Y onto X/M, finishing the proof of the lemma. □

It is worth recording the following simple but useful consequence of (1.3.166):

Theorem 1.3.36. *Let M be a subspace of a vector space X over the field $\mathbb{K}$. Then all algebraic complements of M have the same dimension.*

Exercise 1.3.37. Assume X, Y, Z are three vector spaces over the field $\mathbb{K}$, with the property that $Z \subseteq Y \subseteq X$. Then $\dim_{\mathbb{K}}(X/Z) \geq \dim_{\mathbb{K}}(Y/Z)$. In fact,

$$\dim_{\mathbb{K}}(X/Z) = \dim_{\mathbb{K}}(X/Y) + \dim_{\mathbb{K}}(Y/Z). \tag{1.3.170}$$

As a consequence (corresponding to taking $Z = \{0\}$)

$$\dim_{\mathbb{K}} X = \dim_{\mathbb{K}}(X/Y) + \dim_{\mathbb{K}} Y, \tag{1.3.171}$$

and

$$\begin{aligned} &\text{if } \dim_{\mathbb{K}}(X/Z) < +\infty \text{ then } Z = Y \\ &\text{if and only if } \dim_{\mathbb{K}}(X/Z) = \dim_{\mathbb{K}}(X/Y). \end{aligned} \tag{1.3.172}$$

Hint: Invoke Exercise 1.3.22 for the injection $Y/Z \ni (x + Z) \mapsto (x + Z) \in X/Z$ to conclude that $\dim_{\mathbb{K}}(X/Z) \geq \dim_{\mathbb{K}}(Y/Z)$. As for (1.3.170), observe that if M is an algebraic complement of Z in Y and N is an algebraic complement of Y in X, then $M \boxplus N$ is an algebraic complement of Z in X, so the claim in (1.3.170) follows from

(1.3.168) and Exercise 1.3.28. Next, (1.3.171) is obtained from (1.3.170) by taking $Z := \{0\}$. Finally, (1.3.172) is seen directly from (1.3.170).

Exercise 1.3.38. Suppose X is a vector space over the field $\mathbb{K}$, and consider two linear subspaces A, B of X. Then

$$\dim_{\mathbb{K}}(A/A \cap B) \leq \dim_{\mathbb{K}}(X/B). \tag{1.3.173}$$

As a consequence, if $\dim_{\mathbb{K}}(X/B) < +\infty$ then $\dim_{\mathbb{K}}(A/A \cap B) < +\infty$.

Hint: Again rely on Exercise 1.3.22 for the injection

$$A/A \cap B \ni (a + (A \cap B)) \mapsto (a + B) \in X/B \tag{1.3.174}$$

to conclude that (1.3.173) holds.

Exercise 1.3.39. If X, Y, Z are three vector spaces over the field $\mathbb{K}$ with the property that $Z \subseteq Y \subseteq X$, show that

$$\dim_{\mathbb{K}}\left(\frac{X/Z}{Y/Z}\right) = \dim_{\mathbb{K}}(X/Y). \tag{1.3.175}$$

Hint: Note that $X/Z \ni [x]_{X/Z} \mapsto [x]_{X/Y} \in X/Y$ is a surjective linear map, and use the First Isomorphism Theorem for Vector Spaces (asserting that if $T : V \to W$ is a linear map between two vector spaces, then $V/\ker T$ is isomorphic to $\operatorname{ran} T$).

Exercise 1.3.40. Let X be a vector space over the field $\mathbb{K}$ and suppose Y is a subspace having finite codimension in X. Also, suppose Z is a subspace of X with the property $Z \cap Y = \{0\}$. Then Z is finite dimensional and

$$\operatorname{codim}_{\mathbb{K}}(Y + Z) = \operatorname{codim}_{\mathbb{K}} Y - \dim_{\mathbb{K}} Z. \tag{1.3.176}$$

Hint: Note that $Y + Z = Y \boxplus Z$. Let W be an algebraic complement of $Y \boxplus Z$ in X. Then

$$X = (Y \boxplus Z) \boxplus W = Y \boxplus (Z \boxplus W), \tag{1.3.177}$$

so

$$\operatorname{codim}_{\mathbb{K}} Y = \dim_{\mathbb{K}}(Z \boxplus W) = \dim_{\mathbb{K}} Z + \dim_{\mathbb{K}} W. \tag{1.3.178}$$

On the other hand, $\operatorname{codim}_{\mathbb{K}}(Y + Z) = \dim_{\mathbb{K}} W$.

Exercise 1.3.41. Let X be a vector space over the field $\mathbb{K}$, and let Y, Z be subspaces of X such that $X = Y \boxplus Z$. Then

$$\operatorname{codim}_{\mathbb{K}} Y = \dim_{\mathbb{K}} Z. \tag{1.3.179}$$

Hint: This is a direct consequence of (1.3.166).

Proposition 1.3.42. *Let X, Y be two vector spaces over the field $\mathbb{K}$ and consider $A \subseteq X$ along with $B \subseteq Y$, linear subspaces. Then*

$$\dim_{\mathbb{K}}\Big(\frac{X\times Y}{A\times B}\Big) = \dim_{\mathbb{K}}\Big(\frac{X}{A}\Big) + \dim_{\mathbb{K}}\Big(\frac{Y}{B}\Big). \tag{1.3.180}$$

Proof. Observe that the mapping

$$X\times Y \ni (x,y) \mapsto (x+A, y+B) \in \Big(\frac{X}{A}\Big)\times\Big(\frac{Y}{B}\Big) \tag{1.3.181}$$

is well defined, linear, and surjective. Since its kernel is precisely $A\times B$, the First Isomorphism Theorem for Vector Spaces applies and gives that $\frac{X\times Y}{A\times B}$ and $\big(\frac{X}{A}\big)\times\big(\frac{Y}{B}\big)$ are isomorphic vector spaces. From this, (1.3.180) now follows by virtue of Corollary 1.3.25 and Exercise 1.3.33. □

We continue by establishing a beautiful formula expressing the codimension of a sum in terms of the codimension of one of the summands, the dimension of the other summand, and the dimension of their intersection.

Theorem 1.3.43 (Codimension of a sum). *Let X be a vector space over the field $\mathbb{K}$ and suppose Y, N are subspaces of X. Then*

$$\mathrm{codim}_{\mathbb{K}}\,(Y+N) + \dim_{\mathbb{K}} N = \mathrm{codim}_{\mathbb{K}}\, Y + \dim_{\mathbb{K}}(Y\cap N). \tag{1.3.182}$$

In particular, if N is finite dimensional then

$$\mathrm{codim}_{\mathbb{K}}\,(Y+N) = \mathrm{codim}_{\mathbb{K}}\, Y - \dim_{\mathbb{K}} N + \dim_{\mathbb{K}}(Y\cap N). \tag{1.3.183}$$

Proof. For starters, Theorem 1.3.29 guarantees the existence of two subspaces, $G\subseteq X$ and $H\subseteq N$, satisfying

$$X = (Y+N)\boxplus G \quad\text{and}\quad N = (Y\cap N)\boxplus H. \tag{1.3.184}$$

Observe that $Y\cap H=\{0\}$. Indeed, if $x\in Y\cap H$ then $x\in H\subseteq N$, hence $x\in Y\cap N$, so ultimately $x\in (Y\cap N)\cap H=\{0\}$. Bearing this in mind, by adding Y to the second formula in (1.3.184) we arrive at

$$Y+N = Y\boxplus H, \tag{1.3.185}$$

since $Y+(Y\cap N)=Y$. Hence,

$$X = (Y\boxplus H)\boxplus G = Y\boxplus (H\boxplus G). \tag{1.3.186}$$

From (1.3.184), Exercise 1.3.41, and Exercise 1.3.28 we see that

$$\mathrm{codim}_{\mathbb{K}}\,(Y+N) = \dim_{\mathbb{K}} G \quad\text{and}\quad \dim_{\mathbb{K}} N = \dim_{\mathbb{K}}\,(Y\cap N) + \dim_{\mathbb{K}} H, \tag{1.3.187}$$

while from (1.3.186), Exercise 1.3.41, and Exercise 1.3.28 we see that

$$\mathrm{codim}_{\mathbb{K}}\, Y = \dim_{\mathbb{K}} G + \dim_{\mathbb{K}} H. \tag{1.3.188}$$

Now, (1.3.182) follows by combing (1.3.187) and (1.3.188). □

In view of (1.3.167), the next corollary (cf. (1.3.190) in particular) offers a new proof of Proposition 1.3.31.

Corollary 1.3.44. *Let X be a vector space over the field $\mathbb{K}$ and suppose Y, Z are subspaces of X with the property that $X = Y + Z$. Then*

$$\dim_{\mathbb{K}} Z = \operatorname{codim}_{\mathbb{K}} Y + \dim_{\mathbb{K}}(Y \cap Z). \tag{1.3.189}$$

In particular,

$$\operatorname{codim}_{\mathbb{K}} Y \leq \dim_{\mathbb{K}} Z. \tag{1.3.190}$$

Proof. Formula (1.3.189) is a direct consequence of (1.3.182), keeping in mind that we presently have $\operatorname{codim}_{\mathbb{K}}(Y + Z) = 0$. □

Exercise 1.3.45. Let X be a vector space over the field $\mathbb{K}$ and assume that Y and Z are subspaces of X satisfying $Z \subseteq Y \subseteq X$. Show that if Z has finite codimension in Y and Y has finite codimension in X, then Z has finite codimension in X.

Hint: Let M be an algebraic complement of Z in Y and let N be an algebraic complement of Y in X. We then have $Y = Z \boxplus M$, $X = Y \boxplus N$, as well as $\dim_{\mathbb{K}} M < \infty$ and $\dim_{\mathbb{K}} N < \infty$. It is easy to check then that $X = Z \boxplus (M + N)$ and that the subspace $M + N$ has finite dimension.

We continue by stating and proving two conceptually simple but quite useful propositions.

Proposition 1.3.46. *Let X be a vector space over the field $\mathbb{K}$ and assume that Y and Z are subspaces of X satisfying $Z \subseteq Y \subseteq X$ and that Z has finite codimension in X. Then Z has finite codimension in Y.*

Proof. Let M be an algebraic complement of Z in X. Then M is a subspace of X and we have $Z \boxplus M = X$ and $\dim_{\mathbb{K}} M < \infty$. We claim first that

$$Y = Z \boxplus (Y \cap M). \tag{1.3.191}$$

Clearly, $Z + (Y \cap M) \subseteq Y$. Conversely, if $y \in Y$ then $y \in X$ and there exist $z \in Z$ and $m \in M$ such that $y = z + m$. Since $m = y - z \in Y$ we obtain that $m \in Y \cap M$. Since y was arbitrary in Y this shows that $Y \subseteq Z + (Y \cap M)$ consequently $Y = Z + (Y \cap M)$. However, since $Z \cap M = \{0\}$ we also have $Z \cap (Y \cap M) = \{0\}$, finishing the proof of (1.3.191). Finally $\dim_{\mathbb{K}}(Y \cap M) \leq \dim_{\mathbb{K}} M < \infty$ and in concert with (1.3.191) we obtain that Z has finite codimension in Y. □

Here is the second proposition mentioned above.

Proposition 1.3.47. *Let X be a vector space over the field $\mathbb{K}$ and let Y, Z be subspaces of X such that $Y \cap Z = \{0\}$. Then*

$$\operatorname{codim}_{\mathbb{K}} Y \geq \dim_{\mathbb{K}} Z. \tag{1.3.192}$$

Proof. Let B_Y and B_Z be bases for Y and Z, respectively. Then, according to item *(i)* in Proposition 1.3.26, we have $B_Y \cap B_Z = \varnothing$ and $B_Y \cup B_Z$ is linearly independent. Having established this, Theorem 1.3.9 tells us that it is possible to find a basis B_X for X with the property that $B_X \supseteq B_Y \cup B_Z$. Since $B_Y \cap B_Z = \varnothing$ this entails $(B_X \setminus B_Y) \supset B_Z$, hence $\#(B_X \setminus B_Y) \geq \#B_Z$. Granted this inequality, we then have $\operatorname{codim}_{\mathbb{K}} Y = \#(B_X \setminus B_Y) \geq \#B_Z = \dim_{\mathbb{K}} Z$, as wanted. □

Exercise 1.3.48. Assume X, Y are vector spaces over the field $\mathbb{K}$, and consider a linear operator $T : X \to Y$. Show that

$$\dim_{\mathbb{K}} X = \dim_{\mathbb{K}}(\operatorname{ran} T) + \dim_{\mathbb{K}}(\ker T). \tag{1.3.193}$$

Hint: According to the First Isomorphism Theorem for Vector Spaces, $X/\ker T$ is isomorphic to $\operatorname{ran} T$, hence they have the same dimension. Add $\dim_{\mathbb{K}}(\ker T)$ and use (1.3.171).

The last result in this section further refines formula (1.3.193).

Proposition 1.3.49. *Let X, Y, Z be three vector spaces over the field $\mathbb{K}$ with the property that $Z \subseteq X$, and suppose $T : X \to Y$ is a linear map. Then*

$$\begin{aligned}\dim_{\mathbb{K}}(X/Z) &+ \dim_{\mathbb{K}}(Y/T(X))\\ &= \dim_{\mathbb{K}}(Y/T(Z)) + \dim_{\mathbb{K}}\big(\ker T/(Z \cap \ker T)\big).\end{aligned} \tag{1.3.194}$$

Proof. We will first prove the formula claimed in (1.3.194) under the additional assumption that the linear map $T : X \to Y$ is also surjective. Assuming $Y = T(X)$, consider the vectors $\{y_j\}_{j\in J} \subseteq Y$ and $\{k_i\}_{i\in I} \subseteq \ker T$ such that

$$\begin{aligned}&\{y_j + T(Z)\}_{j\in J} \ \text{ is a basis in } \ Y/T(Z) \ \text{ and}\\ &\{k_i + Z \cap \ker T\}_{i\in I} \ \text{ is a basis in } \ \ker T/(Z \cap \ker T).\end{aligned} \tag{1.3.195}$$

Since T is surjective,

$$\text{for each } j \in J \text{ there exists } x_j \in X \text{ such that } T(x_j) = y_j. \tag{1.3.196}$$

We claim that the equivalence classes

$$\{x_j + Z : \ j \in J\} \cup \{k_i + Z : \ i \in I\} \ \text{ are linearly independent in } \ X/Z. \tag{1.3.197}$$

Indeed, if there exist two finite sets J_* and I_* with $J_* \subseteq J$ and $I_* \subseteq I$ and two families of scalars $\{\lambda_j\}_{j\in J_*}$ and $\{a_i\}_{i\in I_*}$ in $\mathbb{K}$ such that

$$\sum_{j\in J_*} \lambda_j x_j + \sum_{i\in I_*} a_i k_i \in Z, \tag{1.3.198}$$

then after applying T to the expression in (1.3.198) we obtain $\sum_{j\in J_*} \lambda_j T(x_j) \in T(Z)$. This and the first line in (1.3.195) imply $\lambda_j = 0$ for all $j \in J_*$. The latter and (1.3.198) yield $\sum_{i\in I_*} a_i k_i \in Z$, which self improves to $\sum_{i\in I_*} a_i k_i \in Z \cap \ker T$. Recalling the second line in (1.3.195), we necessarily have $a_i = 0$ for all $i \in I_*$. The claim in (1.3.197) is therefore justified.

To proceed, pick $x \in X$ arbitrary. By (1.3.195), there exist a finite set $J_0 \subseteq J$ and a family of scalars $\{\lambda_j\}_{j\in J_0} \subseteq \mathbb{K}$ such that

$$T(x) + T(Z) = \sum_{j\in J_0} \lambda_j (y_j + T(Z)). \tag{1.3.199}$$

Hence, in light of (1.3.196), we have $T(x) - \sum_{j\in J_0} \lambda_j T(x_j) \in T(Z)$, which implies the existence of $z \in Z$ such that

$$x - \sum_{j\in J_0} \lambda_j x_j - z \in \ker T. \tag{1.3.200}$$

Consequently,

$$x - \sum_{j\in J_0} \lambda_j x_j - z + (Z \cap \ker T) \in \ker T/(Z \cap \ker T). \tag{1.3.201}$$

In view of the second line in (1.3.195), there exist a finite set $I_0 \subseteq I$ and a family of scalars $\{a_i\}_{i\in I_0} \subseteq \mathbb{K}$ such that

$$x - \sum_{j\in J_0} \lambda_j x_j - z - \sum_{i\in I_0} a_i k_i \in Z \cap \ker T \subseteq Z. \tag{1.3.202}$$

The latter combined with (1.3.197) and (1.3.195) implies that the equivalence classes in (1.3.197) constitute a basis in the quotient space X/Z and we have

$$\dim_{\mathbb{K}}(X/Z) = \dim_{\mathbb{K}}(Y/T(Z)) + \dim_{\mathbb{K}}\big(\ker T/(Z \cap \ker T)\big). \tag{1.3.203}$$

This concludes the proof of (1.3.194) when $Y = T(X)$.

Next, consider the general case. Based on what we have proved so far, we may write (1.3.203) with $Y := T(X)$ to obtain

$$\dim_{\mathbb{K}}(X/Z) = \dim_{\mathbb{K}}(T(X)/T(Z)) + \dim_{\mathbb{K}}\big(\ker T/(Z \cap \ker T)\big). \tag{1.3.204}$$

In addition, since $T(Z) \subseteq T(X) \subseteq Y$, by (1.3.170) we have

$$\dim_{\mathbb{K}}(Y/T(Z)) = \dim_{\mathbb{K}}(Y/T(X)) + \dim_{\mathbb{K}}(T(X)/T(Z)). \tag{1.3.205}$$

Now (1.3.194) follows by adding $\dim_{\mathbb{K}}(Y/T(X))$ to both sides in (1.3.204) and then making use of (1.3.205). □

1.4 Normed Spaces

In this section we initiate a discussion centered around the notions of p-norms and (genuine) norms.

Definition 1.4.1. *Suppose that* $0 < p \leq 1$ *and that* X *is a vector space over the field* $\mathbb{K}$ *(where* $\mathbb{K}$ *is either* $\mathbb{R}$*, or* $\mathbb{C}$*). A function* $\|\cdot\| : X \to \mathbb{R}$ *is called a* p-`norm` *on* X *provided the following hold.*

(1) For each $x, y \in X$ *one has* $\|x + y\| \leq \|x\| + \|y\|$.
(2) For each $\lambda \in \mathbb{K}$*, and each* $x \in X$*, one has* $\|\lambda x\| = |\lambda|^p \|x\|$.
(3) For each $x \in X$ *the following implication holds* $\|x\| = 0 \Rightarrow x = 0$.

The pair $(X, \|\cdot\|)$*, where* $\|\cdot\|$ *is a p-norm on X, is called a* p-`normed space`.

Finally, refer to a subset A *of* X *as being* $\|\cdot\|$-`bounded` *(or* `norm-bounded`*) if* $\sup_{x \in A} \|x\| < \infty$.

In this context, it should be pointed out that, by making use of assertion *(2)* with $\lambda := 0$, item *(3)* may be expressed as the following if-and-only-if statement:

$$\|x\| = 0 \iff x = 0, \quad \text{for all} \quad x \in X. \tag{1.4.206}$$

Thus, the only difference between a p-norm and a norm is the different (for $p < 1$) notion of homogeneity that occurs in *(2)*. Moreover, every p-norm on a vector space is necessarily nonnegative. Indeed, if x is any vector in X, then

$$2\|x\| = \|x\| + \|(-1)x\| \geq \|x + (-1)x\| = \|0\| = 0, \tag{1.4.207}$$

where the first equality follows from *(2)*, the inequality is a consequence of *(1)*, and the last equality is valid by (1.4.206). Therefore (1.4.207) entails that $\|x\| \geq 0$, as advertised.

From Definition 1.4.1 we also observe that, if X is a vector space and $p \in (0, 1]$, then the following holds:

$$\|\cdot\| \text{ is a } p\text{-norm on } X \implies \|\cdot\|^{q/p} \text{ is a } q\text{-norm on } X \text{ for every } q \in (0, p]. \tag{1.4.208}$$

Given an integrability exponent $p \in (0, 1]$, an important example of a vector space X equipped with a p-norm $\|\cdot\|$ is as follows. Let $(\mathscr{X}, \mu)$ be a measure space, and identify any two μ-measurable functions $f, g : \mathscr{X} \to \mathbb{K}$ such that $f = g$ at μ-a.e. point in $\mathscr{X}$. Denote by $\mathscr{M}(\mathscr{X}, \mu)$ the collection of all equivalence classes of μ-measurable functions thus obtained and, slightly abusing notation, continue to use the symbol f for the equivalence class of every function $f \in \mathscr{M}(\mathscr{X}, \mu)$. Then

$$X := \Big\{ f \in \mathscr{M}(\mathscr{X}, \mu) : \int_{\mathscr{X}} |f|^p \, d\mu < \infty \Big\} \tag{1.4.209}$$

is a vector space over the field $\mathbb{K}$, and

$$\|f\| := \int_{\mathscr{X}} |f|^p \, d\mu \quad \text{for each} \quad f \in X \tag{1.4.210}$$

is a p-norm on X.

We now make two basic definitions, highlighting the role of completeness.

Definition 1.4.2. *Let* $0 < p \leq 1$ *and assume that* $(X, \|\cdot\|)$ *is a p-normed space. Then*

(1) Call $\|\cdot\|$ *a* p-`Banach norm` *if its induced distance d as in* (1.6.308) *is complete (in the sense of Definition 1.2.7).*
(2) Call $(X, \|\cdot\|)$ *a* p-`Banach space` *or, equivalently, a* `complete` p-`normed space`*, if its p-norm is a p-Banach norm.*

In light of this definition, $(X, \|\cdot\|)$ defined in (1.4.209)-(1.4.210) is an example of a p-Banach space, hence, so is every closed subspace of X. In this regard, we note that there is an abundance of p-Banach spaces that are *not* subspaces of some Lebesgue space L^p. For example, this is the case for Besov and Triebel-Lizorkin spaces where the p-norm takes into account the smoothness of the function (in addition to integrability properties).

Exercise 1.4.3. Assume that $(X, \|\cdot\|_X)$ and $(Y, \|\cdot\|_Y)$ are two p-normed spaces, for some $p \in (0, 1]$. Show that

$$\|(x, y)\| := \|x\|_X + \|y\|_Y \quad \text{for all} \quad (x, y) \in X \times Y \tag{1.4.211}$$

is a p-norm on $X \times Y$, and $(X \times Y, \|\cdot\|)$ is actually a p-Banach space if in fact $(X, \|\cdot\|_X)$ and $(Y, \|\cdot\|_Y)$ are p-Banach spaces.

In the second definition alluded to above, we essentially specialize Definition 1.4.1 and Definition 1.4.2 to the case when $p = 1$.

Definition 1.4.4. *Suppose X is a vector space over the field* $\mathbb{K}$ *(where* $\mathbb{K}$ *is either* $\mathbb{R}$*, or* $\mathbb{C}$*). A* `norm` *on X is a* 1*-norm on X. That is, a nonnegative real-valued function* $\|\cdot\|$ *on X is called a norm provided:*

(1) $\|x + y\| \leq \|x\| + \|y\|$*, for each* $x, y \in X$.
(2) $\|\lambda x\| = |\lambda|\|x\|$*, for each* $\lambda \in \mathbb{K}$*, and each* $x \in X$.
(3) $\|x\| = 0 \Rightarrow x = 0$*, for each* $x \in X$.

The ordered pair $(X, \|\cdot\|)$ *is called a* `normed vector space`*, or a* `normed linear space`*, or just a* `normed space`*. In addition, a* 1*-Banach space is simply referred to as a* `Banach space`*, or a* `complete normed space`.

With every p-norm on a vector space one can naturally associate a metric on said space, of the sort described below.

Proposition 1.4.5. *If* $0 < p \leq 1$ *and* $\|\cdot\|$ *is a p-norm on a vector space X, then the mapping* $d : X \times X \longrightarrow \mathbb{R}$*, defined by*

$$d(x, y) := \|x - y\|, \qquad \forall\, x, y \in X, \tag{1.4.212}$$

is a metric on X.

Proof. Let $x, y \in X$ be such that $d(x, y) = 0$. Then $\|x - y\| = 0$, and by property *(3)* from Definition 1.4.1 we obtain $x - y = 0$, i.e., $x = y$. If $x = y$ it is immediate that $d(x, y) = \|0\| = 0$. This shows that d satisfies the non-degeneracy property from the definition of metric in Definition 1.2.1. Using *(2)* from Definition 1.4.1 we also obtain that d is symmetric and the triangle inequality for d is a direct consequence of *(1)* from Definition 1.4.1. This completes the proof that d is a metric on X. □

Definition 1.4.6. *Suppose* $0 < p \leq 1$ *and that* $(X, \|\cdot\|)$ *is a p-normed space. The metric d on X as defined in* (1.4.212) *is called the* `metric`, *or* `distance induced by the` p-`norm` *of X. The topology induced (as in Proposition 1.2.5; cf.* (1.2.55)) *by the distance d defined in* (1.4.212) *will be referred to as the* p-`norm topology` *on X.*

Proposition 1.4.7. *Suppose that* $(X, \|\cdot\|)$ *is a p-normed space, for some* $0 < p \leq 1$. *Equip X with its p-norm topology. Then the mapping* $\|\cdot\| : X \to \mathbb{R}$ *is continuous. Moreover, each ball*

$$B(x, r) := \{y \in X : \|x - y\| < r\}, \quad x \in X, \quad r \in (0, \infty) \tag{1.4.213}$$

is open in the p-norm topology on X.

Proof. Let $x_0 \in X$ be fixed. By property *(1)* of Definition 1.4.1 for each $x \in X$ we have

$$\|x\| - \|x_0\| \leq \|x - x_0\| \text{ and } \|x_0\| - \|x\| \leq \|x_0 - x\|, \qquad \forall\, x \in X. \tag{1.4.214}$$

However,

$$\|x_0 - x\| = \|(-1)(x - x_0)\| = |-1|^p \|x - x_0\| = \|x - x_0\|, \tag{1.4.215}$$

where the first and last equalities are trivial and the second is a consequence of the homogeneity property *(2)* of Definition 1.4.1. Therefore, (1.4.214) and (1.4.215) entail

$$\big|\|x\| - \|x_0\|\big| \leq \|x - x_0\|. \tag{1.4.216}$$

This immediately yields the continuity of the p-norm $\|\cdot\|$ at the point x_0. Since x_0 was arbitrary in X, this shows that the p-norm is indeed continuous on X.

The final claim in the statement is apparent from item *(2)* in Proposition 1.2.5 and the fact that each ball $B(x, r)$ as in (1.4.213) coincides with $B_d(x, r)$ associated with the distance d as in (1.2.51). □

Remark 1.4.8. While, as noticed in Proposition 1.4.7, the balls defined according to (1.4.213), in a p-normed space $(X, \|\cdot\|)$ with $0 < p < 1$, are open (in the p-norm topology on X), they may fail to be convex sets. As an example, fix $p \in (0, 1)$ and take $X := \mathbb{R}^2$ equipped with the p-norm

$$\|(x, y)\| := |x|^p + |y|^p \text{ for each } (x, y) \in \mathbb{R}^2. \tag{1.4.217}$$

Then the boundary of $B((0,0),1) := \{(x,y) \in \mathbb{R}^2 : |x|^p + |y|^p < 1\}$ contained in the first quadrant is the graph of the function $f : (0,1) \to \mathbb{R}$ defined at each $x \in (0,1)$ by $f(x) := (1 - x^p)^{1/p}$. Since f is a convex function, given that

$$f''(x) = (1-p)x^{p-2}(1-x^p)^{1/p-2} > 0 \quad \text{for each} \quad x \in (0,1), \tag{1.4.218}$$

it follows that $B((0,0),1)$ fails to be a convex set. Figure 1.1 shows how $B((0,0),1)$ typically looks like.

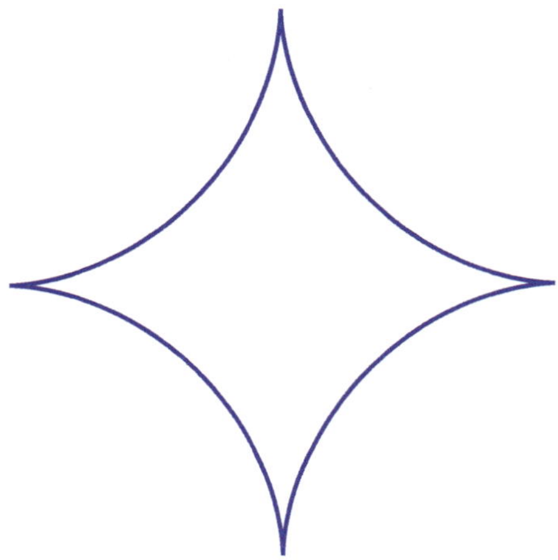

Fig. 1.1 Star-shaped unit ball $B_{\|\cdot\|}(0,1)$ associated with the p-norm $\|\cdot\|$ defined in (1.4.217) with $0 < p < 1$. Although open in the p-norm topology induced by $\|\cdot\|$ in $\mathbb{R}^2$, this fails to be a convex set.

Of course, similar considerations are valid in higher dimensions, i.e., in $\mathbb{R}^n$ with $n \geq 2$ arbitrary, now considering the p-norm

$$\|x\| := |x_1|^p + |x_2|^p + \cdots + |x_n|^p \quad \text{for each} \quad x = (x_1, \ldots, x_n) \in \mathbb{R}^n, \tag{1.4.219}$$

for each given $p \in (0,1)$. Figure 1.2 is a rendering of the unit ball $B(0,1)$ corresponding to $n = 3$.

Exercise 1.4.9. Suppose $0 < p \leq 1$ and that $(X, \|\cdot\|)$ is a p-normed space. Also, fix $q \in (0,p]$. Show that $\|\cdot\|^{q/p}$ is a q-norm on X whose associated q-norm topology on X is the same at the p-norm topology on X associated with the original $\|\cdot\|$.

Hint: Use (1.4.208) and Definition 1.4.6.

Definition 1.4.10. *A* `seminorm` *on a vector space X over the field $\mathbb{K}$ (where $\mathbb{K}$ is either $\mathbb{R}$, or $\mathbb{C}$) is a real-valued function $\|\cdot\|$ on X satisfying:*

(1) $\|x+y\| \leq \|x\| + \|y\|$, *for each* $x, y \in X$.
(2) $\|\lambda x\| = |\lambda|\|x\|$, *for each* $\lambda \in \mathbb{K}$, *and each* $x \in X$.

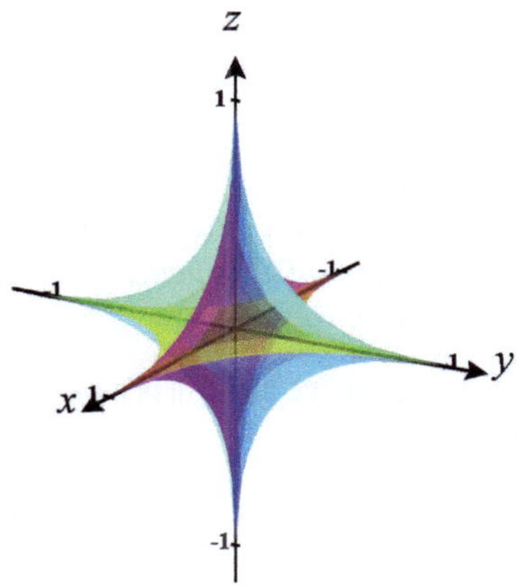

Fig. 1.2 Star-shaped unit ball $B_{\|\cdot\|}(0, 1)$ associated with the p-norm $\|\cdot\|$ defined in (1.4.219) with $n = 3$ and $0 < p < 1$. Even though this is open in the p-norm topology induced by $\|\cdot\|$ in $\mathbb{R}^3$, it fails to be a convex set.

As may be seen from this definition and the exercise below, a seminorm is therefore a real-valued nonnegative function on a vector space satisfying the definition of a norm, except that the value of the seminorm of a nonzero vector is permissible to be zero.

Exercise 1.4.11. If X is a vector space and $\|\cdot\|$ is a seminorm on X, use Definition 1.4.10 to show that $\|0\| = 0$, and $\|x\| \geq 0$ for each $x \in X$. Use (1.3.81) to show that for each $x \in X$ and $r \in (0, \infty)$,

$$\{y \in X : \|y - x\| < r\} \text{ is a convex set.} \tag{1.4.220}$$

Contrast this to Remark 1.4.8.

Here is a companion notion to that introduced in Definition 1.4.1.

Definition 1.4.12. *Let X be a vector space over the field $\mathbb{K}$ (where $\mathbb{K}$ is either $\mathbb{R}$, or $\mathbb{C}$). A mapping $\|\cdot\| : X \to [0, \infty)$ is called a* `quasi-norm` *(on X) provided the following properties hold:*

(i) $\|x\| = 0$ if and only if $x = 0$;
(ii) $\|\lambda x\| = |\lambda|\|x\|$ for each $x \in X$ and $\lambda \in \mathbb{K}$;
(iii) there exists $\kappa \geq 1$ such that $\|x + y\| \leq \kappa(\|x\| + \|y\|)$ for all $x, y \in X$.

The smallest value of κ that works in the last inequality above is occasionally denoted by $\rho(X)$ and is referred to as the `modulus of concavity` *of X.*

Definition 1.4.13. *A* `quasi-normed space` $(X, \|\cdot\|)$ *consists of a vector space* X *and a quasi-norm* $\|\cdot\|$ *on* X*. Occasionally, when the quasi-norm* $\|\cdot\|$ *is understood from the context, or is irrelevant, we shall simply write* X *in place of the quasi-normed space* $(X, \|\cdot\|)$.

Call a quasi-normed space $(X, \|\cdot\|)$ *a* `quasi-Banach space` *if* X *is complete with respect to* $\|\cdot\|$*, in the sense that every sequence* $\{x_n\}_{n\in\mathbb{N}} \subseteq X$ *which is Cauchy with respect to* $\|\cdot\|$ *converges to some vector* $x_* \in X$.

Finally, refer to a given subset A *of* X *as being* $\|\cdot\|$-`bounded` *(or* `quasi-norm-bounded`*) provided* $\sup_{x\in A}\|x\| < \infty$.

Exercise 1.4.14. If X is a vector space and $\|\cdot\|$ is a p-norm on X, for some $p \in (0, 1]$, then $\|\cdot\|^{1/p}$ is a quasi-norm on X.

Definition 1.4.15. *Two quasi-norms,* $\|\cdot\|_1$ *and* $\|\cdot\|_2$*, defined on a common vector space* X *are said to be* `equivalent` *if there exist two real numbers* $c, c' > 0$ *with the property that*

$$c\|x\|_1 \le \|x\|_2 \le c'\|x\|_1, \qquad \forall\, x \in X. \tag{1.4.221}$$

Exercise 1.4.16. Let $(X, \|\cdot\|)$ be a quasi-normed space and suppose $\|\cdot\|' : \mathbb{R}^n \to [0,\infty)$ is a function satisfying the homogeneity condition $\|\lambda x\|' = |\lambda|\|x\|'$ for each $x \in X$ and $\lambda \in \mathbb{K}$, and for which there exist two numbers $C_1 C_2 \in (0, \infty)$ with the property that

$$C_1\|x\| \le \|x\|' \le C_2\|x\|, \qquad \forall\, x \in X. \tag{1.4.222}$$

Show that actually $\|\cdot\|'$ is a quasi-norm on X which is equivalent to $\|\cdot\|$.

It is actually possible to "classify" all quasi-norms on a vector space which are equivalent to a given one, in the manner described below.

Exercise 1.4.17. Assume $(X, \|\cdot\|)$ is a quasi-normed space. Show that for every quasi-norm $\|\cdot\|'$ on X which is equivalent to $\|\cdot\|$ there exists a function

$$f : \{x \in X : \|x\| = 1\} \longrightarrow [a, b] \ \text{ where } \ 0 < a < b < \infty, \tag{1.4.223}$$

with the property that

$$\|x\|' = f\Big(\frac{x}{\|x\|}\Big)\|x\|, \qquad \forall\, x \in X \setminus \{0\}. \tag{1.4.224}$$

Conversely, show that if for every function f as in (1.4.223) one actually defines $\|x\|'$ as in (1.4.224) whenever $x \in X \setminus \{0\}$ and also sets $\|0\|' := 0$, then $\|\cdot\|'$ becomes a quasi-norm on X which is equivalent to $\|\cdot\|$.

Hint: For the first claim, take $f(x) := \|x\|'$. For the converse direction, use Exercise 1.4.16.

Exercise 1.4.18. Given a quasi-normed space $(X, \|\cdot\|)$, denote by $\tau_{\|\cdot\|}$ the collection of all sets $O \subseteq X$ with the property that for every $x \in O$ there exists $r_x > 0$ such that every $y \in X$ satisfying $\|x - y\| < r_x$ belongs to O. Show that $\tau_{\|\cdot\|}$ is a Hausdorff topology on X (called the `topology induced by the quasi-norm` on X).

Exercise 1.4.19. If X is a vector space and $\|\cdot\|_1$ and $\|\cdot\|_2$ are two equivalent quasi-norms on X, show that $\tau_{\|\cdot\|_1}$ coincides with $\tau_{\|\cdot\|_2}$.

We may rephrase the definition of the topology induced by a quasi-norm $\|\cdot\|$ on a vector space X in Exercise 1.4.18 by saying that $\tau_{\|\cdot\|}$ consists of subsets O of X with the property that every point $x \in O$ is the center of a ball

$$B_{\|\cdot\|}(x, r) := \{y \in X : \|x - y\| < r\}, \quad r > 0, \tag{1.4.225}$$

contained in O. The reader should be mindful of the fact that the balls (1.4.225) may *fail* to be themselves open sets in the topology $\tau_{\|\cdot\|}$, even when X is finite dimensional. An example, in which all balls look like *scaled replicas of the planet Saturn*, is offered below:

Proposition 1.4.20. *Fix $n \in \mathbb{N}$ with $n \geq 2$, and define $\|\cdot\| : \mathbb{R}^n \to [0, \infty)$ by setting, for each point $x = (x_1, \dots, x_n) \in \mathbb{R}^n$,*

$$\|x\| := \begin{cases} \sqrt{x_1^2 + \cdots + x_n^2} & \text{if } x_n \neq 0, \\ \frac{1}{2} \cdot \sqrt{x_1^2 + \cdots + x_{n-1}^2} & \text{if } x_n = 0. \end{cases} \tag{1.4.226}$$

Then $\|\cdot\|$ is a quasi-norm on $\mathbb{R}^n$ with the property that

$$\frac{1}{2}\sqrt{x_1^2 + \cdots + x_n^2} \leq \|x\| \leq 2\sqrt{x_1^2 + \cdots + x_n^2} \text{ for each } x = (x_1, \dots, x_n) \in \mathbb{R}^n, \tag{1.4.227}$$

and such that

$$B_{\|\cdot\|}(x, r) \notin \tau_{\|\cdot\|} \text{ for all } x \in \mathbb{R}^n \text{ and all } r \in (0, \infty). \tag{1.4.228}$$

As a consequence,

$$\|\cdot\| : (\mathbb{R}^n, \tau_{\|\cdot\|}) \to [0, \infty) \text{ is not a continuous function.} \tag{1.4.229}$$

Finally, as opposed to (1.4.220), *each ball $B_{\|\cdot\|}(x, r)$ with $x \in \mathbb{R}^n$ and $r \in (0, \infty)$ presently fails to be a convex set* (*take a look at Figure 1.3*).

Proof. It is easy to see from (1.4.226) that $\|\cdot\|$ satisfies the homogeneity condition $\|\lambda x\| = |\lambda|\|x\|$ for each $x \in \mathbb{R}^n$ and $\lambda \in \mathbb{R}$, and for which (1.4.227) holds. Granted these, Exercise 1.4.16 tells us that $\|\cdot\|$ is indeed a quasi-norm on $\mathbb{R}^n$ which is equivalent to the standard Euclidean norm

$$\|x\|_{\text{Eu}} := \sqrt{x_1^2 + \cdots + x_n^2} \text{ for each } x = (x_1, \dots, x_n) \in \mathbb{R}^n. \tag{1.4.230}$$

According to Exercise 1.4.19 the topology $\tau_{\|\cdot\|}$ induced by $\|\cdot\|$ in $\mathbb{R}^n$ is the standard Euclidean topology. However,

$$\begin{aligned} B_{\|\cdot\|}(0,1) &= \{x \in \mathbb{R}^n : \|x\| < 1\} \\ &= \{x \in \mathbb{R}^n : \sqrt{x_1^2 + \cdots + x_n^2} < 1\} \cup \{x \in \mathbb{R}^{n-1} \times \{0\} : \sqrt{x_1^2 + \cdots + x_{n-1}^2} < 2\} \end{aligned} \tag{1.4.231}$$

(which resembles the shape of the planet Saturn). Obviously, this is *not* open in the standard Euclidean topology, hence not open in $\tau_{\|\cdot\|}$. Since

$$B_{\|\cdot\|}(x,r) = x + r \cdot B_{\|\cdot\|}(0,1) \text{ for all } x \in \mathbb{R}^n \text{ and all } r \in (0,\infty), \tag{1.4.232}$$

the claim in (1.4.228) follows. Moreover, since $B_{\|\cdot\|}(0,1)$ is the pre-image under the mapping $\|\cdot\| : \mathbb{R}^n \to \mathbb{R}$ of the interval $[0,1)$, which is a relatively open subset of $[0,\infty)$, the claim in (1.4.229) also follows. □

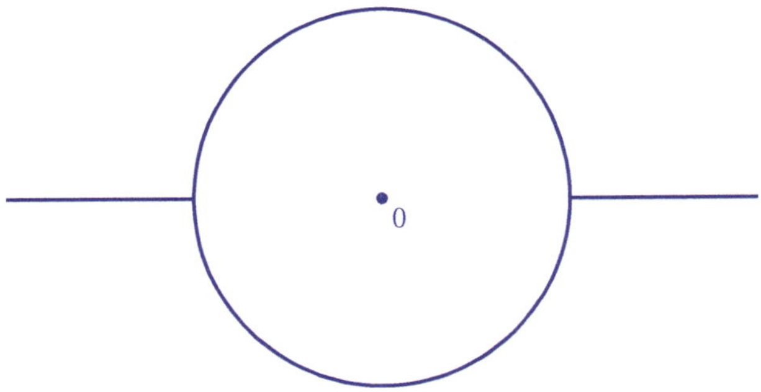

Fig. 1.3 Saturn-like shape of the unit ball $B_{\|\cdot\|}(0,1)$ associated with the quasi-norm $\|\cdot\|$ defined in (1.4.226). This does not contain any miniature version of itself translated so that the center is on one of the flaps.

We can tilt (and change the dimension of) the "ring" of the Saturn-like ball from Proposition 1.4.20 as indicated in the exercise below:

Exercise 1.4.21. Fix $n \in \mathbb{N}$ with $n \geq 2$, and consider a linear subspace V of $\mathbb{R}^n$ with the property that $1 \leq \dim_{\mathbb{R}} V \leq n-1$. Use it to define $\|\cdot\|_V : \mathbb{R}^n \to [0,\infty)$ by setting, for each point $x = (x_1, \dots, x_n) \in \mathbb{R}^n$,

$$\|x\|_V := \begin{cases} \sqrt{x_1^2 + \cdots + x_n^2} & \text{if } x \notin V, \\ \frac{1}{2}\sqrt{x_1^2 + \cdots + x_n^2} & \text{if } x \in V. \end{cases} \tag{1.4.233}$$

Show that $\|\cdot\|_V$ is a quasi-norm on $\mathbb{R}^n$ with the property that

$$\frac{1}{2}\sqrt{x_1^2 + \cdots + x_n^2} \leq \|x\|_V \leq 2\sqrt{x_1^2 + \cdots + x_n^2} \text{ for each } x = (x_1, \dots, x_n) \in \mathbb{R}^n, \tag{1.4.234}$$

and such that

$$B_{\|\cdot\|_V}(x,r) \notin \tau_{\|\cdot\|_V} \text{ for all } x \in \mathbb{R}^n \text{ and all } r \in (0,\infty). \tag{1.4.235}$$

Deduce that

$$\|\cdot\|_V : (\mathbb{R}^n, \tau_{\|\cdot\|_V}) \to [0,\infty) \text{ is not a continuous function.} \tag{1.4.236}$$

We can also add multiple "rings" to the Saturn-like ball from Proposition 1.4.20, as demonstrated by our next exercise.

Exercise 1.4.22. Fix $n, m \in \mathbb{N}$, $n \geq 2$, pick a collection $\mathscr{V} := \{V_1, V_2, \dots, V_m\}$ of one dimensional linear subspaces of $\mathbb{R}^n$, and select an arbitrary family of numbers $c_1, c_2, \dots c_m \in (0,1)$. Use these to define $\|\cdot\|_{\mathscr{V}} : \mathbb{R}^n \to [0,\infty)$ by setting, for each point $x = (x_1, \dots, x_n) \in \mathbb{R}^n$,

$$\|x\|_{\mathscr{V}} := \begin{cases} \sqrt{x_1^2 + \cdots + x_n^2} & \text{if } x \notin V_j \text{ for each } 1 \leq j \leq m, \\ c_j \sqrt{x_1^2 + \cdots + x_n^2} & \text{if there exists } 1 \leq j \leq m \text{ such that } x \in V_j. \end{cases} \tag{1.4.237}$$

Show that $\|\cdot\|_{\mathscr{V}}$ is a quasi-norm on $\mathbb{R}^n$ with the property that

$$\Big(\min_{1\leq j\leq m} c_j\Big)\sqrt{x_1^2 + \cdots + x_n^2} \leq \|x\|_{\mathscr{V}} \leq \Big(\max_{1\leq j\leq m} c_j^{-1}\Big)\sqrt{x_1^2 + \cdots + x_n^2} \tag{1.4.238}$$

for each $x = (x_1, \dots, x_n) \in \mathbb{R}^n$, and such that

$$B_{\|\cdot\|_{\mathscr{V}}}(x,r) \notin \tau_{\|\cdot\|_{\mathscr{V}}} \text{ for all } x \in \mathbb{R}^n \text{ and all } r \in (0,\infty). \tag{1.4.239}$$

Conclude that

$$\|\cdot\|_{\mathscr{V}} : (\mathbb{R}^n, \tau_{\|\cdot\|_{\mathscr{V}}}) \to [0,\infty) \text{ is not a continuous function.} \tag{1.4.240}$$

Moving on, we make a number of important definitions.

Exercise 1.4.23. Define the `quasi-norm` of a given linear operator T from a quasi-normed space $(X, \|\cdot\|_X)$ into a quasi-normed space $(Y, \|\cdot\|_Y)$ as (the possibly infinite quantity)

$$\|T\|_{X\to Y} := \sup_{x\in X\setminus\{0\}} \frac{\|Tx\|_Y}{\|x\|_X} \in [0,\infty] \tag{1.4.241}$$

(one occasionally simply writes $\|T\|$ if the spaces X, Y are clear from the context). Call T, as above, `bounded` if $\|T\|_{X\to Y} < \infty$. Show that the collection of linear and bounded operators from $(X, \|\cdot\|_X)$ to $(Y, \|\cdot\|_Y)$ is a vector space, and that $\|\cdot\|_{X\to Y}$ is a quasi-norm on it. Show that the latter is a genuine norm whenever $(X, \|\cdot\|_X)$ and $(Y, \|\cdot\|_Y)$ are actually normed vector spaces.

Finally, prove that for every linear operator T from a quasi-normed space $(X, \|\cdot\|_X)$ into a quasi-normed space $(Y, \|\cdot\|_Y)$ one has

$$\|T\|_{X\to Y} = \sup\{\|Tx\|_Y : x \in X \text{ and } \|x\|_X \leq 1\}. \tag{1.4.242}$$

Definition 1.4.24. *A linear operator T from a quasi-normed space $(X, \|\cdot\|_X)$ into a quasi-normed space $(Y, \|\cdot\|_Y)$ is said to be an* `isomorphism onto its image` *provided it is bounded, injective, and its inverse mapping T^{-1} (which assigns to each vector y belonging to the range of T the unique vector $x \in X$ such that $Tx = y$) is bounded from the range of T (equipped with the quasi-norm in Y) into X. Simply call such a mapping T an* `isomorphism` *if, in addition, T is also onto. Two quasi-normed spaces X and Y are said to be* `isomorphic` *if there exists an isomorphism of X onto Y.*

We remark that a linear operator T from a quasi-normed space X into a quasi-normed space Y is an isomorphism in the sense of Definition 1.4.24 if and only if it is a vector space isomorphism (in the sense of Definition 1.3.23) with the property that both T and T^{-1} are bounded (in the sense discussed in Exercise 1.4.23).

Proposition 1.4.25. *Let T be a linear operator from a quasi-normed space X into a quasi-normed space Y. Then T is an isomorphism onto its image if and only if there are positive finite constants C_1 and C_2 such that*

$$C_1\|x\| \leq \|Tx\| \leq C_2\|x\|, \qquad \forall\, x \in X. \tag{1.4.243}$$

Proof. The proposition trivially holds when $X = \{0\}$, so it may be assumed that $X \neq \{0\}$. It follows that T and T^{-1} are both nonzero whenever T is an isomorphism onto its image, giving them both positive quasi-norms. Hence if T is an isomorphism onto its image, then

$$\|Tx\| \leq \|T\|\,\|x\| \ \text{ and } \ \|x\| = \|T^{-1}(Tx)\| \leq \|T^{-1}\|\,\|Tx\|, \quad \forall\, x \in X, \tag{1.4.244}$$

ergo letting $C_1 := \|T^{-1}\|^{-1}$ and $C_2 := \|T\|$ entails the desired inequalities in (1.4.243).

In the converse direction, assume there are positive real numbers C_1 and C_2 satisfying (1.4.243). Then the second inequality in (1.4.243) shows that T is bounded while the first inequality yields that T must be injective (since $\|Tx\| = 0$ forces $\|x\| = 0$). Since

$$\|T^{-1}(Tx)\| = \|x\| \leq C_1^{-1}\|Tx\| \ \text{ for each } \ x \in X, \tag{1.4.245}$$

the operator T^{-1} is bounded (from the image of T equipped with the quasi-norm in Y, into X). Thus, T is an isomorphism onto its image, and this completes the proof of Proposition 1.4.25. □

We conclude this section by discussing some related concepts for p-normed spaces.

Definition 1.4.26. *Given a vector space X equipped with a p-norm $\|\cdot\|$ for some $0 < p \leq 1$, and given a linear transformation $T : X \to X$, define (much as in the case of quasi-normed spaces; compare with* (1.4.242)*)*

$$\|T\| := \sup\{\|Tx\| : x \in X \text{ and } \|x\| \leq 1\} \in [0, \infty], \tag{1.4.246}$$

and refer to $\|T\|$ as the `operator` p-`norm` *(or simply* `norm` *if $p = 1$) of T. Call T* `bounded` *(on X) if $\|T\| < \infty$.*

Exercise 1.4.27. In the context of Definition 1.4.26, check that for each $x \in X \setminus \{0\}$ one has

$$\frac{\|Tx\|}{\|x\|} = \left\| T\left(\frac{x}{\|x\|^{1/p}}\right)\right\| \quad \text{and} \quad \left\|\frac{x}{\|x\|^{1/p}}\right\| = \left(\frac{1}{\|x\|^{1/p}}\right)^p \|x\| = 1, \tag{1.4.247}$$

then use this to show that

$$\|T\| = \sup_{x\in X\setminus\{0\}} \frac{\|Tx\|}{\|x\|}. \tag{1.4.248}$$

Conclude from (1.4.248) that

$$\|Tx\| \leq \|T\|\|x\| \text{ for each } x \in X. \tag{1.4.249}$$

Also, prove that that the collection of all linear and bounded operators from the p-normed space $(X, \|\cdot\|)$ into itself is a vector space on which operator p-norm introduced in (1.4.246) is an actual p-norm.

A basic result is that, for linear operators on p-normed spaces, being a continuous function with respect to the p-norm topology (in the sense of Definition 1.1.7) is equivalent with said linear operator being bounded, in the sense of Definition 1.4.26.

Proposition 1.4.28. *Let $(X, \|\cdot\|)$ be a p-normed space for some $0 < p \leq 1$, and suppose T is a linear operator T from X into itself. Then, with the p-norm topology τ on X considered as in Definition 1.4.6, it follows that*

$$\begin{gathered} T : (X, \tau) \to (X, \tau) \ \textit{ is a continuous function (in the sense of Definition 1.1.7)} \\ \textit{if and only if} \\ T : X \to X \textit{ is bounded in the sense of Definition 1.4.26 (i.e., } \|T\| < \infty). \end{gathered} \tag{1.4.250}$$

Proof. The leftward implication in (1.4.250) follows directly from the way τ is introduced in Definition 1.4.6, together with the fact that, as shown in (1.4.249),

$$\|Tx - Ty\| \leq \|T\|\|x - y\| \quad \text{for all } x, y \in X. \tag{1.4.251}$$

To establish the rightward implication in (1.4.250), note from Proposition 1.4.7 that the ball $B(0, 1)$, defined in (1.4.213) with $x = 0$ and $r = 1$, is a neighborhood of 0 in the p-norm topology on X. Since $T0 = 0$, Definition 1.1.7 ensures the existence of some $V \in \mathcal{N}(0; \tau)$ such that $T(V) \subseteq B(0, 1)$. Moreover, by Definition 1.4.6, there exists $r \in (0, \infty)$ with $B(0, r) \subseteq V$, hence

$$T(B(0, r)) \subseteq B(0, 1). \tag{1.4.252}$$

Now, for every $\lambda \in (0, r)$ and every $x \in X \setminus \{0\}$, the vector $\lambda^{1/p}\frac{x}{\|x\|^{1/p}}$ lies in $B(0, r)$, and therefore, by (1.4.252), its image under T belongs to $B(0, 1)$. Consequently,

$$\left\| T\left(\lambda^{1/p}\frac{x}{\|x\|^{1/p}}\right)\right\| < 1, \tag{1.4.253}$$

which implies $\|Tx\| < \lambda^{-1}\|x\|_X$. Combining this with (1.4.248), we deduce

$$\|T\| = \sup_{x\in X\setminus\{0\}} \frac{\|Tx\|}{\|x\|} \leq r^{-1}, \tag{1.4.254}$$

and hence $\|T\| < \infty$, as required. □

Reassuringly, a result similar to Proposition 1.4.28 is also valid in the context of quasi-normed spaces, but this will have to wait until later (see Proposition 1.11.3). We end with the following:

> **Epilogue:** Both quasi-norms and p-norms (with $0 < p < 1$) may be viewed as "defective norms," though each fails in a different respect: a quasi-norm satisfies only a weakened form of the triangle inequality, whereas a p-norm distorts homogeneity through the appearance of the p-th power. Thus, although related, these ways of measuring size in a vector space are rather different in nature. For instance, balls determined by a p-norm are open in the topology induced by that p-norm, whereas balls associated with a quasi-norm need not be open in the topology generated by the quasi-norm (see Proposition 1.4.20). Of course, every genuine norm is simultaneously a quasi-norm and a 1-norm. Moreover, as noted in Exercise 1.4.14, if $\|\cdot\|$ is a p-norm for some $p \in (0, 1]$, then $\|\cdot\|^{1/p}$ is a quasi-norm. The reverse direction (i.e., constructing a p-norm from a given quasi-norm) is considerably more delicate. We shall return to this topic in §1.11, where we will see that, despite not necessarily being an open set, a ball associated with a quasi-norm is always a neighborhood of its center (again, in the topology induced by said quasi-norm).

(1.4.255)

1.5 Hilbert Spaces

This section covers Hilbert spaces and explores their properties, including inner products, orthogonality, and the Pythagorean Theorem. The discussion highlights the importance of the Hilbert space structure in functional analysis, and its role in the study of linear operators. The reader is reminded that $\mathbb{K}$ stands for the fields of real or complex numbers, $\mathbb{R}$ or $\mathbb{C}$, respectively.

Definition 1.5.1. *Let X be a vector space over the field $\mathbb{K}$. An* `inner product` *on X is a mapping*

$$\langle\cdot,\cdot\rangle : X\times X \longrightarrow \mathbb{K} \tag{1.5.256}$$

satisfying the following properties:

(1) $\langle \alpha x + \beta y, z\rangle = \alpha\langle x, z\rangle + \beta\langle y, z\rangle, \quad \forall\, x, y, z \in X$ *and* $\forall\, \alpha, \beta \in \mathbb{K}$.

(2) $\langle y, x\rangle = \overline{\langle x, y\rangle}$, $\forall\, x, y \in X$, *where bar denotes complex conjugation.*
(3) $\langle x, x\rangle \in (0, \infty)$, $\forall\, x \in X \setminus \{0\}$.

The pair $(X, \langle\cdot,\cdot\rangle)$ *of a vector space* X *and an inner product* $\langle\cdot,\cdot\rangle$ *on* X *is called a* `pre-Hilbert space` *over the field* $\mathbb{K}$.

If $(X, \langle\cdot,\cdot\rangle)$ is a pre-Hilbert space over the field $\mathbb{K}$, parts *(1)-(2)* of Definition 1.5.1 readily yield

$$\langle x, \alpha y + \beta z\rangle = \overline{\alpha}\langle x, y\rangle + \overline{\beta}\langle x, z\rangle, \quad \forall\, x, y, z \in X \text{ and } \forall\, \alpha, \beta \in \mathbb{K}. \tag{1.5.257}$$

In particular (corresponding to $\alpha = \beta = 0$),

$$\langle x, 0\rangle = \langle 0, x\rangle = 0, \quad \forall\, x \in X. \tag{1.5.258}$$

Theorem 1.5.2 (Schwarz Inequality). *Suppose that* $(X, \langle\cdot,\cdot\rangle)$ *is a pre-Hilbert space over the field* $\mathbb{K}$. *Then*

$$|\langle x, y\rangle|^2 \leq \langle x, x\rangle \cdot \langle y, y\rangle, \quad \forall\, x, y \in X. \tag{1.5.259}$$

Proof. Let $x, y \in X$. Note that (1.5.259) trivially holds if $\langle x, y\rangle = 0$. Assume next that $\alpha := \langle x, y\rangle \neq 0$. Using properties *(1)* and *(2)* from the definition of the inner product in Definition 1.5.1 (along with the observation made in (1.5.257)), for every $\beta \in \mathbb{K}$, on has

$$0 \leq \langle \beta x + y, \beta x + y\rangle = |\beta|^2\langle x, x\rangle + 2\,\mathrm{Re}\,(\alpha\beta) + \langle y, y\rangle. \tag{1.5.260}$$

Since $\alpha \neq 0$, the observation made in (1.5.258) ensures $x \neq 0$ which, in turn, implies that $\langle x, x\rangle \neq 0$ (by part *(3)* of Definition 1.5.1). Hence (1.5.259) becomes apparent upon substituting β in (1.5.260) by the complex number $-\overline{\alpha}\langle x, x\rangle^{-1}$. □

The next proposition shows that every pre-Hilbert space may be turned into a normed vector space via its inner product.

Proposition 1.5.3. *If* $(X, \langle\cdot,\cdot\rangle)$ *is a pre-Hilbert space over the field* $\mathbb{K}$, *then the mapping*

$$\|\cdot\| : X \to \mathbb{R} \text{ defined by } \|x\| := \sqrt{\langle x, x\rangle}, \qquad \forall\, x \in X, \tag{1.5.261}$$

is a norm on X.

Proof. Parts *(ii)-(iii)* of Definition 1.4.4 of a norm are immediate. Turning to part *(i)* of this definition, for every fixed $x, y \in X$ we may write

$$\|x + y\|^2 = \langle x + y, x + y\rangle = \|x\|^2 + 2\,\mathrm{Re}\langle x, y\rangle + \|y\|^2. \tag{1.5.262}$$

However, $|\mathrm{Re}\langle x, y\rangle| \leq |\langle x, y\rangle| \leq \|x\|\,\|y\|$, where the latter inequality follows from Theorem 1.5.2. This, together with (1.5.262), implies that

$$\|x + y\|^2 \leq \|x\|^2 + 2\,\|x\| \cdot \|y\| + \|y\|^2 = (\|x\| + \|y\|)^2. \tag{1.5.263}$$

This and the fact that $x, y \in X$ were arbitrary, gives the triangle inequality from *(i)* in Definition 1.4.4. This establishes that $\|\cdot\|$ is a norm on X, finishing the proof of Proposition 1.5.3. □

Definition 1.5.4. *A pre-Hilbert space* $(X, \langle\cdot,\cdot\rangle)$ *over the field* $\mathbb{K}$ *is called a* `Hilbert space` *over the field* $\mathbb{K}$ *provided the normed vector space* $(X, \|\cdot\|)$ *is a Banach space, where* $\|\cdot\|$ *is the norm in* (1.5.261).

Proposition 1.5.5. *Assume that* $(H, \langle\cdot,\cdot\rangle)$ *is a Hilbert space over the field* $\mathbb{K}$ *and that* $\{x_n\}_{n\in\mathbb{N}}$ *and* $\{y_n\}_{n\in\mathbb{N}}$ *are two sequences in* H *satisfying*

$$x_n \to x \text{ as } n \to \infty \text{ and } y_n \to y \text{ as } n \to \infty, \tag{1.5.264}$$

for some $x, y \in H$. *Then the sequence* $\{\langle x_n, y_n\rangle\}_{n\in\mathbb{N}}$ *converges to* $\langle x, y\rangle$ *in* $\mathbb{C}$ *as* $n \to \infty$.

Proof. Denote by $\|\cdot\|$ the norm associated with the inner product on H, as in (1.5.261). Since $y_n \to y$ as $n \to \infty$, the continuity of the norm (cf. Proposition 1.4.7) immediately gives that $\|y_n\| \to \|y\|$ as $n \to \infty$. Therefore,

$$\begin{aligned} |\langle x_n, y_n\rangle - \langle x, y\rangle| &= |\langle x_n - x, y_n\rangle + \langle x, y_n - y\rangle| \\ &\le \|x_n - x\|\,\|y_n\| + \|x\|\,\|y_n - y\| \longrightarrow 0, \quad \text{as } n \to \infty, \end{aligned} \tag{1.5.265}$$

as wanted. □

Theorem 1.5.6 (Parallelogram Law). *Let* $(H, \langle\cdot,\cdot\rangle)$ *be a Hilbert space over the field* $\mathbb{K}$ *and denote by* $\|\cdot\|$ *the norm induced by its inner product as in* (1.5.261). *Then, for each* $x, y \in H$ *one has:*

$$\|x + y\|^2 + \|x - y\|^2 = 2(\|x\|^2 + \|y\|^2). \tag{1.5.266}$$

Proof. Fix $x, y \in H$ and using (1.5.261) along with the basic properties of the inner product $\langle\cdot,\cdot\rangle$ write

$$\begin{aligned} \|x + y\|^2 + \|x - y\|^2 &= \langle x + y, x + y\rangle + \langle x - y, x - y\rangle \\ &= \langle x, x\rangle + \langle x, y\rangle + \langle y, x\rangle + \langle y, y\rangle + \langle x, x\rangle - \langle x, y\rangle - \langle y, x\rangle + \langle y, y\rangle \\ &= 2\langle x, x\rangle + 2\langle y, y\rangle \\ &= 2(\|x\|^2 + \|y\|^2), \end{aligned} \tag{1.5.267}$$

yielding (1.5.266), as desired. □

Definition 1.5.7. *Assume that* $(H, \langle\cdot,\cdot\rangle)$ *is a Hilbert space over the field* $\mathbb{K}$.

(i) An element $x \in H$ *is said to be* `orthogonal`, *or* `perpendicular`, *to another element* $y \in Y$ *provided* $\langle x, y\rangle = 0$. *If* $x, y \in H$ *are such that* x *is orthogonal to* y, *this shall be denoted by* $x \perp y$.

(2) If $E \subseteq H$, define the `orthogonal complement` *of E, $E^\perp$, as the collection of all elements in H that are orthogonal to every point in E; i.e.,*

$$E^\perp := \{x \in H \,:\, \langle x, y\rangle = 0, \ \forall\, y \in E\}. \tag{1.5.268}$$

It is straightforward to show that $E^\perp$ is a closed subspace of H in the topology induced by the norm (1.5.261), for every subset E of a Hilbert space H. Indeed, this follows from Proposition 1.5.5 and the linearity of the inner product on H in its first argument.

Exercise 1.5.8. Let $(H, \langle\cdot,\cdot\rangle)$ be a Hilbert space.

(1) Show that if $E \subseteq H$ then $\overline{E} = (E^\perp)^\perp$, where the closure of E is taken in the topology induced by the norm defined in (1.5.261).
(2) Show that if E is a closed subspace of H then $E = (E^\perp)^\perp$.

Theorem 1.5.9 (Pythagorean Theorem). *Let $(H, \langle\cdot,\cdot\rangle)$ be a Hilbert space and consider a collection $x_1, \dots, x_n \in H$ of mutually orthogonal vectors, i.e.,*

$$x_j \perp x_k \ \textit{for every} \ \ j, k \in \{1, \dots, n\} \ \ \textit{such that} \ \ j \neq k. \tag{1.5.269}$$

Then the following identity holds:

$$\Big\|\sum_{i=1}^n x_i\Big\|^2 = \sum_{i=1}^n \|x_i\|^2. \tag{1.5.270}$$

Proof. Let $x_1, \dots, x_n \in H$ satisfying (1.5.269). We may therefore write

$$\begin{aligned}\Big\|\sum_{i=1}^n x_i\Big\|^2 &= \Big\langle \sum_{j=1}^n x_j, \sum_{k=1}^n x_k \Big\rangle \\ &= \sum_{j,k=1}^n \langle x_j, x_k\rangle = \sum_{j=1}^n \langle x_j, x_j\rangle = \sum_{j=1}^n \|x_j\|^2, \end{aligned} \tag{1.5.271}$$

where the second-to-last equality uses (1.5.269). This completes the proof of the theorem. □

Theorem 1.5.10. *If M is a closed subspace of a Hilbert space $(H, \langle\cdot,\cdot\rangle)$ over the field $\mathbb{K}$, then*

$$H = M + M^\perp \ \ \textit{and} \ \ M \cap M^\perp = \{0\}. \tag{1.5.272}$$

Proof. Let M be a closed subspace of H and x an arbitrary point in H. Set

$$\delta := \inf\,\{\|x - y\| : y \in M\}, \tag{1.5.273}$$

and let $\{y_n\}_{n\in\mathbb{N}} \subseteq M$ such that

$$\|x - y_n\| \to \delta \ \text{as} \ n \to \infty. \tag{1.5.274}$$

We claim that the sequence $\{y_n\}_{n\in\mathbb{N}}$ is Cauchy. To see this, start by using Theorem 1.5.6 to write for each $n, m \in \mathbb{N}$

$$2(\|y_n - x\|^2 + \|y_m - x\|^2) = \|y_n - y_m\|^2 + \|y_n + y_m - 2x\|^2. \tag{1.5.275}$$

Also, observe that for each $n, m \in \mathbb{N}$ we have $\frac{1}{2}(y_n + y_m) \in M$ hence

$$\left\|\tfrac{1}{2}(y_n + y_m) - x\right\| \geq \delta \tag{1.5.276}$$

by the definition of δ. Therefore, by (1.5.275)-(1.5.276), for each $n, m \in \mathbb{N}$ we have

$$\begin{aligned}\|y_n - y_m\|^2 &= 2\|y_n - x\|^2 + 2\|y_m - x\|^2 - 4\left\|\tfrac{1}{2}(y_n + y_m) - x\right\|^2 \\ &\leq 2\|y_n - x\|^2 + 2\|y_m - x\|^2 - 4\delta^2.\end{aligned} \tag{1.5.277}$$

Thus, (1.5.277) and (1.5.274) entail that $\{y_n\}_{n\in\mathbb{N}}$ is a Cauchy sequence in M, as desired. Consequently, since M is closed, there exists $y \in M$ such that $\lim_{n\to\infty} y_n = y$. This, together with (1.5.274) and Proposition 1.5.5 yield

$$\|x - y\| = \sqrt{\langle x - y, x - y\rangle} = \lim_{n\to\infty} \sqrt{\langle x - y_n, x - y_n\rangle} = \delta. \tag{1.5.278}$$

Introduce next $z := x - y \in H$, where $y \in M$ is the limit of the sequence $\{y_n\}_{n\in\mathbb{N}}$. Our goal is to show that

$$z \in M^\perp. \tag{1.5.279}$$

We start by letting let $\lambda_u := \langle z, u\rangle \in \mathbb{K}$ for each $u \in M$. Then

$$\langle z, \lambda_u u\rangle = \overline{\lambda_u}\langle z, u\rangle = |\langle z, u\rangle|^2 \in \mathbb{R}, \quad \forall\, u \in M. \tag{1.5.280}$$

Thus, if $u \in M$ is fixed, the function $f_u : \mathbb{R} \longrightarrow \mathbb{K}$ given by

$$f_u(t) := \|z + t(\lambda_u u)\|^2 = \|z\|^2 + 2t\langle z, \lambda_u u\rangle + t^2\|\lambda_u u\|^2, \quad \forall\, t \in \mathbb{R}, \tag{1.5.281}$$

is in fact real valued. Note that since $z + t(\lambda_u u) = x - (y - t(\lambda_u u))$ and $y - t(\lambda_u u) \in M$, then, by (1.5.273),

$$f_u(t) = \left\|x - (y - t(\lambda_u u))\right\| \geq \delta^2 \qquad \forall\, u \in M \text{ and } \forall\, t \in \mathbb{R}. \tag{1.5.282}$$

However, using (1.5.278), we have that $\delta^2 = f_u(0)$. Thus, using (1.5.282), the quadratic function f_u attains its minimum at $t = 0$ and therefore $f_u'(0) = 0$. Differentiating in (1.5.281) this yields

$$0 = 2\langle z, \lambda_u u\rangle = 2|\langle z, u\rangle|^2, \tag{1.5.283}$$

where the last equality comes from (1.5.280). Hence $\langle z, u\rangle = 0$, proving (1.5.279). As a consequence,

$$x = y + (x - y) = y + z \in M + M^\perp. \tag{1.5.284}$$

Since $x \in H$ was arbitrary this shows that $H = M + M^{\perp}$, proving the first part in (1.5.272).

At this point, it remains to verify that $M \cap M^{\perp} = \{0\}$. Indeed, if $x \in M \cap M^{\perp}$, then $\langle x, x \rangle = 0$, hence $x = 0$, as desired. □

Theorem 1.5.11. *Let $(H, \langle \cdot, \cdot \rangle)$ be a Hilbert space over the field $\mathbb{K}$ and $f : H \to \mathbb{K}$ be a continuous linear functional. Then there exists a unique vector $y \in H$ with the property that*

$$f(x) = \langle x, y \rangle \ \text{for each} \ \ x \in H. \tag{1.5.285}$$

Proof. Let $f : H \to \mathbb{K}$ be a continuous linear functional. Start with the case when $f \equiv 0$. In such a scenario take $y := 0$, for which (1.5.285) clearly is satisfied. In addition, if $y \in H$ is such that (1.5.285) holds then, in particular,

$$0 = f(y) = \langle y, y \rangle, \tag{1.5.286}$$

which, by property *(3)* from Definition 1.5.1, forces $y = 0$.

Consider next the case when $f : H \to \mathbb{K}$ is a non-zero continuous linear functional. Then by Proposition 1.3.18 and the continuity of f it follows that $\ker f$ is a closed proper subspace of H. Using Theorem 1.5.10 then $\ker f$ has an orthogonal complement $(\ker f)^{\perp}$ which is non-trivial. Pick some $z \in (\ker f)^{\perp} \setminus \{0\}$ and by normalizing this appropriately, we may assume that $\|z\| = 1$. For each $x \in H$ set $u(x) := f(x)z - f(z)x \in H$ and note that $u(x) \in \ker f$, by the linearity of f. Thus, using (1.5.257), we obtain

$$0 = \langle u(x), z \rangle = f(x) \cdot \|z\|^2 - f(z) \cdot \langle x, z \rangle = f(x) - \langle x, \overline{f(z)} z \rangle \qquad \forall\, x \in H. \tag{1.5.287}$$

This shows that if one sets $y := \overline{f(z)} z$ then (1.5.285) holds.

As far as the uniqueness of y as in (1.5.285) is concerned, assume the vectors $y_1, y_2 \in H$ are such that $\langle x, y_1 \rangle = \langle x, y_2 \rangle$ for each $x \in H$. Then take $x := y_1 - y_2$ and write

$$0 = \langle y_1 - y_2, y_1 \rangle - \langle y_1 - y_2, y_2 \rangle = \langle y_1 - y_2, y_1 - y_2 \rangle = \|y_1 - y_2\|^2. \tag{1.5.288}$$

Consequently $y_1 = y_2$, showing the uniqueness of $y \in H$ satisfying (1.5.285), and completing the proof of the theorem. □

Exercise 1.5.12. Let $(H, \langle \cdot, \cdot \rangle)$ be a Hilbert space, and consider a linear and continuous operator $T : H \to H$. Show that there exists a (unique) linear and continuous operator $T^* : H \to H$, called the `adjoint` of T, with the property that

$$\langle Tx, y \rangle = \langle x, T^* y \rangle \ \text{ for each } \ x, y \in H. \tag{1.5.289}$$

Hint: Use Theorem 1.5.11.

Definition 1.5.13. *Let $(H, \langle \cdot, \cdot \rangle)$ be a Hilbert space over $\mathbb{K}$. A subset $\{u_n\}_{n \in \mathbb{N}} \subseteq H$ is said to be an* `orthonormal sequence` *provided $\|u_n\| = 1$ for each $n \in \mathbb{N}$, and $u_n \perp u_m$ whenever $n, m \in \mathbb{N}$ are such that $n \neq m$.*

Theorem 1.5.14 (Bessel's Inequality). *Suppose that* $(H, \langle\cdot,\cdot\rangle)$ *is a Hilbert space over the field* $\mathbb{K}$ *and that* $\{u_n\}_{n\in\mathbb{N}}$ *is an orthonormal sequence in* H. *Then*

$$\sum_{n\in\mathbb{N}} |\langle x, u_n\rangle|^2 \le \|x\|^2, \qquad \forall\, x \in H. \tag{1.5.290}$$

Proof. Fix $x \in H$ and $k \in \mathbb{N}$ and write

$$\begin{aligned}
0 &\le \Big\| x - \sum_{n=1}^{k} \langle x, u_n\rangle u_n \Big\|^2 \qquad (1.5.291)\\
&= \|x\|^2 - 2\,\mathrm{Re}\,\langle x, \sum_{n=1}^{k} \langle x, u_n\rangle u_n\rangle + \Big\| \sum_{n=1}^{k} \langle x, u_n\rangle u_n \Big\|^2\\
&= \|x\|^2 - 2 \sum_{n=1}^{k} |\langle x, u_n\rangle|^2 + \sum_{n=1}^{k} |\langle x, u_n\rangle|^2\\
&= \|x\|^2 - \sum_{n=1}^{k} |\langle x, u_n\rangle|^2,
\end{aligned}$$

where the second equality is justified by Theorem 1.5.9. Thus

$$\sum_{n=1}^{k} |\langle x, u_n\rangle|^2 \le \|x\|^2, \qquad \forall\, x \in H, \quad \forall\, k \in \mathbb{N}. \tag{1.5.292}$$

Letting $k \to \infty$ in (1.5.292) immediately yields (1.5.290). □

Proposition 1.5.15. *Let* $\mathbb{K}$ *be the field* $\mathbb{R}$ *or* $\mathbb{C}$ *and assume that* $n \in \mathbb{N}$ *is given. Show that any two norms on the vector space* $\mathbb{K}^n$ *(over the field* $\mathbb{K}$*) are equivalent.*

Proof. Denote by $\|\cdot\|$ the canonical norm on $\mathbb{K}^n$, i.e.,

$$\|x\| := \sqrt{|x_1|^2 + \cdots + |x_n|^2}, \qquad \forall\, x = (x_1, \ldots, x_n) = x_1 e_1 + \ldots x_n e_n \in \mathbb{K}^n, \tag{1.5.293}$$

where $\{e_1, \ldots, e_n\}$ denotes the canonical basis in $\mathbb{K}^n$ as a vector field over $\mathbb{K}$. Then, if $|||\cdot|||$ is another norm on $\mathbb{K}^n$ then, for each $x = (x_1, \ldots, x_n) = x_1 e_1 + \ldots x_n e_n \in \mathbb{K}^n$ the following holds

$$\begin{aligned}
|||x||| = |||x_1 e_1 + \ldots x_n e_n||| &\le \sum_{j=1}^{n} |x_j| \cdot |||e_j||| \qquad (1.5.294)\\
&\le \Big(\sum_{j=1}^{n} |||e_j|||^2\Big)^{1/2} \cdot \Big(\sum_{j=1}^{n} |x_j|^2\Big)^{1/2}\\
&= C\|x\|,
\end{aligned}$$

where

$$C := \Big(\sum_{j=1}^{n} |||e_j|||^2\Big)^{1/2} \in (0, \infty). \tag{1.5.295}$$

Indeed, the first inequality in (1.5.294) is a consequence of the triangle inequality and homogeneity properties of the norm $|||\cdot|||$, the second inequality follows from Schwarz' inequality from Theorem 1.5.2 in the Hilbert space $\mathbb{R}^n$. All in all, we have shown that

$$|||x||| \leq C\|x\|, \quad \forall\, x \in \mathbb{K}^n, \tag{1.5.296}$$

where C is as in (1.5.295). Our goal is to show next that

$$\exists\, c \in (0, \infty) \text{ such that } \|x\| \leq c|||x|||, \quad \forall\, x \in \mathbb{K}^n, \tag{1.5.297}$$

and our first step in this direction is to note that on grounds of (1.5.296) it follows immediately that the identity function id : $(\mathbb{K}^n, \|\cdot\|) \longrightarrow (\mathbb{K}^n, |||\cdot|||)$ is continuous. Combining this with the continuity property of the norm $|||\cdot|||$, cf. Proposition 1.4.7, we may further conclude that the mapping

$$T : (\mathbb{K}^n, \|\cdot\|) \longrightarrow \mathbb{R} \text{ given by } T(x) := |||x|||, \quad \forall\, x \in \mathbb{K}^n, \tag{1.5.298}$$

is continuous. Since the set

$$S := \{x \in \mathbb{K}^n : \|x\| = 1\} \text{ is compact in } (\mathbb{K}^n, \|\cdot\|), \tag{1.5.299}$$

the continuity of T from (1.5.298) entails that $T\big|_S$ attains its minimum, i.e.,

$$\exists\, x_o \in S \text{ such that } |||x||| \geq |||x_o||| \quad \forall\, x \in S. \tag{1.5.300}$$

Note that since $|||\cdot|||$ is a norm on $\mathbb{K}^n$ and the set S does not contain the zero vector, then $|||x_o||| > 0$. Thus, it is meaningful to set

$$c := \frac{1}{|||x_o|||} \in (0, \infty). \tag{1.5.301}$$

Then, for each $x \in \mathbb{K}^n \setminus \{0\}$ we have that $x/\|x\| \in S$ and as such, using (1.5.300), we have

$$|||(x/\|x\|)||| \geq |||x_o||| = 1/c, \tag{1.5.302}$$

from which it immediately follows that

$$c|||x||| \geq \|x\|, \quad \forall\, x \in \mathbb{K}^n \setminus \{0\}. \tag{1.5.303}$$

Since the inequality in (1.5.303) trivially holds when $x = 0$, this and (1.5.303) give (1.5.297).

Now, (1.5.295) and (1.5.297) show that the arbitrary norm $|||\cdot|||$ on $\mathbb{K}^n$ is equivalent with the canonical one given in (1.5.293). Finally, the claim in the statement of the proposition follows from the transitivity property of equivalence of norms. □

Exercise 1.5.16. Let X be a vector space over the field $\mathbb{K}$ and assume that X has finite dimension. Show that any two norms on X are equivalent.

Hint: Reason as in the proof of Proposition 1.5.15.

In principle, each norm induces a specific topology on a given vector space, though any two such "norm topologies" coincide if the corresponding norms are equivalent. We therefore have:

Corollary 1.5.17. *For each $n \in \mathbb{N}$, the vector space $\mathbb{K}^n$ has a unique norm topology, namely, the topology induced by the canonical norm (i.e., the Euclidean topology).*

The first of the next two corollaries is immediately obtained upon realizing that each normed vector space with finite dimension n must be isomorphic to the Euclidean n-space $\mathbb{K}^n$, itself a Banach space. The second then follows from the first, since each complete subset of a metric space is closed in that space.

Corollary 1.5.18. *Every finite-dimensional normed vector space is a Banach space.*

Corollary 1.5.19. *Every finite-dimensional subspace of a normed vector space is a closed subset of the space.*

1.6 The Rudiments of Linear Topological Spaces

This section, comprising eight subsections, may be regarded as a fast-track introduction to linear topological spaces. These are vector spaces equipped with a topology making the vector addition and scalar multiplication continuous operations. Here we review and develop some key concepts and results pertaining to this topic, which in turn are used as foundational material throughout the monograph. We begin with the following basic definition.

Definition 1.6.1. *Call a topology τ on a vector space X a* `vector topology` *if vector addition and scalar multiplication are continuous, i.e.,*

(1) the vector addition map $(x, y) \mapsto x + y$ maps $X \times X$ continuously into X;
(2) the scalar multiplication map $(\lambda, x) \mapsto \lambda x$ maps $\mathbb{K} \times X$ continuously into X.

Above, the product spaces $X \times X$ and $\mathbb{K} \times X$ are equipped with the natural product topologies (cf. Definition 1.1.15).

If τ is a vector topology on the vector space X, the pair (X, τ) (or just X itself when τ is irrelevant or understood) is said to be a `topological vector space`*, or a* `linear topological space`*.*

Note that if (X, τ) is a linear topological space and $x \in X$, $\lambda \in \mathbb{K} \setminus \{0\}$ are fixed, then the mappings $X \ni y \mapsto x + y \in X$ and $X \ni y \mapsto \lambda y \in X$ are homeomorphisms. (As a side-note, to contrast the continuity properties of the aforementioned maps from those in items *(1)-(2)* of Definition 1.6.1 we shall occasionally refer to the latter as *joint* continuity properties.) In particular,

$$A \in \tau \implies x + A \in \tau \quad \text{and} \quad \lambda A \in \tau. \tag{1.6.304}$$

Here, as usual, we have set

$$x + A := \{x + y : y \in A\} \quad \text{and} \quad \lambda A := \{\lambda y : y \in A\}. \tag{1.6.305}$$

Furthermore, for each $x \in X$

$$V \in \mathcal{N}(x;\tau) \iff -x + V \in \mathcal{N}(0;\tau), \tag{1.6.306}$$

and for each $\alpha \in \mathbb{K} \setminus \{0\}$ we have

$$V \in \mathcal{N}(0;\tau) \iff \alpha V \in \mathcal{N}(0;\tau). \tag{1.6.307}$$

We next discuss several basic examples of linear topological spaces.

Proposition 1.6.2. *Let* $0 < p \leq 1$. *Then every p-norm* $\|\cdot\|$ *on a vector space X induces a translation-invariant distance d on X, defined by*

$$d(x,y) := \|x-y\|, \qquad \forall\, x,y \in X. \tag{1.6.308}$$

The vector space X, equipped with the induced topology by the metric d, becomes a Hausdorff linear topological space.

Proof. The fact that d is a metric on X has been established in Proposition 1.4.5. Also, it is clear from its definition in (1.6.308) that d is translation invariant.

There remains to deal with the last part of the statement in Proposition 1.6.2. We first recall the induced topology τ_d from Proposition 1.2.5 and claim that the vector addition is continuous from $(X \times X, \tau_d \times \tau_d)$ into (X, τ_d). To see this it suffices to show that

$$\forall\, O \in \tau_d \;\text{ we have }\; f_+^{-1}(O) \in \tau_d \times \tau_d, \tag{1.6.309}$$

where $f_+(x,y) := x + y$ for each $x, y \in X$. To see this, pick $(y_1,y_2) \in f_+^{-1}(O)$. Thus $y_1 + y_2 \in O$ and, since $O \in \tau_d$, there exists $r > 0$ such that $B_d(y_1 + y_2, r) \subseteq O$. It is then an immediate application of the triangle inequality property for d to see that

$$B_d(y_1, r/2) \times B_d(y_2, r/2) \subseteq f_+^{-1}(O). \tag{1.6.310}$$

Now (1.6.310) combined with the fact that balls are open in τ_d (cf. item *(2)* in Proposition 1.2.5) shows that $f_+^{-1}(O) \in \mathcal{N}((y_1,y_2);\tau_d \times \tau_d)$. Since $(y_1,y_2) \in f_+^{-1}(O)$ was arbitrary, this completes the argument that $f_+^{-1}(O)$ is open. Using Proposition 1.1.8 this shows that vector addition is continuous.

Consider next the scalar multiplication $f_* : \mathbb{K} \times X \longrightarrow X$, given by $f_*(\alpha, x) = \alpha x$ for each $(\alpha, x) \in \mathbb{K} \times X$. In order to show that f_* is continuous we appeal again to Proposition 1.1.8 and reduce matters to proving that

$$\forall\, O \in \tau_d \ \text{ we have } \ f_*^{-1}(O) \in \tau_{\mathbb{K}} \times \tau_d, \tag{1.6.311}$$

where $\tau_{\mathbb{K}}$ stands for the Euclidean topology on $\mathbb{K}$ (recall that throughout $\mathbb{K}$ is either $\mathbb{R}$ or $\mathbb{C}$). To establish (1.6.311) fix $O \in \tau_d$ and $(\alpha, x) \in f_*^{-1}(O)$. In particular $\alpha x \in O$ and since $O \in \tau_d$ there exists $r > 0$ such that $B_d(\alpha x, r) \subseteq O$. Since for each pair $(\beta, y) \in \mathbb{K} \times X$ one may write that, as a consequence of the triangle inequality for d,

$$\begin{aligned} d(\alpha x, \beta y) &\leq d(\alpha x, \alpha y) + d(\alpha y, \beta y) \\ &= |\alpha|^p \cdot d(x, y) + |\alpha - \beta|^p \cdot \|y\| \\ &\leq |\alpha|^p \cdot d(x, y) + |\alpha - \beta|^p \cdot d(x, y) + |\alpha - \beta|^p \cdot \|x\|, \end{aligned} \tag{1.6.312}$$

where the last inequality above follows from the sub-additivity of $\|\cdot\|$ (cf., part *(1)* in Definition 1.4.1) and the definition of d from (1.6.308). Next, set

$$r_1 := \min\Big\{\frac{r}{3|\alpha|^p}, 1\Big\} \ \text{ and } \ r_2 := \min\Big\{\Big(\frac{r}{3}\Big)^{1/p}, \Big(\frac{r}{3\|x\|}\Big)^{1/p}\Big\}, \tag{1.6.313}$$

with the convention that $r/(3|\alpha|^p)$ is set to be $+\infty$ in the case when $\alpha = 0$ and, similarly, $r/(3\|x\|)$ is set to be $+\infty$ when $x = 0$. Then $r_1, r_2 \in (0, \infty)$ and (1.6.313) together with (1.6.312) yield the implication

$$(\beta, y) \in B_{\mathbb{K}}(\alpha, r_2) \times B_d(x, r_1) \implies \beta y \in B_d(\alpha x, r). \tag{1.6.314}$$

Here for $\eta \in \mathbb{K}$ and $t > 0$ we denoted by $B_{\mathbb{K}}(\eta, t)$ the Euclidean ball in $\mathbb{K}$ of center η and radius t. As a consequence of (1.6.314) we have

$$(\alpha, x) \in B_{\mathbb{K}}(\alpha, r_2) \times B_d(x, r_1) \in f_*^{-1}(O). \tag{1.6.315}$$

Denote by $\tau_{\mathbb{K}}$ the canonical topology on the field of scalars $\mathbb{K}$. Since by item *(2)* in Proposition 1.2.5 we have $B_{\mathbb{K}}(\alpha, r_2) \times B_d(x, r_1) \in \tau_{\mathbb{K}} \times \tau_d$ and the pair (α, x) was arbitrarily chosen in $f_*^{-1}(O)$, this ultimately gives (1.6.311). Thus the scalar multiplication f_* is continuous, as desired. Since the topology induced by any metric is Hausdorff (cf. Exercise 1.2.8), this completes the proof of Proposition 1.6.2. □

Example 1.6.3. From Proposition 1.6.2 we know that every normed vector space, with the metric topology induced by its norm, is a linear topological space. In particular, every Banach space, and hence every Hilbert space, is a Hausdorff linear topological space.

Example 1.6.4. The goal of this example is to describe a vector topology on $C(\mathbb{D})$, the space of continuous complex-valued functions on the unit disk in $\mathbb{C}$. To this end let $\mathbb{D}$ stand for the `unit disk` in $\mathbb{C}$, i.e.,

$$\mathbb{D} := \{z \in \mathbb{C} : |z| < 1\}, \tag{1.6.316}$$

and for each $n \in \mathbb{N}$ and each $f \in C(\mathbb{D})$ set

$$p_n(f) := \max\,\{|f(z)| : |z| \le 1 - 1/n\}. \tag{1.6.317}$$

Then the following hold.

(1) For each $n \in \mathbb{N}$, the function p_n is the *seminorm* on $C(\mathbb{D})$ (i.e., p_n is subadditive and homogeneous, but not necessarily non-degenerate; cf. Definition 1.4.10) which describes uniform convergence on the closed disk $\overline{\mathbb{D}}_n$, where we have denoted $\mathbb{D}_n := \{z \in \mathbb{D} : |z| < 1 - 1/n\}$. That is, given any integer $n \in \mathbb{N}$ along with a sequence $\{f_j\}_{j\in\mathbb{N}} \subseteq C(\mathbb{D})$ and $f \in C(\mathbb{D})$, then

$$\lim_{j\to\infty} p_n(f_j - f) = 0, \tag{1.6.318}$$

if and only if the sequence $\left\{f_j\big|_{\overline{\mathbb{D}}_n}\right\}_{j\in\mathbb{N}}$ converges to $f\big|_{\overline{\mathbb{D}}_n}$ uniformly.

(2) The function $d : C(\mathbb{D}) \times C(\mathbb{D}) \longrightarrow [0,\infty)$ given by

$$d(f,g) := \sum_{n=1}^{\infty} 2^{-n} \frac{p_n(f-g)}{1 + p_n(f-g)}, \qquad \forall\, f, g \in C(\mathbb{D}), \tag{1.6.319}$$

induces a complete metric on $C(\mathbb{D})$ for which a sequence converges precisely when it converges uniformly on compact subsets of $\mathbb{D}$. The resulting topology is a vector topology on $C(\mathbb{D})$ (called the topology of uniform convergence on compact sets of $\mathbb{D}$).

Proof. Clearly if $f, g \in C(\mathbb{D})$ and $|z| \le 1 - 1/n$ for some $n \in \mathbb{N}$, then

$$|f(z) + g(z)| \le |f(z)| + |g(z)| \le p_n(f) + p_n(g), \tag{1.6.320}$$

from which it immediately follows that $p_n(f+g) \le p_n(f) + p_n(g)$, i.e., p_n is subadditive. Also if $\alpha \in \mathbb{C}$ and $f \in C(\mathbb{D})$, then for each $n \in \mathbb{N}$,

$$\begin{aligned} p_n(\alpha f) &= \max\,\{|\alpha \cdot f(z)| : |z| \le 1 - 1/n\} \\ &= |\alpha| \cdot \max\,\{|f(z)| : |z| \le 1 - 1/n\} \\ &= |\alpha| \cdot p_n(f), \end{aligned} \tag{1.6.321}$$

showing that p_n is homogeneous. Thus p_n is a semi-norm on $C(\mathbb{D})$, for each $n \in \mathbb{N}$. Regarding the lack of non-degeneracy property of p_n for some fixed $n \in \mathbb{N}$, it is immediate that

$$f, g \in C(\mathbb{D}) \text{ are such that } p_n(f-g) = 0 \iff f\big|_{\overline{\mathbb{D}}_n} = g\big|_{\overline{\mathbb{D}}_n}, \tag{1.6.322}$$

without necessarily requiring that $f = g$ on $\mathbb{D}$.

Next fix $n \in \mathbb{N}$ and let $\{f_j\}_{j\in\mathbb{N}} \subseteq C(\mathbb{D})$ and $f \in C(\mathbb{D})$ be such that (1.6.318) holds. Fix an arbitrary $\varepsilon > 0$. Then there exists $J_{\varepsilon,n} \in \mathbb{N}$ such that $p_n(f_j - f) \le \varepsilon$ for each $j \in \mathbb{N}$ with $j \ge J_{\varepsilon,n}$. Using the definition of p_n from (1.6.317) this immediately yields

$$|f_j(z) - f(z)| \leq \varepsilon, \qquad \forall\, z \in \overline{\mathbb{D}}_n \text{ provided } j \geq J_{\varepsilon,n}. \tag{1.6.323}$$

In turn, (1.6.323) implies that $\{f_j\big|_{\overline{\mathbb{D}}_n}\}_{j\in\mathbb{N}}$ converges to $f\big|_{\overline{\mathbb{D}}_n}$ uniformly, as desired. The converse implication is proved similarly, finishing the proof of *(1)*.

Turning attention to *(2)* we start with the observation that since for each pair of functions $f, g \in C(\mathbb{D})$ and each $n \in \mathbb{N}$ one has

$$0 \leq \frac{p_n(f-g)}{1 + p_n(f-g)} < 1, \tag{1.6.324}$$

then the series in the right hand-side of (1.6.319) is controlled by the geometric series with ratio $1/2$ and, as such, is absolutely convergent. Clearly $d(f, g) = 0$ only if $p_n(f - g) = 0$ for each $n \in \mathbb{N}$. Employing (1.6.322) we obtain that

$$f\big|_{\overline{\mathbb{D}}_n} = g\big|_{\overline{\mathbb{D}}_n}, \qquad \forall\, n \in \mathbb{N}, \tag{1.6.325}$$

which in turn allows us to conclude that $f = g$ on $\mathbb{D}$. Conversely, it is immediate that if $f = g$ on $\mathbb{D}$ then $d(f, g) = 0$. Thus d satisfies the non-degeneracy property of distances. Also, one can see directly from definitions (1.6.319) and (1.6.317) that d is symmetric.

To complete the proof that d is a distance on $C(\mathbb{D})$ it remains to show that d satisfies the triangle inequality. However, based on (1.6.317), each p_n satisfies the triangle inequality, and it is an elementary task (see Exercise 1.2.3) to see that this further implies that

$$\frac{p_n(\cdot)}{1 + p_n(\cdot)} \quad \text{satisfies the triangle inequality on } C(\mathbb{D}). \tag{1.6.326}$$

Next, (1.6.326) and (1.6.319) permit us to conclude that d satisfies the triangle inequality. Combined with the non-degeneracy and symmetry properties established earlier, this shows that d is a distance on $C(\mathbb{D})$.

Next we prove that the metric space $(C(\mathbb{D}), d)$ is complete. With this goal in mind, consider a Cauchy sequence $\{f_j\}_{j\in\mathbb{N}} \subseteq (C(\mathbb{D}), d)$. Thus,

$$\begin{gathered} \text{for each } \varepsilon > 0 \text{ there exists } J_\varepsilon \in \mathbb{N} \text{ such that} \\ j, k \geq J_\varepsilon \implies d(f_j, f_k) < \varepsilon. \end{gathered} \tag{1.6.327}$$

Our first claim is that for each $n \in \mathbb{N}$ the sequence $\{f_j\}_{j\in\mathbb{N}} \subseteq (C(\mathbb{D}), d)$ is Cauchy in the seminorm p_n, i.e.,

$$\begin{gathered} \text{for each } n \in \mathbb{N} \text{ and each } \varepsilon > 0 \text{ there exists } J_{\varepsilon,n} \in \mathbb{N} \text{ such that} \\ j, k \geq J_{\varepsilon,n} \implies p_n(f_j, f_k) < \varepsilon. \end{gathered} \tag{1.6.328}$$

To justify this claim, let us first notice that, since $[0, 1) \ni t \mapsto F(t) := t/(1 - t)$ is an increasing function, given any $\varepsilon > 0$ there exists $\delta_\varepsilon \in (0, 1)$ such that the following implication holds

$$[0,\delta_\varepsilon) \ni t \implies F(t) = t/(1-t) \in [0,\varepsilon). \tag{1.6.329}$$

Fix $n \in \mathbb{N}$ and $\varepsilon > 0$ and let $\delta_\varepsilon > 0$ be such that the implication in (1.6.329) is valid. Employing (1.6.327) with $2^{-n} \cdot \delta_\varepsilon$ in place of ε yields the existence of $J_{\varepsilon,n} \in \mathbb{N}$ such that the following implication holds

$$j,k \in \mathbb{N},\ j,k \geq J_{\varepsilon,n} \implies d(f_j,f_k) < 2^{-n}\delta_\varepsilon. \tag{1.6.330}$$

In particular, using the definition of d from (1.6.319) we have

$$d(f,g) \geq 2^{-n} \cdot \frac{p_n(f,g)}{1+p_n(f,g)} \quad \text{for each} \quad f,g \in C(\mathbb{D}). \tag{1.6.331}$$

Consequently

$$\frac{p_n(f_j,f_k)}{1+p_n(f_j,f_k)} \in [0,\delta_\varepsilon) \ \text{ for each } \ j,k \geq J_{\varepsilon,n} \tag{1.6.332}$$

and, combining this with (1.6.329), we may conclude that

$$p_n(f_j,f_k) = F\left(\frac{p_n(f,g)}{1+p_n(f,g)}\right) < \varepsilon \qquad \forall\, j,k \geq J_{\varepsilon,n}, \tag{1.6.333}$$

completing the proof of the claim made in (1.6.328).

An immediate consequence of (1.6.328) is the fact that, for each $z \in \mathbb{D}$, the sequence of complex numbers $\{f_j(z)\}_{j\in\mathbb{N}}$ is Cauchy. Indeed, for each $z \in \mathbb{D}$ there exists $n \in \mathbb{N}$ such that $z \in \overline{\mathbb{D}_n}$ and thus $|f_j(z) - f_k(z)| \leq p_n(f_j,f_k)$ for each $j,k \in \mathbb{N}$. Then (1.6.328) grants that $\{f_j(z)\}_{j\in\mathbb{N}}$ is Cauchy and therefore, since $\mathbb{C}$ equipped with the Euclidean distance is complete, we have that

$$\forall\, z \in \mathbb{D} \ \text{ the limit } \ f(z) := \lim_{j\to\infty} f_j(z) \ \text{ exists.} \tag{1.6.334}$$

Our second claim is that the function $f : \mathbb{D} \longrightarrow \mathbb{C}$ given in (1.6.334) is continuous. Indeed, note first that for each $K \subseteq \mathbb{D}$ fixed compact set, there exists $n_K \in \mathbb{N}$ such that $K \subseteq \overline{\mathbb{D}_n}$ for each $n \geq n_K$. On grounds of (1.6.328) and the definition of p_n from (1.6.317) , fixing some $n = n_K$, we may write that

$$\forall\, \varepsilon > 0\ \exists\, J_{\varepsilon,K} \in \mathbb{N} \ \text{ such that } \ j,k \geq J_{\varepsilon,K} \implies |f_j(z) - f_k(z)| < \varepsilon, \quad \forall z \in K. \tag{1.6.335}$$

Passing to the limit as $j \to \infty$ in the right hand side of (1.6.335) (using (1.6.334)) then yields

$$\forall\, \varepsilon > 0\ \exists\, J_{\varepsilon,K} \in \mathbb{N} \ \text{ such that } \ k \geq J_{\varepsilon,K} \implies |f(z) - f_k(z)| \leq \varepsilon, \quad \forall z \in K, \tag{1.6.336}$$

i.e., the sequence $\{f_j\}_{j\in\mathbb{N}}$ converges uniformly on K to the function f. Since K was an arbitrary compact set in $\mathbb{D}$ and since continuity is preserved by uniform convergence, it follows that f belongs to $C(\mathbb{D})$, as desired.

Our third claim is that the sequence $\{f_j\}_{j\in\mathbb{N}}$ converges to f in the distance d, i.e.,

$$\lim_{j\to\infty} d(f_j, f) = 0. \tag{1.6.337}$$

To this end, let $\varepsilon > 0$ and let $n_\varepsilon \in \mathbb{N}$ be such that $\sum_{n\geq n_\varepsilon+1} 2^{-n} < \varepsilon/2$, and for each $j \in \mathbb{N}$ write, based on the definition of d,

$$\begin{aligned} d(f_j, f) &\leq \sum_{n=1}^{n_\varepsilon} \frac{1}{2^n} \cdot \frac{p_n(f_j, f)}{1 + p_n(f_j, f)} + \sum_{n=n_\varepsilon+1}^{\infty} \frac{1}{2^n} \\ &\leq \sum_{n=1}^{n_\varepsilon} \frac{1}{2^n} \cdot \frac{p_n(f_j, f)}{1 + p_n(f_j, f)} + \varepsilon/2. \end{aligned} \tag{1.6.338}$$

Next, notice that since the function $[0,\infty) \ni t \mapsto G(t) := t/(1+t)$ is increasing, there exists $\delta_\varepsilon > 0$ such that the following implication holds

$$[0, \delta_\varepsilon] \ni t \implies G(t) \in [0, \varepsilon/2). \tag{1.6.339}$$

Letting $K := \overline{\mathbb{D}}_{n_\varepsilon}$ and employing the uniform convergence result established in (1.6.336) allows us to conclude that there exists $J_\varepsilon \in \mathbb{N}$ such that for each $j \in \mathbb{N}$ such that $j \geq J_\varepsilon$ we have $p_{n_\varepsilon}(f_j, f) \leq \delta_\varepsilon$. Since $\overline{\mathbb{D}}_m \subseteq \overline{\mathbb{D}}_k$ for each $m, k \in \mathbb{N}$ with $m \leq k$, this yields

$$p_n(f_j, f) \leq \delta_\varepsilon, \qquad \forall\, j \geq J_\varepsilon \text{ and } \forall\, n \in \{1, \dots, n_\varepsilon\}. \tag{1.6.340}$$

Combined with (1.6.339) this yields that

$$G(p_n(f_j, f)) = \frac{p_n(f_j, f)}{1 + p_n(f_j, f)} < \varepsilon/2, \qquad \forall\, j \geq J_\varepsilon \text{ and } \forall\, n \in \{1, \dots, n_\varepsilon\}. \tag{1.6.341}$$

Taking this back in (1.6.338) allows us to finally conclude that for each $j \in \mathbb{N}$ such that $j \geq J_\varepsilon$ we have

$$d(f_j, f) \leq (\varepsilon/2) \cdot \sum_{n=1}^{n_\varepsilon} \frac{1}{2^n} + \varepsilon/2 < \varepsilon, \tag{1.6.342}$$

finishing the proof of (1.6.337). Finally, a combination of (1.6.337) and of our earlier result to the effect that f belongs to $C(\mathbb{D})$, shows that the Cauchy sequence $\{f_j\}_{j\in\mathbb{N}}$ in $(C(\mathbb{D}), d)$ is convergent to $f \in C(\mathbb{D})$, i.e., the metric space $(C(\mathbb{D}), d)$ is complete.

To finish the treatment of item *(2)*, there remains to show that a sequence $\{f_j\}_{j\in\mathbb{N}}$ in $(C(\mathbb{D}), d)$ is convergent to $f \in C(\mathbb{D})$ if and only if the sequence $\{f_j\}_{j\in\mathbb{N}}$ converges to f uniformly on all compact sets $K \subset \mathbb{D}$. However this is already implicit in what we have done so far. For the direct implication the sequence $\{f_j\}_{j\in\mathbb{N}} \subseteq (C(\mathbb{D}), d)$ is Cauchy in $(C(\mathbb{D}), d)$ and uniform convergence to f follows much like (1.6.336).

Conversely, if $\{f_j\}_{j\in\mathbb{N}} \subseteq C(\mathbb{D})$ converges to f uniformly on all compact sets $K \subset \mathbb{D}$, then (1.6.337) holds as shown above. □

Example 1.6.5. Let $\mathbb{D}$ be the unit disk as in (1.6.316) and consider $\mathscr{O}(\mathbb{D})$, the vector space of holomorphic functions in $\mathbb{D}$, and recall the space $C(\mathbb{D})$ and the distance d from Example 1.6.4. Then the following properties are true:

(1) $\mathscr{O}(\mathbb{D})$ is a subspace of the vector space $C(\mathbb{D})$.
(2) When equipped with the topology induced by d from (1.6.319), $\mathscr{O}(\mathbb{D})$ becomes a linear topological space where the topology is that induced by uniform convergence on compact sets.
(3) $\mathscr{O}(\mathbb{D})$ is closed in $(C(\mathbb{D}), d)$, and is therefore a complete metric space in its own right.

Proof. Property *(1)* is immediate due to basic properties of holomorphic functions, while property *(2)* is a consequence of *(2)* from Example 1.6.4. Finally, since analyticity is preserved by uniform convergence on compact sets (a consequence of Morera's Theorem from Complex Analysis), property *(3)* immediately follows. □

The category of F-spaces, which we recall next (cf., e.g., [39, Chapter 1]), offers additional examples of topological linear spaces.

Definition 1.6.6. *An F-*`space` *is a linear topological space (X, τ) whose topology τ is induced by a complete and translation-invariant metric.*

It is clear from Exercise 1.2.8 and definitions that

$$\begin{gathered}\text{every } F\text{-space is Hausdorff, and}\\ \text{every } p\text{-Banach space is an } F\text{-space (in particular, Hausdorff).}\end{gathered} \tag{1.6.343}$$

The spaces $C(\mathbb{D})$ and $\mathscr{O}(\mathbb{D})$ just discussed are F-spaces, as are their counterparts defined on general open subsets of the complex plane, or of $\mathbb{C}^n$ for $n \in \mathbb{N}$.

It turns out that F-spaces provide a remarkably fertile setting for functional analysis. For example, the Open Mapping Theorem remains valid for F-spaces, and a formal statement is given below (see e.g., [39, Corollary 2.12, part (a), p. 49]; for related material, see also [23] and [33]).

Theorem 1.6.7 (Open Mapping Theorem for F-spaces). *Suppose X and Y are two F-spaces, and let*

$$T : X \longrightarrow Y \tag{1.6.344}$$

be a continuous linear surjection. Then T is an open mapping (that is, the image $T(U)$ is open in Y for every open set $U \subseteq X$).

Additional examples of topological vector spaces are generated by the proposition below, which asserts that being a topological vector space is a hereditary property.

Proposition 1.6.8. *Suppose (X, τ) is a topological vector space, and Y is a linear subspace of X. Then $(Y, \tau|_Y)$ is itself a topological vector space. Moreover, if (X, τ) is Hausdorff, then so is $(Y, \tau|_Y)$. Also, the canonical inclusion $\iota_Y : (Y, \tau|_Y) \hookrightarrow (X, \tau)$ is a continuous function.*

Proof. Consider the addition map in X,

$$\begin{aligned} &f_+ : (X \times X, \tau \times \tau) \longrightarrow (X, \tau) \ \text{ given by} \\ &f_+(x, y) := x + y \ \text{ for every } \ x, y \in X. \end{aligned} \tag{1.6.345}$$

Since (X, τ) is a topological vector space, it follows that f_+ is continuous. Based on this, the fact that $f_+(Y \times Y) \subseteq Y$, Exercise 1.1.11, and Proposition 1.1.16 we conclude that

$$\begin{aligned} &g_+ : \big(Y \times Y, (\tau|_Y) \times (\tau|_Y)\big) \longrightarrow (Y, \tau|_Y) \ \text{ given by} \\ &g_+(x, y) := x + y \ \text{ for every } \ x, y \in Y \end{aligned} \tag{1.6.346}$$

is a continuous function. This is one of the requirements needed to conclude that $(Y, \tau|_Y)$ is a topological vector space (cf. item *(1)* in Definition 1.6.1). Similarly, the continuity of the multiplication map $(\mathbb{K}, \tau_{\mathbb{K}}) \times (X, \tau) \ni (\lambda, x) \mapsto \lambda x \in (X, \tau)$ ensures that the multiplication map $(\mathbb{K}, \tau_{\mathbb{K}}) \times (Y, \tau|_Y) \ni (\lambda, x) \mapsto \lambda x \in (Y, \tau|_Y)$ is also continuous. Hence all requirements in Definition 1.6.1 are satisfied for the vector space Y equipped with the topology $\tau|_Y$. This renders $(Y, \tau|_Y)$ a topological vector space. That $(Y, \tau|_Y)$ is Hausdorff when (X, τ) is so is seen directly from Exercise 1.1.6. The final claim in the statement follows from Exercise 1.1.11 (applied with f the identity on X). □

Exercise 1.6.9. Given a linear topological space X, show that $K_0 + K_1$ is compact in X whenever K_0, K_1 are compact in X.

Hint: Use the continuity of the addition map (cf. item *(1)* in Definition 1.6.1), Exercise 1.1.29, and Exercise 1.1.32.

Given a vector space X along with a linear subspace Y of X, the quotient map

$$\pi : X \to X/Y, \qquad \pi(x) := [x]_{X/Y} = x + Y \in X/Y \ \text{ for each } \ x \in X, \tag{1.6.347}$$

is a well-defined linear operator. Assume next that (X, τ_X) is a Hausdorff topological vector space and that Y is a closed linear subspace of X. Then

$$\tau_{X/Y} := \{A \subseteq X/Y : \pi^{-1}(A) \in \tau_X\} \tag{1.6.348}$$

is a topology on X/Y, called the `quotient topology`. Some of the most basic properties of $\tau_{X/Y}$ are listed in the lemma below (for a proof, see [39, Theorem 1.41, p. 31]). Before stating this, we clarify some terminology in relation of a topological vector space X, with topology τ. First,

> call a linear topological space (X, τ) `locally convex` if there exists a local base at zero for τ consisting of zero-neighborhoods which are convex sets (cf. (1.3.81)). (1.6.349)

Second,

call a linear topological space (X, τ) `normable` if τ is the topology induced by some norm on X. (1.6.350)

Third,

call a topological space (X, τ) `metrizable` if τ is the topology induced by some metric on X. (1.6.351)

Finally, (cf. Definition 1.6.6; see also [39, §1.8, items *(e)* and *(f)*, p. 9]):

a Hausdorff topological vector space X is said to be a `Fréchet space` if X is a locally convex F-space. (1.6.352)

Lemma 1.6.10. *Let Y be a closed linear subspace of a Hausdorff topological vector space (X, τ_X). Then, in relation to the quotient topology on X/Y, defined as in* (1.6.348), *the following statements are true.*

(1) When equipped with $\tau_{X/Y}$, the quotient space X/Y becomes a Hausdorff topological vector space, and the quotient map $\pi : (X, \tau_X) \to (X/Y, \tau_{X/Y})$ is linear, continuous, surjective, open, with $\ker \pi = Y$.
(2) If $\mathscr{B}_X$ is a local base at zero for τ_X, then $\{\pi(U)\}_{U \in \mathscr{B}_X}$ is a local base at zero for $\tau_{X/Y}$.
(3) Each of the following properties of (X, τ_X) is inherited by $(X/Y, \tau_{X/Y})$: local convexity (cf. (1.6.349)*), normability (cf.* (1.6.350)*), metrizability (cf.* (1.6.351)*), local boundedness (cf. Definition 1.6.46).*
(4) If X is an F-space (cf. Definition 1.6.6), or a Fréchet space (cf. (1.6.352)*), or a Banach space, then so is X/Y.*

The proof we have in mind for our next result, to the effect that the quality of being a closed subspace of finite codimension is preserved under pre-images of linear continuous mappings, makes use of quotient spaces.

Lemma 1.6.11. *Let X, Y be Hausdorff linear topological spaces over the field $\mathbb{K}$ and assume that $T : X \to Y$ is a linear and continuous mapping. If Z is a closed subspace of Y having finite codimension, then $T^{-1}Z$ is a closed subspace of finite codimension in X.*

Proof. Since T is continuous and Z is a closed set, it follows that $T^{-1}Z$ is closed as well. Next, define

$$\widehat{T} : X/T^{-1}Z \longrightarrow Y/Z, \qquad \widehat{T}[x] := [Tx] \ \text{ for each } \ x \in X, \tag{1.6.353}$$

where for each $x \in X$, $[x]$ stands for the class of x in $X/T^{-1}Z$, and $[Tx]$ stands for the class of Tx in Y/Z. Clearly, $\widehat{T}$ is well defined, linear, and one-to-one, which then entails

$$\dim_{\mathbb{K}}\big(X/T^{-1}Z\big) \leq \dim_{\mathbb{K}}\big(Y/Z\big) < +\infty. \tag{1.6.354}$$

Thus, $T^{-1}Z$ is a closed subspace of finite codimension in X. □

We conclude this section with a brief foray into duality for linear topological spaces.

Definition 1.6.12. *Let X be a linear topological space X over the field $\mathbb{K}$ (on X which, as usual, is either $\mathbb{R}$, or $\mathbb{C}$). Any linear map $\Lambda : X \to \mathbb{K}$ is referred to as* `functional` *(on X). The (topological)* `dual` *of X, denoted by X^*, consists of all continuous linear functionals on X, i.e.,*

$$X^* := \{\Lambda : X \to \mathbb{K} : \Lambda \ \text{ linear and continuous}\}. \tag{1.6.355}$$

Say that X has enough continuous linear functionals to `separate points` *if for every $x_0, x_1 \in X$ with $x_0 \neq x_1$ there exists a functional $\Lambda \in X^*$ with the property that $\Lambda(x_0) \neq \Lambda(x_1)$.*

Alternatively, call a linear topological space X `dual-rich` if X has enough continuous linear functionals to separate points. Note that the latter property may be equivalently reformulated as the demand that for each $x \in X \setminus \{0\}$ there exists $\Lambda \in X^*$ so that $\Lambda(x) \neq 0$. The classical Hahn-Banach Theorem guarantees that:

$$\text{every normed vector space is dual-rich.} \tag{1.6.356}$$

More generally, in Theorem 1.12.2 we shall show that every Hausdorff locally convex topological vector space is dual-rich.

The reader is also referred to Exercises 1-10 and Exercise 14 in §1.9 which deal, in one form or another, with linear functionals on a topological vector space.

1.6.1 Local Bases for Vector Topologies

So far we have only discussed *metrizable* examples of linear topological spaces. In order to provide non-metrizable examples we need to focus a bit on the special properties possessed by neighborhoods in vector topologies. Because of the continuity of addition on a linear topological space (X, τ), it is easy to see that the mapping of *translation by a fixed vector* $x_0 \in X$ (that is, the map $X \ni x \mapsto x + x_0 \in X$) is a homeomorphism of (X, τ) onto itself (see Proposition 1.6.13). For this reason the neighborhood structure at any point x is the same as the neighborhood structure at the origin. As such our goal in this subsection is to try to understand how the continuity of vector operations in a linear topological space (X, τ) influences the structure of $\mathcal{N}(0; \tau)$, the collection of neighborhoods of the zero-vector, henceforth referred to as `zero-neighborhoods`.

We start with formally stating the preliminary result we already advertised whose proof is elementary.

Proposition 1.6.13. *Let (X, τ) be linear topological space and $x_0 \in X$. Then the translation by x_0 denoted by f_{x_0} and given by*

$$f_{x_0} : (X, \tau) \longrightarrow (X, \tau), \quad f_{x_0}(x) := x_0 + x, \quad \forall\, x \in X, \tag{1.6.357}$$

is a homeomorphism of (X, τ).

Proof. It is trivial to see that for each $x_0 \in X$ the function f_{x_0} is a bijection with inverse f_{-x_0}. Both f_{x_0} and f_{-x_0} are continuous from (X, τ) onto itself as a consequence of the continuity of scalar addition in the linear topological space (X, τ). □

We proceed by making the following definition.

Definition 1.6.14. *Let X be a vector space over the field $\mathbb{K}$. Then*

(a) $A \subseteq X$ is called `circled` *or* `balanced` *if $\lambda x \in A$ for every $x \in A$ and every $\lambda \in \mathbb{K}$ with $|\lambda| \leq 1$, i.e., the following implication holds*

$$\lambda \in \mathbb{K} \text{ and } |\lambda| \leq 1 \implies \lambda A \subseteq A. \tag{1.6.358}$$

(b) A set $A \subseteq X$ is said to `absorb` *a set $B \subseteq X$ (or B is said to be* `absorbed` *by A) if there exists $\rho > 0$ such that $rA \supseteq B$ for every $r \in [\rho, \infty)$. Equivalently, $A \subseteq X$ absorbs $B \subseteq X$ if there exists $\varepsilon_o > 0$ such that $A \supseteq \varepsilon B$ for each $\varepsilon \in (0, \varepsilon_o]$.*

(c) $A \subseteq X$ is called `absorbing` *provided A absorbs all singletons of X; that is, for every $x \in X$ there exists $\delta_x > 0$ such that the following implication holds*

$$t \in (0, \delta_x) \implies tx \in A. \tag{1.6.359}$$

(d) $A \subseteq X$ is called `discretely-absorbing` *if for every $x \in X$ there exists $n \in \mathbb{N}$ such that $x \in nA$, i.e., if*

$$X = \bigcup_{n \in \mathbb{N}} nA. \tag{1.6.360}$$

(e) If $A \subseteq X$, the `circled hull of` *A denoted by $\operatorname{circ}(A)$ is defined by setting*

$$\operatorname{circ}(A) := \bigcup_{|\lambda| \leq 1} \lambda A. \tag{1.6.361}$$

It is worth noting that every nontrivial vector space contains a nonempty circled set which is not absorbing; namely, the singleton $\{0\}$. Alternative denominations for circled hull of a set are `balanced hull`, or `balanced envelope`. Also, alternative terms used for an absorbing set are `radial set`, or `absorbent set` (or, simply, `absorbent`).

Exercise 1.6.15. Let X be a vector space over $\mathbb{R}$ and $A \subseteq X$. Then

$$\operatorname{circ}(A) = \bigcup_{a \in A} \big([0, a] \cup [0, -a]\big), \tag{1.6.362}$$

where for every $x, y \in X$ the closed segment in X joining x and y is denoted by $[x, y]$, i.e.,

$$[x, y] := \big\{z \in X : \exists t \in [0, 1] \text{ such that } z = (1 - t)x + ty\big\}. \tag{1.6.363}$$

Proof. Let $a \in A$ and suppose $\lambda \in \mathbb{R}$ is such that $|\lambda| \le 1$. It is then immediate that $\lambda a \in [0, a] \cup [0, -a]$ from which we obtain the left-to-right inclusion in (1.6.363). To establish the converse inclusion, it is enough to notice that, using (1.6.363), for each $x \in [0, a] \cup [0, -a]$ we have $x = ta$ for some $t \in \mathbb{R}$ with $|t| \le 1$. □

Some of the most basic properties of the classes of sets introduced above are collected in the next proposition.

Proposition 1.6.16. *Assume that X is a vector space over the field $\mathbb{K}$. Then the following hold.*

(1) Every absorbing subset of X is discretely-absorbing and contains the zero vector.
(2) Every circled set $A \subseteq X$ is invariant under multiplication with -1, i.e., $A = -A$.
(3) For every $A \subseteq X$ the set circ (A) *is circled and contains A.*
(4) If $A, B \subseteq X$ are such that for each $b \in B$ there exists $a \in A$ with the property that b belongs to the line segment with endpoints $-a$ and a (i.e., $b = (2t-1)a$, for some $t \in [0, 1]$), then $\mathrm{circ}(B) \subseteq \mathrm{circ}(A)$. *In particular, one has*

$$B \subseteq A \subseteq X \Longrightarrow \mathrm{circ}\,(B) \subseteq \mathrm{circ}\,(A). \tag{1.6.364}$$

(5) For every $A \subseteq X$ the set circ (A) *is the smallest circled set in X containing A.*
(6) For every $A \subseteq X$,

$$\mathrm{circ}\,(\alpha A) = \alpha \cdot \mathrm{circ}\,(A), \quad \forall\, \alpha \in \mathbb{K}. \tag{1.6.365}$$

(7) For every $A \subseteq X$,

$$A = \mathrm{circ}\,(A) \iff A \text{ is circled.} \tag{1.6.366}$$

(8) A finite intersection of absorbing sets of X is absorbing. Also, any positive dilate of an absorbing set is absorbing.
(9) Circled subsets of X satisfy the following expansion property:

$$A \subseteq X \text{ is circled} \iff \forall\, \lambda, \gamma \in \mathbb{K} \text{ with } |\lambda| \le |\gamma| \text{ one has } \lambda A \subseteq \gamma\, A. \tag{1.6.367}$$

(10) If $X \neq \{0\}$, it follows that

$$\text{every circled subset of } X \text{ not equal to } \{0\} \text{ is absorbing} \iff \dim_{\mathbb{K}} X = 1. \tag{1.6.368}$$

(11) If (X, τ) is a linear topological space then

$$A \in \tau \text{ and } 0 \in A \Longrightarrow \mathrm{circ}\,(A) \in \tau. \tag{1.6.369}$$

(12) If (X, τ) is a linear topological space and $A \subseteq X$ is circled then its closure $\overline{A}$ is circled.
(13) If (X, τ) is a linear topological space and $A \subseteq X$ is circled then its interior Int(A) *is circled provided A is a neighborhood of the zero vector, i.e.,*

$$A = \mathrm{circ}(A) \text{ and } A \in \mathcal{N}(0; \tau) \Longrightarrow \mathrm{Int}(A) = \mathrm{circ}(\mathrm{Int}(A)). \tag{1.6.370}$$

Proof. All properties are straightforward consequences of definitions. We include the elementary proof for the sake of completeness.

To justify *(1)*, let $A \subseteq X$ be an absorbing set and select $x \in X$. By *(c)* in Definition 1.6.14 there exists δ_x such that implication (1.6.359) is true. In particular, since there exists $n \in \mathbb{N}$ such that $1/n < \delta_x$, we have $x \in nA$ and, consequently, the left-to-right inclusion in (1.6.360) is true. Since the reverse inclusion is trivial, the equality in (1.6.360) is valid, showing that A is discretely absorbing. Finally, since A absorbs the singleton $\{0\}$ it is immediate that $0 \in A$ as desired.

Turning our attention to the statement in *(2)*, let $A \subseteq X$ be circled. Then (1.6.358) gives $-A \subseteq A$. Using this inclusion, for each $a \in A$ we may write

$$a = -(-a) \in -(-A) \subseteq -A, \tag{1.6.371}$$

thus showing that $A \subseteq -A$, and finishing the proof of *(2)*.

Consider next the statement made in *(3)*. Let $A \subseteq X$ and using definition (1.6.361) is immediate that $A \subseteq \mathrm{circ}(A)$. Pick next $x \in \mathrm{circ}(A)$ and let $\lambda \in \mathbb{K}$ be such that $|\lambda| \leq 1$ and $x \in \lambda A$. Then, for every $\lambda_o \in \mathbb{K}$ satisfying $|\lambda_o| \leq 1$, we may write

$$\lambda_o x \in \lambda_o(\lambda A) = (\lambda_o\lambda)A \subseteq \mathrm{circ}(A), \tag{1.6.372}$$

where the inclusion above is a consequence of the fact that $|\lambda_o\lambda| \leq 1$ and the definition of $\mathrm{circ}(A)$ from (1.6.361). Since $x \in \mathrm{circ}(A)$ was arbitrary, this show that $\mathrm{circ}(A)$ is circled, as desired.

Next, let $A, B \subseteq X$ be such that

$$\begin{aligned} &\text{for each } b \in B \text{ there exists } a \in A \text{ such that} \\ &b \in [-a, a] := \{(2t-1)a : t \in [0,1]\}. \end{aligned} \tag{1.6.373}$$

Then, for each $\lambda \in \mathbb{K}$ such that $|\lambda| \leq 1$ we have

$$\lambda b \in [-\lambda a, \lambda a] \subseteq \bigcup_{t \in [0,1]} ((2t-1)\lambda)A \subseteq \mathrm{circ}(A), \tag{1.6.374}$$

where the last inclusion follows from the definition of $\mathrm{circ}(A)$ from (1.6.361) and the fact that for each $t \in [0,1]$ we have $|(2t-1)\lambda| \leq 1$. The validity (1.6.374) for each $b \in B$ and each $\lambda \in \mathbb{K}$ such that $|\lambda| \leq 1$ finally yields that $\mathrm{circ}(B) \subseteq \mathrm{circ}(A)$. In the scenario when $B \subseteq A$, property (1.6.373) trivially holds. Indeed for each $b \in B$ let $a := b$ and $t := 1$. Consequently, the implication in (1.6.364) follows from what we have proved so far. This completes the proof of *(4)*.

As for the claim in item *(5)*, using *(4)* matters reduce to showing that if $B \subseteq X$ is circled and $A \subseteq B$ then $\mathrm{circ}(A) \subseteq B$. Indeed,

$$\text{for each } \lambda \in \mathbb{K} \text{ such that } |\lambda| \leq 1 \text{ we have } \ \lambda A \subseteq \lambda B \subseteq B, \tag{1.6.375}$$

where the last inclusion in (1.6.375) is a consequence of the implication (1.6.358) written for the circled set B. Consequently, using (1.6.375) and (1.6.361) we obtain

$$\operatorname{circ}(A) \subseteq B, \tag{1.6.376}$$

as desired.

Moving on, fix $A \subseteq X$ and $\alpha \in \mathbb{K}$. For each $\lambda \in \mathbb{K}$ satisfying $|\lambda| \le 1$, using (1.6.361), write

$$\lambda(\alpha A) = \alpha(\lambda A) \subseteq \alpha \cdot \operatorname{circ}(A). \tag{1.6.377}$$

Hence, once again availing ourselves of (1.6.361) (with αA now playing the role of A) and (1.6.377) we may conclude that $\operatorname{circ}(\alpha A) \subseteq \alpha \cdot \operatorname{circ}(A)$. Conversely, if $x \in \alpha \cdot \operatorname{circ}(A)$ there exists $\lambda \in \mathbb{K}$ such that

$$x \in \alpha(\lambda A) = \lambda(\alpha A) \subseteq \operatorname{circ}(\alpha A). \tag{1.6.378}$$

Since the vector x was arbitrary chosen in $\alpha \cdot \operatorname{circ}(A)$, the membership (1.6.378) immediately gives that $\alpha \cdot \operatorname{circ}(A) \subseteq \operatorname{circ}(\alpha A)$. Thus $\operatorname{circ}(\alpha A) = \alpha \cdot \operatorname{circ}(A)$, proving *(6)*.

The proof of the equivalence in (1.6.366) in *(7)* is immediate since the direct implication follows from *(3)* while the reverse implication is a consequence of *(5)*. Consider next $n \in \mathbb{N}$ along with $F_1, \dots, F_n \subseteq X$ absorbing sets and set

$$F := \bigcap_{j=1}^{n} F_j. \tag{1.6.379}$$

Consider $x \in X$ and notice that since each of the sets F_j is absorbing, we have that

$$\begin{gathered} \text{for each } j \in \{1, \dots, n\} \text{ there exists } \delta_{x,j} > 0 \text{ such that} \\ t \in (0, \delta_{x,j}) \Longrightarrow tx \in F_j. \end{gathered} \tag{1.6.380}$$

Then, setting $\delta_x := \min\{\delta_{x,j} : j \in \{1, \dots, n\}\} > 0$, it follows from (1.6.380) that for each $t \in (0, \delta_x)$ one has $tx \in F$, showing that F is absorbing. This takes care of the first claim in item *(8)*. The second claim in item *(8)* follows by unraveling definitions.

We shall consider next the expansion property of circled sets from (1.6.367). For the direct implication let $A \subseteq X$ be circled and let $\lambda, \gamma \in \mathbb{K}$ be such that $|\lambda| \le |\gamma|$. Using *(1)* we have that $0 \in A$, thus if $\lambda = 0$ then $\lambda A = \{0\}$ and the inclusion $\{0\} \subseteq \gamma A$ trivially holds. If $\lambda \neq 0$ then $\gamma \neq 0$ and we shall proceed to write that

$$\lambda A = \gamma \cdot \left(\frac{\lambda}{\gamma} A\right) \subseteq \gamma \operatorname{circ}(A) = \gamma\, A, \tag{1.6.381}$$

where the first equality above is trivial, the inclusion follows from the definition of $\operatorname{circ}(A)$ in (1.6.361) on grounds that $|\lambda/\gamma| \le 1$, and the last equality is a consequence of the fact that A is circled and *(7)*. This establishes the left-to-right implication in (1.6.367). To establish the reverse implication, assume $A \subseteq X$ is such that $\lambda A \subseteq \gamma A$ for each $\lambda, \gamma \in \mathbb{K}$ with $|\lambda| \le |\gamma|$. Then, $\lambda A \subseteq A$ for each $\lambda \in \mathbb{K}$ such that $|\lambda| \le 1$ and thus $\operatorname{circ}(A) \subseteq A$. This, combined with the fact that $\operatorname{circ}(A)$ contains A (cf. *(3)*), yields that $A = \mathrm{circ}(A)$. Using *(7)* we may conclude that A is circled, as desired. This shows the right-to-left implication in (1.6.367), finishing the proof of *(9)*.

Consider now the equivalence from *(10)*. To show the direct implication, we shall proceed by contradiction and assume that $\dim_{\mathbb{K}} X > 1$. In particular, there exist vectors $x_1, x_2 \in X \setminus \{0\}$ which are linearly independent. Consider next the segment with end points $-x_1$ and x_1, i.e.,

$$I_{x_1} := (-x_1, x_1) = \{(2t-1)x_1 : t \in (0,1)\} \tag{1.6.382}$$

and set

$$A := \mathrm{circ}(I_{x_1}) \subseteq X. \tag{1.6.383}$$

Since $I_{x_1} \subseteq \mathrm{circ}(I_{x_1}) = A$, it follows that $A \neq \{0\}$. Also, A is circled by *(3)*. By hypothesis, A is absorbing, so there exists $\delta > 0$ such that for each $t \in (0, \delta)$ one has $tx_2 \in A$. In particular, by the definition of $\mathrm{circ}(I_{x_1})$ and the fact that $x_2 \neq 0$ there exists $\lambda \in \mathbb{K} \setminus \{0\}$ with $|\lambda| \leq 1$ such that $tx_2 \in \lambda I_{x_1}$ for each $t \in (0, \delta)$. However this implies that x_1 and x_2 are linearly dependent over the field $\mathbb{K}$, reaching a contradiction. This concludes the proof of the left-to-right implication in (*10*).

To show the converse implication in (*10*), assume X is a vector field over $\mathbb{K}$ with $\dim_{\mathbb{K}} X = 1$. Then $X = \{\alpha x : \alpha \in \mathbb{K}\}$ for some $x \in X \setminus \{0\}$. Fix next $A \subseteq X$ circled and satisfying $A \neq \{0\}$. Our goal is to show that A is absorbing. To this end, select an arbitrary $a \in A \setminus \{0\}$ and consider the line segment with end points $-a$ and a, given by

$$I_a := (-a, a) = \{(2t-1)a : t \in (0,1)\}. \tag{1.6.384}$$

Then

$$I_a \subseteq \mathrm{circ}(A) = A \text{ and thus } \mathrm{circ}(I_a) \subseteq \mathrm{circ}(A) = A, \tag{1.6.385}$$

where we have used that A is circled and (1.6.364) from *(4)*. Pick next $y \in X$. If $y = 0$ this is clearly absorbed by A which contains $\{0\}$, since A is circled. If $y \neq 0$, there exists $\gamma_o \in \mathbb{K} \setminus \{0\}$ such that $y = \gamma_o x$. However, since $a \in X \setminus \{0\}$ we also have that $a = \alpha_o x$ for some $\alpha_o \in \mathbb{K} \setminus \{0\}$ and, using (1.6.385) we may write that

$$ty = (t\gamma_o)x = \frac{t\gamma_o}{\alpha_o} \cdot a \in \mathrm{circ}(I_a) \subseteq A, \tag{1.6.386}$$

whenever

$$t \in \left(0, \frac{\alpha_o}{\gamma_o}\right). \tag{1.6.387}$$

In concert, (1.6.386) and (1.6.387) show that A absorbs $\{y\}$ as desired. This completes the treatment of item *(10)*.

For the remaining portion of the proof, (X, τ) is assumed to be a linear topological space. As far as *(11)* is concerned, to prove the implication in (1.6.369) start by fixing $A \in \tau$ such that $0 \in A$. Then

$$\mathrm{circ}(A) = \bigcup_{|\lambda| \leq 1} \lambda A = \bigcup_{0 < |\lambda| \leq 1} \lambda A, \tag{1.6.388}$$

where for the second equality in (1.6.388) we have used that $0 \in A$. However, since (X, τ) is a linear topological space

$$\forall\, \lambda \in \mathbb{K} \setminus \{0\} \text{ the function } f_\lambda : X \longrightarrow X \text{ given by } f_\lambda(x) := \lambda x, \quad \forall\, x \in X$$
$$\text{is a homeomorphism.} \tag{1.6.389}$$

Thus, using Proposition 1.1.8, it follows that $f_\lambda(A) = \lambda A \in \tau$. From this and (1.6.388) it then follows that $\operatorname{circ}(A) \in \tau$, as desired.

Next, we turn our attention to the statement made in *(12)*. Consider $A \subseteq X$ circled set. By *(3)* we have that $\overline{A} \subseteq \operatorname{circ}(\overline{A})$ and, since A is circled, $0 \in A \subseteq \overline{A}$. Our goal is to show that the reverse inclusion

$$\operatorname{circ}(\overline{A}) \subseteq \overline{A} \tag{1.6.390}$$

also holds, yielding that $\overline{A} = \operatorname{circ}(\overline{A})$ and thus by (7), that $\overline{A}$ is circled.

Let then $y \in \operatorname{circ}(\overline{A})$ and pick $\lambda \in \mathbb{K} \setminus \{0\}$ such that $|\lambda| \le 1$ and $y \in \lambda\overline{A} = f_\lambda(\overline{A})$. Using that multiplication by λ is continuous, that $\overline{A}$ is closed, and that $A \subseteq \overline{A}$, it follows that

$$\lambda\overline{A} \text{ is a closed set in } (X, \tau) \text{ which contains } \lambda A. \tag{1.6.391}$$

This and the discussion following Definition 1.1.21 imply that

$$\overline{\lambda A} \subseteq \lambda \overline{A}. \tag{1.6.392}$$

If $z \in \overline{A}$ then there exist $\{z_j\}_{j\in\mathbb{N}} \subseteq A$ such that $\lim\limits_{j\to\infty} z_j = z$. Appealing again to the continuity of the scalar multiplication by λ in the linear topological space (X, τ), this further yields that $\lim\limits_{j\to\infty} \lambda z_j = \lambda z$ and ultimately that $\lambda z \in \overline{\lambda A}$. This shows that

$$\lambda \overline{A} \subseteq \overline{\lambda A}, \tag{1.6.393}$$

and, combined with (1.6.392), that $\lambda\overline{A} = \overline{\lambda A}$. Thus, since A is circled and $|\lambda| \le 1$,

$$y \in \overline{\lambda A} \subseteq \overline{A}, \tag{1.6.394}$$

completing the proof of (1.6.390), and hence the proof of *(12)*.

Finally, we shall prove the statement made in *(13)*. To this end let $A \subseteq X$ be circled and $A \in \mathcal{N}(0; \tau)$. Using *(3)* and *(7)* matters reduce to showing that

$$\operatorname{circ}(\operatorname{Int}(A)) \subseteq \operatorname{Int}(A). \tag{1.6.395}$$

Further employing the definition of the circled hull (in (1.6.361)) it suffices to show that

$$\forall\, \lambda \in \mathbb{K} \text{ such that } |\lambda| \le 1 \text{ one has } \lambda \cdot \operatorname{Int}(A) \subseteq \operatorname{Int}(A). \tag{1.6.396}$$

With this in mind, fix $\lambda \in \mathbb{K}$ such that $|\lambda| \le 1$ and $x \in \operatorname{Int}(A)$. If $\lambda = 0$ then $\lambda x = 0$ and since $A \in \mathcal{N}(0; \tau)$ it follows that $0 \in \operatorname{Int}(A)$, proving (1.6.396) when $\lambda = 0$.

In the case when $\lambda \neq 0$, note that since $x \in \text{Int}(A)$ there exists $O \in \tau$ such that $x \in O$ and $O \subseteq A$. Then $\lambda x \in \lambda O$ and $\lambda O \subseteq \lambda A \subseteq A$, where the last inclusion follows since A is circled. Since $\lambda O = f_{1/\lambda}^{-1}(O)$, where $f_{1/\lambda}$ is the homeomorphism of scalar multiplication by $1/\lambda$ (as in (1.6.389) with $1/\lambda$ in place of λ), it follows that $\lambda O \in \tau$ by Proposition 1.1.8. This shows that λx belongs to Int(A), ultimately giving (1.6.396). This finishes the argument for *(13)*. The proof of Proposition 1.6.16 is now complete. □

Comment 1.6.17 *With regard to* (1.6.369)*, it is worth noting that in general, the circled hull of an open set is not necessarily open if that set fails to contain the origin.*

To see this, consider the linear topological space $\mathbb{R}^2$ over the field $\mathbb{R}$, and consider the set $A := B((1,0), 1) \subseteq \mathbb{R}^2$, i.e., A is the open ball in $\mathbb{R}^2$ centered at the point $(1,0)$ with radius 1. We claim that

$$\text{circ}\,(A) = (-A) \cup \{(0,0)\} \cup A. \tag{1.6.397}$$

Consequently, circ (A) is not a neighborhood of the origin and hence it cannot be open.

To check (1.6.397), simple considerations based on the definition of the circled hull from (1.6.361) yield that $(-A) \cup \{(0,0)\} \cup A \subseteq \text{circ}(A)$. For the reverse inclusion, appeal again to (1.6.361) and pick $\lambda \in \mathbb{R}$ such that $\lambda \in [-1,1]$. Clearly

$$\lambda \in \{-1,0,1\} \implies \lambda A \subseteq (-A) \cup \{(0,0)\} \cup A. \tag{1.6.398}$$

Assume next that $\lambda \in (0,1)$ and select $(x_1, x_2) \in A$. Then

$$x_1 > 0 \text{ and } (x_1 - 1)^2 + x_2^2 < 1, \tag{1.6.399}$$

and $\lambda(x_1, x_2)$ satisfies

$$\begin{aligned} |\lambda(x_1,x_2) - (1,0)|^2 &= (\lambda x_1 - 1)^2 + (\lambda x_2)^2 \\ &= \lambda^2 \left[(x_1-1)^2 + x_2^2\right] + (1-\lambda^2) + 2x_1\lambda(\lambda-1) \\ &< \lambda^2 + (1-\lambda^2) + 2x_1\lambda(\lambda-1) \\ &< 1. \end{aligned} \tag{1.6.400}$$

Here the first and second equality follow from the definition of the distance in $\mathbb{R}^2$ and from trivial algebraic manipulations, the first inequality is a consequence of the second inequality in (1.6.399), while the second inequality follows from the first inequality in (1.6.399) and the fact that $\lambda \in (0,1)$. This shows that in this case $\lambda(x_1, x_2) \in A$ and hence

$$\lambda \in (0,1) \implies \lambda A \subseteq A. \tag{1.6.401}$$

One can establish in a similar way that

$$\lambda \in (-1,0) \implies \lambda A \subseteq -A. \tag{1.6.402}$$

This establishes the left-to-right inclusion in (1.6.397), completing the proof of the claim made in (1.6.397).

Another way to see that (1.6.397) holds is to appeal to Exercise 1.6.15 and to notice that in the case when $A = B((1,0),1) \subseteq \mathbb{R}^2$ then

$$\bigcup_{a\in A}\Big([0,a] \cup [0,-a]\Big) = (-A) \cup \{0\} \cup A. \tag{1.6.403}$$

Indeed, (1.6.403) is immediately seen geometrically. ■

We are now ready to begin the study of the geometric consequences of the continuity of vector operations in a linear topological space.

Theorem 1.6.18. *Let (X,τ) be a linear topological space over the field $\mathbb{K}$ and pick $U \in \mathcal{N}(0;\tau)$, where 0 is the zero vector in X. Then the following properties are true:*

(a) One has

$$\forall\, x \in X \quad \exists\, \delta_x > 0 \;\text{ such that }\; \lambda x \in U \;\text{ if } \lambda \in \mathbb{K} \text{ satisfies } |\lambda| < \delta_x. \tag{1.6.404}$$

In particular U is absorbing.

(b) The set U contains a circled zero-neighborhood, i.e.,

$$\exists\, W \in \mathcal{N}(0;\tau) \;\text{ such that }\; W = \operatorname{circ}(W) \;\text{ and }\; W \subseteq U. \tag{1.6.405}$$

(c) There exists $V \in \mathcal{N}(0;\tau)$ such that $V + V \subseteq U$.

Proof. To establish *(a)* start by fixing $x \in X$. The joint continuity of scalar multiplication implies that the map $\mathbb{K} \ni \lambda \mapsto \lambda x \in X$ is continuous and maps the scalar zero into the zero vector in X. Using Proposition 1.1.8 and Proposition 1.2.5 along with the fact that $U \in \mathcal{N}(0;\tau)$, this further yields the existence of $\delta_x > 0$ such that

$$\lambda \in B_{\mathbb{K}}(0,\delta_x) \implies \lambda x \in U, \tag{1.6.406}$$

establishing (1.6.404). The fact that U is absorbing is now a direct consequence of (1.6.404) and the definition of absorbing sets from (c) in Definition 1.6.14.

The proof of statement *(b)* requires the full strength of joint continuity of scalar multiplication

$$f_* : \mathbb{K} \times X \longrightarrow X, \quad f_*(\alpha,x) := \alpha x, \quad \forall\, (\alpha,x) \in \mathbb{K} \times X. \tag{1.6.407}$$

Since $U \in \mathcal{N}(0;\tau)$ and $f_*(0,0) = 0$, Definition 1.1.7 (cf. (1.1.4)) and Proposition 1.2.5 guarantee the existence of a set $V \in \mathcal{N}(0;\tau)$ and a number $\delta > 0$ such that

$$f_*\big(B_{\mathbb{K}}(0,\delta) \times V\big) \subseteq U. \tag{1.6.408}$$

With these in hand, let

$$W := \bigcup_{|\alpha|\leq\delta} \alpha V = \bigcup_{0<|\alpha|\leq\delta} \alpha V, \tag{1.6.409}$$

where the second equality follows form the fact that $0 \in V$. Since for each $\alpha \in \mathbb{K}\setminus\{0\}$ the left-to-right implication in (1.6.307) holds, it follows that $W \in \mathcal{N}(0;\tau)$. Invoking (1.6.408) we also obtain that $W \subseteq U$. To finish, we claim that W defined in (1.6.409) is circled. Indeed, using the definition of the circled hull and (1.6.409) we may write

$$\begin{aligned}\operatorname{circ}(W) &= \bigcup_{|\lambda|\leq 1} \lambda\Big(\bigcup_{|\alpha|\leq\delta} \alpha V\Big) = \bigcup_{|\lambda|\leq 1}\bigcup_{|\alpha|\leq\delta} (\lambda\alpha)V \\ &= \bigcup_{|\gamma|\leq\delta} \gamma V = W.\end{aligned} \tag{1.6.410}$$

Thus, W is circled by *(7)* in Proposition 1.6.16.

To deal with part *(c)*, recall that the vector addition, viewed as a map from the product space $X\times X$ into X is continuous. Its continuity at the point $(0,0)$ guarantees the existence of a product neighborhood $V_1 \times V_2$, where $V_1, V_2 \in \mathcal{N}(0;\tau)$, that the addition map carries into U. Consequently, we have $V_1 + V_2 \subseteq U$. As a result, $V := V_1 \cap V_2 \in \mathcal{N}(0;\tau)$ has the desired property that $V + V \subseteq U$. □

It is easy to check that given a linear topological space (X,τ), a set $U \in \mathcal{N}(0;\tau)$, and a number $n \in \mathbb{N}$, repeated applications of property *(c)* in Theorem 1.6.18 give that there exists $V \in \mathcal{N}(0;\tau)$ such that

$$\underbrace{V + V + \cdots + V}_{2^n \text{ times}} \subseteq U. \tag{1.6.411}$$

Since for each $j,k \in \mathbb{N}$ with $j \leq k$ the j-fold sum of V with itself is a subset of the k-fold sum of V with itself, this ultimately shows that

$$\forall\, n \in \mathbb{N} \quad \exists\, V \in \mathcal{N}(0;\tau) \text{ such that } \underbrace{V + V + \cdots + V}_{n \text{ times}} \subseteq U. \tag{1.6.412}$$

1.6.2 Characterizing Local Bases

We begin by recording a particularly useful result.

Proposition 1.6.19. *In a linear topological space, the following hold:*

(1) The circled zero-neighborhoods form a local base at the origin.
(2) The open circled zero-neighborhoods form a local base at the origin.

Proof. The statement in *(1)* is just a rephrasing of part *(b)* of Theorem 1.6.18 while the statement in *(2)* is a direct consequence of part *(13)* of Proposition 1.6.16 and the current item *(1)*. □

Comment 1.6.20 *It is instructive to note that if X is a vector space over the field $\mathbb{K}$ then*

$$\begin{array}{c}\textit{a family of subsets of } X \textit{ may be a local base at the origin} \\ \textit{for at most one vector topology.}\end{array} \tag{1.6.413}$$

Indeed, suppose $\mathcal{U}$ is a local base at zero for two vector topologies τ_1 and τ_2 on X. Fix $i, j \in \{1,2\}$ such that $i \neq j$ and $x \in X$ and let $N \in \mathcal{N}(x;\tau_i)$. Since $\mathcal{U}$ is a local base at the origin for τ_i and τ_i is a vector topology on X, using (1.6.306) it follows that there exists $U \in \mathcal{U}$ such that $U \subseteq N - x$. However, $U \in \mathcal{N}(0;\tau_j)$, for $j \in \{1,2\} \setminus \{i\}$. Thus, $N - x$ becomes a neighborhood of the origin relative to τ_j, $j \in \{1,2\}$. Appealing again to (1.6.306) this further yields that $N \in \mathcal{N}(x;\tau_j)$ for each $j \in \{1,2\}$. Consequently,

$$\mathcal{N}(x;\tau_i) \subseteq \mathcal{N}(x;\tau_j). \tag{1.6.414}$$

Switching the roles of i and j in (1.6.414) and since $x \in X$ was arbitrary, (1.6.414) further gives that

$$\mathcal{N}(x;\tau_1) = \mathcal{N}(x;\tau_2) \qquad \forall\, x \in X, \tag{1.6.415}$$

from which it immediately follows that $\tau_1 = \tau_2$, finishing the proof of (1.6.413). ■

Theorem 1.6.18 provides the foundation for the following important characterization of local bases at the origin in linear topological spaces.

Theorem 1.6.21 (Local Base Characterization). *Suppose (X,τ) is a linear topological space and assume that $\mathcal{U}$ is a local base at the origin in (X,τ). Then the following properties hold.*

(a) For every $U, V \in \mathcal{U}$ there exists $W \in \mathcal{U}$ with the property that $W \subseteq U \cap V$.
(b) If $U \in \mathcal{U}$ then there exists $V \in \mathcal{U}$ such that $V + V \subseteq U$.
(c) Each member of $\mathcal{U}$ is absorbing.
(d) If $U \in \mathcal{U}$ then there exists $V \in \mathcal{U}$ such that $\mathrm{circ}\,(V) \subseteq U$.

Conversely, if a nonempty collection $\mathcal{U}$ of subsets of a vector space X over the field $\mathbb{K}$ satisfies conditions (a)-(d), then there exists a unique topology τ on X with the property that (X,τ) is a linear topological space and $\mathcal{U}$ is a local base at the origin for τ.

Proof. Suppose that (X,τ) is a linear topological space and that $\mathcal{U}$ is a local base at zero for τ. Property *(a)* follows from Definition 1.1.19 and the fact that $\mathcal{N}(0;\tau)$ is closed under finite intersections.

Moving on, statement *(b)* follows from part *(c)* of Theorem 1.6.18 and the assumption that $\mathcal{U}$ is a local base at the origin. Turning our attention to *(c)*, this is an immediate consequence of part *(a)* in Theorem 1.6.18. As for *(d)*, this follows from part *(2)* in Proposition 1.6.19 in conjunction with (1.6.366).

Turning to the converse, suppose that $\mathcal{U}$ is a collection of subsets of X that satisfies properties *(a)*-*(d)* above. Set then

$$\tau := \{A \subseteq X : \ \forall\, a \in A \ \ \exists\, U_a \in \mathcal{U} \ \text{ such that } \ a + U_a \subseteq A\}. \tag{1.6.416}$$

It is straightforward to verify that τ is a topology on X. Indeed, clearly $\varnothing \in \tau$ and $X \in \tau$. Pick next $O_1, O_2 \in \tau$ and fix $a \in O_1 \cap O_2$. Then, by (1.6.416) there exist $U_{a,1}, U_{a,2} \in \mathcal{U}$ such that $a + U_{a,j} \subseteq O_j$ for each $j \in \{1,2\}$. Appealing to property *(a)* we may further conclude that there exists $W_a \in \mathcal{U}$ such that $W_a \subseteq U_{a,1} \cap U_{a,2}$. Then $a + W_a \subseteq O_1 \cap O_2$. Since $a \in O_1 \cap O_2$ was arbitrary, this and (1.6.416) show that $O_1 \cap O_2 \in \tau$, as desired. Finally, consider a collection $\{O_i\}_{i\in I} \subseteq \tau$ and fix some $a \in \bigcup_{i\in I} O_i$. Thus, there exists $i_o \in I$ such that $a \in O_{i_o} \in \tau$. In particular, there exists $U_a \in \mathcal{U}$ such that $a + U_a \in O_{i_o}$. However this immediately gives that $a + U_a \in \bigcup_{i\in I} O_i$, showing that $\bigcup_{i\in I} O_i \in \tau$, and finishing the proof of the fact that τ from (1.6.416) is a topology on X.

We next claim that

$$U \in \mathcal{N}(0;\tau) \ \text{ for each } \ U \in \mathcal{U}. \tag{1.6.417}$$

To justify this claim, fix $U \in \mathcal{U}$. Our goal is to show that $U \in \mathcal{N}(0;\tau)$, i.e., to identify a set O such that

$$O \in \tau \ \text{ and } \ 0 \in O \subseteq U. \tag{1.6.418}$$

With this goal in mind, set

$$O := \{x \in X : \exists\, U_x \in \mathcal{U} \ \text{ such that } \ x + U_x \subseteq U\}, \tag{1.6.419}$$

and observe that since $0 + U = U \in \mathcal{U}$ we have $0 \in O$. Next, by property *(d)* each $U \in \mathcal{U}$ contains a circled set, which in turn contains the zero vector by (1.6.361). Thus, for each $x \in O$ we have

$$x = x + 0 \in x + U_x \subseteq U, \tag{1.6.420}$$

where $U_x \in \mathcal{U}$ is as in (1.6.419). This shows that $O \subseteq U$.

To complete the proof of (1.6.418) there remains to show that $O \in \tau$. To see this, start by fixing $a \in O$. On grounds of (1.6.419) choose $U_a \in \mathcal{U}$ such that $a + U_a \subseteq U$. Since the collection $\mathcal{U}$ satisfies property *(b)*, there exists $W_a \in \mathcal{U}$ with the property that $W_a + W_a \subseteq U_a$. Thus if $y \in a + W_a$, one has

$$W_a + y \subseteq W_a + (W_a + a) \subseteq U_a + a \subseteq U. \tag{1.6.421}$$

In particular $y \in O$ and since $y \in a + W_a$ was arbitrary, this implies $a + W_a \subseteq O$. Consequently, by (1.6.416), we have $O \in \tau$, as desired. This completes the proof of (1.6.417).

At this stage, we shall observe that the collection of all zero-neighborhoods in X, relative to the topology τ, is precisely

$$\mathcal{N}(0;\tau) = \{N \subseteq X : \exists\, U \in \mathcal{U} \ \text{ such that } \ U \subseteq N\}. \tag{1.6.422}$$

Indeed, if $N \in \mathcal{N}(0;\tau)$ then there exists a set $O \in \tau$ such that $0 \in O$ and $O \subseteq N$. Hence, using the definition of the topology τ from (1.6.416) it follows that there exists some $U_o \in \mathcal{U}$ with the property that $U_o = 0 + U_o \subseteq O$. In particular, $U_o \subseteq N$

as desired. This establishes the left-to-right inclusion in (1.6.422). The right-to-left inclusion is an immediate consequence (1.6.417), thus the equality in (1.6.422) is proven. Collectively (1.6.417) and (1.6.422) show that

$$\text{the collection } \mathcal{U} \text{ constitutes a local base at the origin for the topology } \tau. \tag{1.6.423}$$

In this context, it is worth pointing out several properties. The first is that

$$x + O \in \tau \quad \text{for each} \quad x \in X \quad \text{and} \quad O \in \tau. \tag{1.6.424}$$

Indeed if $a \in x + O$ there exists $x_o \in O$ such that $x + x_o = a$. Using the definition of τ from (1.6.416) in concert with the fact that $O \in \tau$ yields a set $U \in \mathcal{U}$ such that $x_o + U \subseteq O$. Thus, we have $a + U = x + (x_o + U) \subseteq x + O$, establishing (1.6.424). As a consequence of (1.6.424) and (1.6.417) we obtain that for

$$x + U \in \mathcal{N}(x;\tau) \quad \text{for all} \quad x \in X \quad \text{and all} \quad U \in \mathcal{U}. \tag{1.6.425}$$

On the other hand, if $x \in X$ is fixed and $N \in \mathcal{N}(x;\tau)$ then there exists some set $O \in \tau$ such that $x \in O \subseteq N$. By (1.6.424) we have $\widetilde{O} := -x + O \in \tau$ and clearly $0 \in \widetilde{O}$ and $\widetilde{O} \subseteq -x + N$. This shows that $-x + N \in \mathcal{N}(0;\tau)$, i.e.,

$$-x + N \in \mathcal{N}(0;\tau) \quad \text{for all} \quad x \in X \quad \text{and} \quad N \in \mathcal{N}(x;\tau). \tag{1.6.426}$$

The next order of business is to show the (joint) continuity of addition on $X \times X$ and that of scalar multiplication on $\mathbb{K}\times X$. To this end, fix first a point $(x_0, y_0) \in X\times X$. To prove the continuity of the addition map

$$f_+ : X \times X \longrightarrow X, \qquad f_+(x,y) := x + y, \quad \forall\, x, y \in X, \tag{1.6.427}$$

at the point (x_0, y_0), fix an arbitrary $N \in \mathcal{N}(x_0 + y_0;\tau)$ with the goal of identifying two neighborhoods, $A \in \mathcal{N}(x_0;\tau)$ and $B \in \mathcal{N}(y_0;\tau)$, such that $f_+(A \times B) \subseteq N$. Since $N \in \mathcal{N}(x_0 + y_0;\tau)$ it follows that $-x_0 - y_0 + N \in \mathcal{N}(0;\tau)$, by (1.6.426). In particular, using (1.6.423) there exists $U \in \mathcal{U}$ such that $U \subseteq -x_0 - y_0 + N$. Invoking property *(b)* from the hypotheses satisfied by the collection $\mathcal{U}$, there exists $V \in \mathcal{U}$ such that

$$V + V \subseteq U \subseteq -x_0 - y_0 + N. \tag{1.6.428}$$

Thus if we set $A := x_0 + V$ and $B := y_0 + V$, by (1.6.425) we have $A \in \mathcal{N}(x_0;\tau)$ and $B \in \mathcal{N}(y_0;\tau)$ and

$$f_+(A \times B) = A + B = (x_0 + V) + (y_0 + V) = x_0 + y_0 + V + V \subseteq N, \tag{1.6.429}$$

where the last inclusion is a result of (1.6.428). This completes the proof of the continuity of f_+ at (x_0, y_0) and since this point was arbitrary in $X \times X$ it follows that f_+ is continuous on $X \times X$, as desired.

We consider next the scalar multiplication map

$$f_* : \mathbb{K} \times X \longrightarrow X, \qquad f_*(\lambda, x) := \lambda x, \quad \forall\, \lambda \in \mathbb{K} \text{ and } \forall\, x \in X. \tag{1.6.430}$$

Fix $\lambda_0 \in \mathbb{K}$ and $x_0 \in X$ and with the goal in mind of establishing the continuity of f_* from (1.6.430) at the point (λ_0, x_0), select $N \in \mathcal{N}(\lambda_0 x_0; \tau)$. Using Proposition 1.1.8 matters reduce to identifying $\delta > 0$ and $A \in \mathcal{N}(x_0; \tau)$ such that

$$f_*(B_{\mathbb{K}}(\lambda_0, \delta) \times A) \subseteq N. \tag{1.6.431}$$

Since $N \in \mathcal{N}(\lambda_0 x_0; \tau)$, from (1.6.426) we see that $-\lambda_0 x_0 + N \in \mathcal{N}(0; \tau)$. Using (1.6.423), this gives that there exists $U \in \mathcal{U}$ such that $\lambda_0 x_0 + U \subseteq N$. Appealing to properties *(b)* and *(d)* that collection $\mathcal{U}$ satisfies, this further implies

$$\exists\, V \in \mathcal{U} \text{ such that } \lambda_0 x_0 + \mathrm{circ}(V) + \mathrm{circ}(V) \subseteq N. \tag{1.6.432}$$

At this point, invoking property *(c)* of $\mathcal{U}$, the set V is absorbing, and as such

$$\exists\, \delta > 0 \text{ such that } t x_0 \in V \text{ for each } t \in (0, \delta). \tag{1.6.433}$$

In particular,

$$\gamma x_0 \in \mathrm{circ}(V) \text{ for each } \gamma \in \mathbb{K} \text{ such that } |\gamma| \in (0, \delta). \tag{1.6.434}$$

Indeed, if $\eta := \gamma|\gamma|^{-1}$ then $\gamma x_0 = \eta(|\gamma| x_0) \in \eta V \subseteq \mathrm{circ}(V)$ where, the membership follows from (1.6.433), and the inclusion is a consequence of the definition of the circled hull of a set and the fact that η has magnitude equal to 1.

Note now that for each $\lambda \in \mathbb{K}$ and each $x \in X$ we have

$$\lambda x - \lambda_0 x_0 = (\lambda - \lambda_0) x_0 + \lambda(x - x_0). \tag{1.6.435}$$

Thus, whenever $\lambda \in B_{\mathbb{K}}(\lambda_0, \delta)$ by (1.6.434) we have $(\lambda - \lambda_0) x_0 \in \mathrm{circ}(V)$. Using this and (1.6.435) we may therefore write

$$|\lambda - \lambda_0| < \delta \implies \lambda x \in \lambda_0 x_0 + \mathrm{circ}(V) + \lambda(x - x_0). \tag{1.6.436}$$

For the remaining part of the argument fix $\lambda \in B_{\mathbb{K}}(\lambda_0, \delta)$ and observe that

$$|\lambda| \leq |\lambda - \lambda_0| + |\lambda_0| < \delta + |\lambda_0| < 2^n, \tag{1.6.437}$$

for some $n \in \mathbb{N}$ large enough, which depends only on δ and λ_0, and which we also fix for the rest of the discussion. Next, with $V \in \mathcal{U}$ as in (1.6.432), we shall repeatedly apply property *(b)* and then property *(d)* to conclude

$$\exists\, W \in \mathcal{U} \text{ such that } \underbrace{\mathrm{circ}(W) + \mathrm{circ}(W) + \cdots + \mathrm{circ}(W)}_{2^n \text{ times}} \subseteq V \subseteq \mathrm{circ}(V), \tag{1.6.438}$$

where the last inclusion is a consequence of part *(3)* in Proposition 1.6.16.

We shall set

$$A := x_0 + W \in \mathcal{N}(x_0; \tau), \tag{1.6.439}$$

where the membership above is a consequence of (1.6.425). Observe that whenever $(\lambda, x) \in B_{\mathbb{K}}(\lambda_0, \delta) \times A$, by (1.6.436) we have

$$\lambda x \in \lambda_0 x_0 + \mathrm{circ}(V) + \lambda(x - x_0) \quad \text{and} \quad x - x_0 \in W. \tag{1.6.440}$$

Thus,

$$\lambda(x - x_0) \in \lambda W \subseteq \lambda \mathrm{circ}(W) \subseteq 2^n \mathrm{circ}(W) \subseteq \mathrm{circ}(V), \tag{1.6.441}$$

where the membership in (1.6.441) follows from the second part of (1.6.440), the first inclusion is a consequence of part *(3)* in Proposition 1.6.16, the second inclusion uses the expansion property of circled sets (cf. the direct implication in part *(9)* in Proposition 1.6.16) since $|\lambda| < 2^n$, and the third inclusion follows from the fact that W is as in (1.6.438). Taking this back in (1.6.440) yields

$$\lambda x \in \lambda_0 x_0 + \mathrm{circ}(V) + \mathrm{circ}(V) \subseteq N, \tag{1.6.442}$$

where the inclusion above is a consequence of the fact that V is as in (1.6.432). This shows that (1.6.431) is satisfied, completing the proof of the continuity of the scalar multiplication function f_* at the point (λ_0, x_0). Since this point was arbitrary in $\mathbb{K} \times X$, it follows that f_* is continuous on $\mathbb{K} \times X$, as desired.

Combining the continuity results proved so far we may conclude that the vector space X equipped with the topology τ from (1.6.416) becomes a linear topological space. Finally, the uniqueness of the topology τ follows readily from (1.6.413), completing the proof of Theorem 1.6.21. □

Moving on, the following proposition shows that even within the friendly confines of a Hilbert space, there lurk non-metrizable vector topologies.

Proposition 1.6.22 (Weak topology of Hilbert space). *Let $(H, \langle \cdot, \cdot \rangle)$ be a Hilbert space (cf. Definition 1.5.4). For each $\varepsilon > 0$ and each finite set $F \subseteq H$ define*

$$N(F, \varepsilon) := \Big\{ x \in H : |\langle x, f \rangle| < \varepsilon, \ \ \forall f \in F \Big\}. \tag{1.6.443}$$

Then the collection

$$\mathcal{U} = \Big\{ N(F, \varepsilon) : \varepsilon > 0 \ \ \textit{and} \ \ F \subseteq H \ \ \textit{finite} \Big\} \tag{1.6.444}$$

satisfies properties (a)-(d) of Theorem 1.6.21, and is therefore a local base at the origin for a vector topology on H. This topology, denoted by τ_w, is called the `weak topology` *of H.*

Proof. We start with the proof of the fact that the collection $\mathcal{U}$ from (1.6.444) satisfies property *(a)* from Theorem 1.6.21. For this, consider two arbitrary elements of $\mathcal{U}$, i.e., two sets $N(F_1, \varepsilon_1)$ and $N(F_2, \varepsilon_2)$ as in (1.6.443), with $\varepsilon_1, \varepsilon_2 > 0$ and $F_1, F_2 \subseteq H$ finite sets. Choose then

$$\varepsilon := \min\{\varepsilon_1, \varepsilon_2\} \ \text{ and } \ F := F_1 \cup F_2, \tag{1.6.445}$$

and notice that $\varepsilon > 0$ and $F \subseteq H$ is finite. Then, it is immediate that $N(F, \varepsilon) \in \mathcal{U}$ and

$$N(F, \varepsilon) \subseteq N(F_1, \varepsilon_1) \cap N(F_2, \varepsilon_2). \tag{1.6.446}$$

Turning to property *(b)* in Theorem 1.6.21, pick $N(F, \varepsilon) \in \mathcal{U}$. Thus $\varepsilon > 0$ and $F \subseteq H$ is finite. Then $N(F, \varepsilon/2) \in \mathcal{U}$ and for each $x, y \in N(F, \varepsilon/2)$ and each $f \in F$ we have

$$|\langle x + y, f\rangle| = |\langle x, f\rangle + \langle y, f\rangle| \leq |\langle x, f\rangle| + |\langle y, f\rangle| \leq \varepsilon/2 + \varepsilon/2 = \varepsilon, \tag{1.6.447}$$

which shows that $N(F, \varepsilon/2) + N(F, \varepsilon/2) \subseteq N(F, \varepsilon)$, and finishes the proof of property *(b)*.

Next we shall check that the collection $\mathcal{U}$ in (1.6.444) satisfies property *(c)* in Theorem 1.6.21. To this end, pick $N(F, \varepsilon) \in \mathcal{U}$, for some $\varepsilon > 0$ and some finite set $F \subseteq H$. Our goal is to show that $N(F, \varepsilon)$ is absorbing. By part *(c)* of Definition 1.6.14 this amounts to showing that for each $x \in H$ there exists $\delta_x > 0$ such that the following implication holds

$$t \in (0, \delta_x) \implies tx \in N(F, \varepsilon). \tag{1.6.448}$$

Consider first the case when $x \in N(F, \varepsilon)$ and observe that in this case one can take $\delta_x := 1$ in (1.6.448). Indeed, for each $t \in (0, 1)$ we have

$$|\langle tx, f\rangle| = t|\langle x, f\rangle| \leq |\langle x, f\rangle| \leq \varepsilon, \tag{1.6.449}$$

where the first equality is a consequence of property (1.5.257) of the inner product, the first inequality uses $t \in (0, 1)$ and the last inequality uses that $x \in N(F, \varepsilon)$ and (1.6.443). All in all, (1.6.449) shows that (1.6.448) holds for $x \in N(F, \varepsilon)$ and $\delta_x := 1$.

Going further, fix $x \in H \setminus N(F, \varepsilon)$. Using (1.6.443) and the finiteness of F, this guarantees that

$$\varepsilon \leq \max\Big\{|\langle x, f\rangle| : f \in F\Big\} < \infty. \tag{1.6.450}$$

In view of (1.6.450), it is meaningful to set

$$\delta_x := \frac{\varepsilon}{\max\Big\{|\langle x, f\rangle| : f \in F\Big\}} \in (0, 1]. \tag{1.6.451}$$

Then, for each $t \in (0, \delta_x)$ and each $g \in F$ we have

$$|\langle tx, g\rangle| = t|\langle x, g\rangle| \leq \delta_x \cdot \max\Big\{|\langle x, f\rangle| : f \in F\Big\} = \varepsilon. \tag{1.6.452}$$

Indeed, the first equality uses again property (1.5.257) of the inner product, the first inequality follows from the fact that $t \in (0, \delta_x)$ and the fact that $g \in F$, while the last equality uses the definition of δ_x from (1.6.451). Thus, (1.6.452) establishes the implication (1.6.448) for $x \in H \setminus N(F, \varepsilon)$ and δ_x as in (1.6.451). Combining this conclusion with the one in the case $x \in N(F, \varepsilon)$ which we already analyzed, shows that the implication (1.6.448) holds for each $x \in H$. This finishes the proof of the fact that $N(F,\varepsilon)$ is absorbing, thus establishes that collection $\mathcal{U}$ satisfies property *(c)* in Theorem 1.6.21.

Finally, property *(d)* in Theorem 1.6.21 follows as soon as we show that each element of $\mathcal{U}$ is in fact a circled set. With this goal in mind, pick $N(F,\varepsilon) \in \mathcal{U}$, for some $\varepsilon > 0$ and $F \subseteq H$ finite set. Then, for each $x \in N(F,\varepsilon)$, each $f \in F$, and each $\lambda \in \mathbb{K}$ such that $|\lambda| \leq 1$ we have

$$\left|\langle \lambda x, f\rangle\right| = |\lambda| \cdot \left|\langle x, f\rangle\right| < \varepsilon, \tag{1.6.453}$$

i.e., $\lambda x \in N(F,\varepsilon)$. In conclusion,

$$\lambda \cdot N(F,\varepsilon) \subseteq N(F,\varepsilon) \;\text{ whenever }\; \lambda \in \mathbb{K} \;\text{ is such that }\; |\lambda| \leq 1. \tag{1.6.454}$$

As a consequence of this and *(e)* in Definition 1.6.14 it follows that we have $\operatorname{circ}(N(F,\varepsilon)) \subseteq N(F,\varepsilon)$. However, by part *(3)* in Proposition 1.6.16 the inclusion $N(F,\varepsilon) \subseteq \operatorname{circ}(N(F,\varepsilon))$ also holds and, as such, $N(F,\varepsilon) = \operatorname{circ}(N(F,\varepsilon))$, showing that $N(F,\varepsilon)$ is circled (by appealing again to the statement *(3)* in Proposition 1.6.16). This completes the proof of the fact that the collection $\mathcal{U}$ from (1.6.444) satisfies properties *(a)-(d)* from Theorem 1.6.21. Consequently, $\mathcal{U}$ is a local base at the origin for a vector topology on H, as desired. □

Let us mention here that according to results proved later (cf. Corollary 1.6.37), for the weak topology τ_w on H (introduced in Proposition 1.6.22) we have:

$$\text{the space } (H,\tau_w) \text{ is a Hausdorff topological space.} \tag{1.6.455}$$

Exercise 1.6.23. Let $(H,\langle\cdot,\cdot\rangle)$ be a Hilbert space. Then a sequence $\{x_n\}_{n\in\mathbb{N}}$ in H converges to zero in the weak topology τ_w if and only if for any $y \in H$ it follows that $\langle x_n, y\rangle \to 0$ as $n \to \infty$.

For this reason the weak topology on a Hilbert space H is often called the `topology of pointwise convergence` on H. Exercise 1.6.23 together with the Uniform Boundedness Principle show that *every weakly convergent sequence in H is bounded in norm*. This is *not*, however, true in general for *nets* (this may be deduced from Exercise 9 in §1.7 specialized to the case when X is an infinite-dimensional Hilbert space).

Going further, the goal is to establish that, in any Hilbert space, every zero-neighborhood in the weak topology is also a zero-neighborhood in the norm topology.

Proposition 1.6.24. *Let $(H,\langle\cdot,\cdot\rangle)$ be a Hilbert space. Recall from Proposition 1.6.2 that H is a linear topological space when equipped with the metric topology τ (induced by its norm $\|\cdot\|$). In addition, consider the weak topology τ_w of H introduced in Proposition 1.6.22. Then*

$$\mathcal{N}(0;\tau_w) \subseteq \mathcal{N}(0;\tau). \tag{1.6.456}$$

Proof. Pick an arbitrary $U \in \mathcal{N}(0;\tau_w)$. According to Proposition 1.6.2, there exist a finite subset F of H along with $\varepsilon > 0$ such that $N(F,\varepsilon) \subseteq U$. If $F = \{0\}$ then $N(F,\varepsilon) = H$ and this forces $U = H$. Thus, in this case we have $U \in \mathcal{N}(0;\tau)$. Assume next that $F \neq \{0\}$. In this scenario, we claim that $N(F,\varepsilon)$ contains the open ball in H centered at the origin and of radius $\varepsilon/\max\{\|f\| : f \in F\}$, where $\|\cdot\|$ is the norm induced by the inner product as in (1.5.261). Indeed, first notice that if $F \neq \{0\}$ is finite then $\max\{\|f\| : f \in F\}$ is meaningful and, by property *(3)* from Definition 1.5.1, we have $\max\{\|f\| : f \in F\} \neq 0$. Thus, if $g \in H$ is such that

$$\|g\| < \frac{\varepsilon}{\max\{\|f\| : f \in F\}} \in (0,\infty), \tag{1.6.457}$$

then

$$|\langle g,h\rangle| \leq \|g\| \cdot \|h\| < \frac{\varepsilon}{\max\{\|f\| : f \in F\}} \cdot \|h\| \leq \varepsilon, \qquad \forall\, h \in F. \tag{1.6.458}$$

Above, the first inequality is a consequence of the Schwarz inequality from Theorem 1.5.2, the second follows from (1.6.457), while the last uses that $h \in F$. Together, (1.6.457) and (1.6.458) guarantee that the open ball centered at the origin with radius $\varepsilon/\max\{\|f\| : f \in F\}$ is contained in $N(F,\varepsilon)$, as desired. This places $N(F,\varepsilon)$ in $\mathcal{N}(0;\tau)$, and since U contains $N(F,\varepsilon)$ it follows that $U \in \mathcal{N}(0;\tau)$. □

Proposition 1.6.24 shows that the weak topology of H is possibly weaker than the topology induced on H by the norm $\|\cdot\|$. In fact, this can be substantially weaker! The following result sheds light in this regard.

Proposition 1.6.25. *Suppose $\{e_n\}_{n\in\mathbb{N}}$ is an orthonormal sequence in a Hilbert space $(H,\langle\cdot,\cdot\rangle)$. Then $\{e_n\}_{n\in\mathbb{N}}$ converges to zero in the weak topology on H, but does not converge to any vector in H in the norm topology.*

Proof. That $\{e_n\}_{n\in\mathbb{N}}$ converges to the zero vector as $n \to \infty$ in the weak topology is seen using Bessel's inequality from Theorem 1.5.14. Indeed, given $U \in \mathcal{N}(0;\tau_w)$, where as before τ_w denotes the weak topology on H, then the results in Proposition 1.6.22 ensure that there exists $\varepsilon > 0$ and $F \subseteq H$ finite such that

$$N(F,\varepsilon) \subseteq U. \tag{1.6.459}$$

Appealing to Bessel's inequality (1.5.290) and the fact that the set F is finite we may therefore infer that there exists $n(F,\varepsilon) \in \mathbb{N}$ such that

$$\sum_{n\geq n(F,\varepsilon)} |\langle f,e_n\rangle|^2 \leq \varepsilon^2, \qquad \forall\, f \in F. \tag{1.6.460}$$

In particular (1.6.460) yields that $e_n \in N(F,\varepsilon) \subseteq U$ for each $n \geq n(F,\varepsilon)$. This completes the proof of the fact that $\{e_n\}_{n\in\mathbb{N}}$ converges to the zero vector in the weak topology on H.

Next, we reason by contradiction and assume that there exists $h \in H$ such that

$$\lim_{n\to\infty} \|e_n - h\| = 0. \tag{1.6.461}$$

Since by the discussion preceding this example the weak topology τ_w is coarser than the norm topology on H, then the sequence $\{e_n\}_{n\in\mathbb{N}}$ converges to h in the weak topology. Using (1.6.455) it follows that h must be the zero vector of H. In light of (1.6.461), the numerical sequence $\{\|e_n\|\}_{n\in\mathbb{N}}$ converges to $0 \in \mathbb{R}$, which contradicts the fact that each e_n is a unit vector in H. Thus, $\{e_n\}_{n\in\mathbb{N}}$ is not convergent in the norm topology on H. □

Before continuing, we recall below the Uniform Boundedness Principle (aka, the Banach-Steinhaus Theorem):

> Let X be a Banach space and Y a normed vector space. Suppose $\{T_\alpha\}_{\alpha\in A}$ is a family of bounded linear operators from X into Y. If $\sup_{\alpha\in A} \|T_\alpha\, x\|_Y < \infty$ for all $x \in X$, then the family $\{T_\alpha\}_{\alpha\in A}$ is uniformly bounded, i.e., $\sup_{\alpha\in A} \|T_\alpha\|_{X\to Y} < \infty$. (1.6.462)

For a more general version see [39, Theorem 2.5].

Proposition 1.6.26. *The weak topology of an infinite-dimensional Hilbert space $(H, \langle\cdot,\cdot\rangle)$ does not have a countable local base at the origin.*

Proof. We reason by contradiction. Assume that the weak topology τ_w of the infinite-dimensional Hilbert space H has a countable local base at the origin

$$\mathfrak{U} := \{U_n : n \in \mathbb{N}\}. \tag{1.6.463}$$

Bring in the collection $\mathcal{U}$ from (1.6.444), which by Proposition 1.6.22 is a local base at the origin for τ_w. As such, for each $n \in \mathbb{N}$ there exist a finite set $F_n \subseteq H$ and some $\varepsilon_n > 0$ such that

$$N(F_n, \varepsilon_n) \subseteq U_n. \tag{1.6.464}$$

In view of the fact that for any two finite sets $F, G \subseteq H$ such that $F \subseteq G$ and each $\varepsilon > 0$ one has $N(G, \varepsilon) \subseteq N(F, \varepsilon)$, we may, without loss of generality, assume that the sequence of sets $\{F_n\}_{n\in\mathbb{N}}$ in (1.6.464) is increasing, i.e.,

$$F_n \subseteq F_{n+1}, \quad \forall\, n \in \mathbb{N}. \tag{1.6.465}$$

Given that H is infinite dimensional, for each $n \in \mathbb{N}$ there exists a vector $x_n \in H$ with $\|x_n\| = n$ and $x_n \perp F_n$. Now suppose $x \in H$ and $\varepsilon > 0$ are given. Since $N(\{x\}, \varepsilon)$ is a zero-neighborhood in τ_w and the family $\mathfrak{U}$ is a local base at the origin, there exists $n_o \in \mathbb{N}$ such that $U_{n_o} \subseteq N(\{x\}, \varepsilon)$. Consequently, using (1.6.464), we obtain

$$N(F_{n_o}, \varepsilon_{n_o}) \subseteq N(\{x\}, \varepsilon). \tag{1.6.466}$$

Thus, if $n \in \mathbb{N}$ is such that $n \geq n_o$, then x_n is perpendicular to F_n which in turn contains F_{n_o}. Consequently, for each such n we have $x_n \in N(F_{n_o}, \varepsilon_{n_o})$. This and (1.6.466) then yield

$$n \geq n_o \implies x_n \in N(F_{n_o}, \epsilon_{n_o}) \subseteq N(\{x\}, \epsilon). \tag{1.6.467}$$

As a result, $|\langle x_n, x\rangle| < \varepsilon$ for all $n \in \mathbb{N}$ such that $n \geq n_o$, and therefore, since $\varepsilon > 0$ and $x \in H$ are arbitrary, we obtain

$$\lim_{n\to\infty} \langle x_n, x\rangle = 0, \quad \forall\, x \in H. \tag{1.6.468}$$

The Uniform Boundedness Principle (cf. (1.6.462)) applied to $\{\langle\,\cdot\,, x_n\rangle\}_{n\in\mathbb{N}}$, the family of bounded linear functionals from H into $\mathbb{C}$, guarantees that $\sup_n \|x_n\| < \infty$. But this contradicts the fact that we have specified $\|x_n\| = n$ for each $n \in \mathbb{N}$. Thus the original sequence of sets $\{U_n : n \geq 1\}$ could not have been a local base at zero for the weak topology of H. □

Theorem 1.6.27. *The weak topology of a Hilbert space is metrizable if and only if the space is finite dimensional.*

Proof. Every metrizable linear topological space has a countable local base at the origin, namely the sequence $\{B_n : n = 1, 2, \dots\}$ where B_n is the ball of radius $1/n$ centered at the origin. In concert, this and Proposition 1.6.26 give that the weak topology of an infinite-dimensional Hilbert space is not metrizable. The converse is left as an exercise (see Exercise 6 in §1.7). □

Proposition 1.6.22 may be generalized considerably. Let X be a Banach space, and let X^* be its dual space (all continuous linear functionals on X). Define the `weak-star` topology on X^* in a manner similar to Proposition 1.6.22 (see Exercise 7 in §1.7). Much as in Theorem 1.6.27, if X is infinite dimensional, then the weak-star topology on X^* is not metrizable (again, see Exercise 7 in §1.7).

Exercise 1.6.28. Let $(H, \langle\cdot,\cdot\rangle)$ be a Hilbert space. In particular, H is Banach and its dual, H^*, is canonically identified with H (via $H \ni x \equiv \langle x, \cdot\rangle \in H^*$; cf. Theorem 1.5.11). Show that, under this identification, the weak-star topology on H^* coincides with the weak topology τ_w on H.

Next, recall that the Banach-Alaoglu Theorem (cf. [39, Theorem 3.15, p. 68]) asserts that:

given a topological vector space X and a neighborhood V of the origin in X, then $\{\Lambda \in X^* : |\Lambda x| \leq 1 \text{ for every } x \in V\}$ is a compact set in the weak-star topology on X^* (for the latter, see Exercise 7 in §1.7). (1.6.469)

We close by recording a useful consequence of the Banach-Alaoglu Theorem and Exercise 1.6.28.

Theorem 1.6.29. *If H is a Hilbert space, then its closed unit ball $\{h \in H : \|h\| \leq 1\}$ is compact in the weak topology.*

1.6.3 Closed Neighborhoods

In the Euclidean space endowed with the topology induced by the Euclidean norm, the collection of open balls centered at the origin forms a local base at zero. However, the closed balls about the origin serve just as well. Using Theorem 1.6.18 we can show that something similar happens in every linear topological space. We have already seen that the *circled* neighborhoods of zero form a local base at the origin. We now show that the same is true for the *closed* circled zero-neighborhoods. The key to this is an amusing arithmetic property of circled sets.

Lemma 1.6.30 (Circled Addition). *Let X be a vector space, and suppose $C \subseteq X$ is symmetric, in the sense that $C = -C$. Then, for every subset $A \subseteq X$, one has*

$$A + C = \{x \in X : (x + C) \cap A \neq \varnothing\}. \tag{1.6.470}$$

Proof. This follows by unraveling definitions. □

Remark 1.6.31. By property *(2)* in Proposition 1.6.16, every circled set in a vector space X is symmetric, hence (1.6.470) holds for every circled set C in X.

Lemma 1.6.30 has the following important corollary which provides an efficient way of dealing with the notion of closure in a linear topological setting.

Corollary 1.6.32 (Closure). *Given a linear topological space (X, τ), for every subset A of X one has*

$$\overline{A} = \bigcap_{U \in \mathcal{N}(0;\tau)} (A + U). \tag{1.6.471}$$

Proof. The proof proceeds by listing the following statements about an element $x \in X$, each of which is equivalent to the previous one (thanks to the result invoked in parentheses):

(1) x belongs to the closure of A.
(2) each neighborhood of x intersects A (definition of $\overline{A}$).
(3) $(x + U) \cap A \neq \varnothing$ for every circled zero-neighborhood U (item *(b)* in Theorem 1.6.18).
(4) $x \in A + U$ for every circled zero-neighborhood U (Lemma 1.6.30).
(5) $x \in A + U$ for every zero-neighborhood U (item *(b)* in Theorem 1.6.18).

The equivalence of statements (1) and (5) then proves (1.6.471). □

Corollary 1.6.33. *In any linear topological space, every zero-neighborhood contains a closed zero-neighborhood. Equivalently, the closed zero-neighborhoods form a local base at zero.*

Proof. Fix an arbitrary zero-neighborhood U and use part *(c)* of Theorem 1.6.18 to choose a zero-neighborhood V whose sum with itself is contained in U. Now, thanks to Corollary 1.6.32, the closure of V, which is clearly a zero-neighborhood, lies in $V + V$, and therefore in U. □

Corollary 1.6.34. *In a linear topological space X, the closed and circled zero-neighborhoods form a local base at the origin.*

Proof. Start with an arbitrary zero-neighborhood U in X. By part *(c)* of Theorem 1.6.18, there exists a zero-neighborhood V in X such that $V + V \subseteq U$. By part *(b)* of Theorem 1.6.18 there exists a circled zero-neighborhood W in X which is contained in V. Part *(12)* of Proposition 1.6.16 guarantees that $\overline{W}$ is a closed circled zero-neighborhood in X. In addition, Corollary 1.6.32 tells us that $\overline{W}$ is contained in $W + W$ which, in turn, is contained in $V + V$, and therefore in U. The desired conclusion now follows. □

1.6.4 Hausdorff Linear Topological Spaces

In every Hausdorff topological space the singletons are closed, but this property does not in general characterize Hausdorffness.

Exercise 1.6.35. Consider an infinite set X endowed with the topology

$$\tau := \{\mathcal{O} \subseteq X : \mathcal{O}^c := X \setminus \mathcal{O} \text{ is finite}\}. \tag{1.6.472}$$

Show that the closed subsets of X, with respect to the topology τ, are precisely X and all finite sets, and yet the topological space (X, τ) fails to be Hausdorff.

The next result shows that linear topological spaces do not suffer from this affliction.

Theorem 1.6.36. *In a linear topological space the following are equivalent:*

(a) The space is Hausdorff.
(b) The intersection of all the zero-neighborhoods is the singleton $\{0\}$.
(c) The set $\{0\}$ *is closed.*
(d) Every singleton is closed.

Proof. Suppose X is a linear topological space. If X is Hausdorff, then for every non-zero point there is a zero-neighborhood that does not contain that point. In other words, *(a)* implies *(b)*.

We have just seen that the *closed* zero-neighborhoods in X form a local base at the origin, hence the intersection of all zero-neighborhoods is closed. Thus, *(b)* implies *(c)*. Moreover, since translation by any fixed vector is a homeomorphism, *(d)* follows from *(c)*.

Finally, suppose *(d)* holds. To show X is Hausdorff, it suffices to show that if x and y are distinct vectors in X then there is a zero-neighborhood V with $x+V$ disjoint from $y+V$. To produce V, observe that since the singleton $\{x-y\}$ is closed, and does not contain the origin, there is a zero-neighborhood U that does not contain $x - y$. Choose a circled zero-neighborhood V contained in U with $V + V \subseteq U$. We claim that V is the desired zero-neighborhood.

Indeed, if $x + V$ intersected $y + V$, then $x - y + V$ would intersect V. In this case, Lemma 1.6.30, or rather Remark 1.6.31 (used with $x - y$ in place of x and V used in place of both C and A), would guarantee that $x - y \in V + V \subseteq U$, contradicting the fact that $x - y \notin U$. This shows that *(d)* implies *(a)*, completing the proof of the theorem. □

Corollary 1.6.37. *The weak topology on a Hilbert space is Hausdorff.*

Proof. Suppose that H is a Hilbert space. Let a vector $x \in H$ be in every weak neighborhood of the local base at the origin; i.e., x belongs to the neighborhood $N(F, \epsilon)$ introduced in (1.6.443), for each $\epsilon > 0$ and every finite subset F of H. Then in particular it belongs to $N(\{x\}, \epsilon)$ for every $\epsilon > 0$, and is therefore orthogonal to itself. Therefore $x = 0$. Thus, the intersection of all the neighborhoods in a local base at zero is $\{0\}$, hence the same is true for the intersection of all the zero-neighborhoods. The desired result now follows from Theorem 1.6.36. □

1.6.5 Boundedness

Recall that given a Banach space, a set is called bounded if it is bounded in norm. However when dealing with spaces with no norm, the appropriate generalization of this concept is not obvious. Even in metric spaces problems can arise, and a case in point is the linear topological space $C(\mathbb{D})$ introduced in Example 1.6.4. If we try to mimic the Banach space situation and say that a set is bounded if and only if it is contained in some metric ball then we have a problem: the whole space is bounded, since it is contained in every ball having radius bigger than one! The real story is, in fact, quite absorbing.

By way of a preamble, the reader is referred to part *(b)* of Definition 1.6.14.

Definition 1.6.38. *Suppose X is a vector space equipped with a topology. A subset A of X is called* `bounded` *in X provided every zero-neighborhood in X absorbs A.*

That is,

$$A \subseteq X \text{ is bounded if for each zero-neighborhood } U \text{ in } X \text{ there exists } \varepsilon_U > 0 \text{ such that } U \supseteq \varepsilon A \text{ for each } \varepsilon \in (0, \varepsilon_U]. \tag{1.6.473}$$

Proposition 1.6.39. *Assume that X is a vector space and let τ be a topology on X. Then the union of two bounded sets of X is a bounded set of X, every subset of a bounded set in X is itself a bounded set in X, and every positive scalar multiple of a bounded set in X is itself a bounded set in X.*

Also, if T is a linear mapping of X into a vector space Y, and τ_Y is a topology on Y, whenever T is τ-to-τ_Y continuous it follows that T maps bounded subsets of X into bounded subsets of Y.

Proof. Let $A, B \subseteq X$ be bounded sets and U be a zero-neighborhood in X. By boundedness of A and B, for all $r > 0$ sufficiently large $A \subseteq rU$ and $B \subseteq rU$. Thus U

absorbs $A \cup B$, ultimately proving that $A \cup B$ is bounded. The fact that every subset of a bounded set in X is itself a bounded set in X is clear from definitions, and so is the fact that every positive scalar multiple of a bounded set is itself bounded.

Turning our attention to the last part in the statement, let B be a bounded subset of X and assume U is an arbitrary zero-neighborhood (relative to τ_Y) in Y. The goal is to show that U absorbs $T(B)$. Using the continuity of T and that $T(0) = 0$, there exists a zero-neighborhood V (relative to τ) in X with $T(V) \subseteq U$. By the boundedness of B, for all sufficiently large positive numbers r we have $B \subseteq rV$, hence by linearity,

$$T(B) \subseteq T(rV) = rT(V) \subseteq rU. \tag{1.6.474}$$

This shows that U absorbs $T(B)$, hence $T(B)$ is bounded. □

The new notion of boundedness introduced in Definition 1.6.38 is motivated by the fact that in a normed space (or even a p-normed space) a set is bounded if and only if it is absorbed by some multiple (and hence all sufficiently large multiples) of the unit ball. Thus, sets in p-normed spaces are "norm-bounded" if and only if they satisfy the new linear topological definition of boundedness from Definition 1.6.38.

By part *(a)* of Theorem 1.6.18 we have that

$$\begin{array}{c}\text{in a linear topological space, every singleton is bounded}\\ \text{and, in fact, so is every finite set.}\end{array} \tag{1.6.475}$$

Here are some further properties of bounded sets. To get a real feeling for what boundedness is all about, cover up the proofs and do them as exercises.

Theorem 1.6.40 (Boundedness basics). *Suppose (X, τ) is a linear topological space. Then the following properties are valid.*

(a) The algebraic sum of two bounded sets of X is bounded in X. In particular, in such a scenario, every translate of a bounded set by a fixed vector is bounded. Also, every scalar multiple of a bounded set is itself a bounded set.
(b) If $B \subseteq X$ is bounded, then so is $\overline{B}$.
(c) Every compact subset of X is bounded. Also, every relatively compact subset of X is bounded.
(d) The circled hull of a bounded subset of X is bounded. Consequently, if U is a bounded zero-neighborhood of X then circ (U) *is a bounded, circled and absorbing zero-neighborhood of X.*

Proof. Consider the claims made in *(a)*. First, suppose A and B are bounded subsets of the linear topological space X, and suppose U is a zero-neighborhood in X. We must show that U absorbs $A + B$. For this, using part *(c)* in Theorem 1.6.18, choose a zero-neighborhood V with $V + V \subseteq U$. Since V absorbs both A and B, for all sufficiently large positive numbers r we have

$$rU \supseteq r(V + V) = rV + rV \supseteq A + B. \tag{1.6.476}$$

Thus, U absorbs $A+B$, as desired. The second assertion in *(a)* becomes apparent by invoking property (1.6.475). The very last claim in *(a)* is clear from definitions.

Pressing on, consider the statement in part *(b)*. Suppose B is bounded in the linear topological space X and fix U, a zero-neighborhood in X. We want to show that $\overline{B}$ is absorbed by U. Upon recalling property *(c)* in Theorem 1.6.18, fix a zero-neighborhood V in X such that $V+V\subseteq U$. Then V absorbs B, so for all sufficiently large positive numbers r we have $rV\supseteq B$, and hence

$$rU\supseteq rV+rV\supseteq B+rV. \tag{1.6.477}$$

Since rV is also a zero-neighborhood, (1.6.477) entails

$$rU\supseteq \bigcap_{W\in\mathcal{N}(0;\tau)}(B+W). \tag{1.6.478}$$

By (1.6.471) in Corollary 1.6.32, $rU\supseteq\overline{B}$, showing that $\overline{B}$ is absorbed by U, as desired.

As far as the first claim in part *(c)* is concerned, suppose K is a compact subset of the linear topological space X, and let U be a zero-neighborhood in X. We wish to show that U absorbs K. Appealing to part *(c)* in Theorem 1.6.18, choose an open zero-neighborhood V with $V+V\subseteq U$, and observe that the family $\{x+V:\ x\in K\}$ covers K. Since translations in X are homeomorphisms, this family of sets becomes an open cover of K. Invoking the compactness of K, there exists a finite subcover $\{x_j+V:\ x_j\in K,\ 1\le j\le n\}$. As V is a zero-neighborhood, part *(a)* in Theorem 1.6.18 implies that for all $r>0$ sufficiently large, $x_j\in rV$ for each $1\le j\le n$. Hence for all $r>0$ sufficiently large and $j\in\{1,\dots,n\}$,

$$x_j+rV\subseteq rV+rV=r(V+V)\subseteq rU, \tag{1.6.479}$$

which shows that $K\subseteq rU$. Hence, K is bounded. The last claim in part *(c)* then follows from what we have just proved and Proposition 1.6.39.

To treat part *(d)*, assume B is a bounded set of a linear topological space X, and pick a zero-neighborhood U. From part *(2)* of Proposition 1.6.19 we know that there exists a zero-neighborhood V which is circled and such that $V\subseteq U$. The fact that B is bounded then guarantees that $B\subseteq rV$, for all $r>0$ sufficiently large. In this context, for each such r we have

$$\operatorname{circ}(B)\subseteq\operatorname{circ}(rV)=r\operatorname{circ}(V)=rV\subseteq rU, \tag{1.6.480}$$

by (1.6.364), (1.6.365) and (1.6.366). This shows that $\operatorname{circ}(B)$ is bounded. As regards the second part of *(d)*, let $U\subseteq X$ be a bounded zero-neighborhood. In virtue of the foregoing discussion, $\operatorname{circ}(U)$ is a bounded and circled zero-neighborhood. The fact that $\operatorname{circ}(U)$ is also absorbing is an immediate consequence of part *(a)* in Theorem 1.6.18, completing the proof of *(d)*. □

Exercise 1.6.41. Prove that every subset of a relatively compact set in a linear topological space is bounded.

Hint: Use Proposition 1.6.39 and item *(c)* in Theorem 1.6.40.

Shortly, we shall need the following result, of a purely topological nature.

Lemma 1.6.42. *Let* (X, τ) *be a linear topological space. Then*

$$\bigcap_{N \in \mathcal{N}(0;\tau)} N \tag{1.6.481}$$

is a subspace of X*, which actually is the largest bounded subspace of* X*. In particular, if* Y *is a bounded subspace of* X *then* Y *must be contained in* $\bigcap_{N \in \mathcal{N}(0;\tau)} N$*.*

Proof. First, we show that

$$W := \bigcap_{N \in \mathcal{N}(0;\tau)} N \tag{1.6.482}$$

is a subspace of X. Clearly, $0 \in W \subseteq X$. Let $x, y \in W$ and $V \in \mathcal{N}(0; \tau)$. By part *(c)* of Theorem 1.6.18, we may choose $U \in \mathcal{N}(0; \tau)$ with the property that $U + U \subseteq V$. Thus $x + y \in U + U \subseteq V$, and hence $x + y \in V$ for all $V \in \mathcal{N}(0; \tau)$; that is, $x + y \in W$. Going further, let $\alpha \in \mathbb{K}$ and $x \in W$. In the case when $\alpha = 0$, it is immediate to see that $\alpha x \in W$. If $\alpha \neq 0$, then $\alpha^{-1} N \in \mathcal{N}(0; \tau)$, for every $N \in \mathcal{N}(0; \tau)$, by (1.6.307). Hence,

$$\bigcap_{N \in \mathcal{N}(0;\tau)} N = \bigcap_{N \in \mathcal{N}(0;\tau)} \alpha^{-1} N. \tag{1.6.483}$$

Since $x \in W$, it follows that $x \in \alpha^{-1} N$ for all $N \in \mathcal{N}(0; \tau)$, thus $\alpha x \in W$. To recapitulate, we have just shown that W is a genuine subspace of X. One moment's thought shows that W is bounded in X (cf. Definition 1.6.38). The last issue that remains to be clarified is that W is indeed the largest bounded subspace of X. For this reason, suppose that Y is a bounded subspace of X. Then

$$\forall\, N \in \mathcal{N}(0; \tau),\ \ \exists\, \varepsilon_N > 0 \text{ such that } tY \subseteq N \ \text{ for all } t \in (0, \varepsilon_N] \tag{1.6.484}$$

(cf. by item *(b)* in Definition 1.6.14). Since Y is a vector space, $Y = tY$ for all t. As a result, $Y \subseteq W$, as wanted. This concludes the proof of the lemma. □

Proposition 1.6.43. *In any linear topological space* X*, one has*

$$X \text{ is Hausdorff} \iff \text{every bounded subspace of } X \text{ is the trivial subspace.} \tag{1.6.485}$$

Proof. Since by Lemma 1.6.42 every bounded subspace of X is contained in

$$W := \bigcap_{N \in \mathcal{N}(0;\tau)} N, \tag{1.6.486}$$

and since W coincides with the space $\{0\}$ by Theorem 1.6.36, the left-to-right implication is valid. Reciprocally, Lemma 1.6.42 warrants that W is a bounded subspace of X, and thus, by our assumption, it must be the trivial space. At this stage, the desired conclusion becomes apparent upon appealing once more to Theorem 1.6.36.
□

1.6.6 Weak Boundedness in Hilbert Spaces

The main goal of this subsection is to identify the bounded subsets of a Hilbert space H endowed with its weak topology (cf. Proposition 1.6.22). Recall from Proposition 1.6.24 that every weak zero-neighborhood in H is also a zero-neighborhood for the norm topology in H. Hence, in this context, the norm-bounded sets are all bounded (where boundedness is understood in the norm topology and, respectively, weak topology on H—cf. Definition 1.6.38).

Conversely, suppose B is a weakly bounded subset of H. Then given any vector $f \in H$, for all sufficiently large $r > 0$ we have

$$B \subseteq r \cdot N(\{f\}, 1) = N(\{f\}, r). \tag{1.6.487}$$

Thus, for all $r > 0$ sufficiently large, $|\langle b, f\rangle| < r$ for each $b \in B$, i.e., $\{\langle \cdot, b\rangle\}_{b\in B}$ is "pointwise bounded." Therefore, using the Uniform Boundedness Principle (cf. (1.6.462)) we obtain that B is indeed a norm-bounded set. Accordingly, this argument (perhaps unexpected in view of the fact that the weak topology of a Hilbert space is much weaker than the norm topology) yields:

Theorem 1.6.44. *A subset of a Hilbert space is weakly bounded if and only if it is a norm-bounded set.*

Recall Theorem 1.6.29 and note that this guarantees that norm-bounded subsets of a Hilbert space are relatively compact in the weak topology. This, along with Theorem 1.6.44 show that the Heine-Borel Theorem persists in some infinite-dimensional linear topological spaces—Hilbert space endowed with its weak topology being one of them. (For another, see Exercise 12 in §1.7.) Specifically, weak bounded sets are relatively compact in the weak topology. Let us point out, however, that in general boundedness does not imply relative compactness. Indeed, recalling Example 1.6.4, the space $C(\mathbb{D})$ contains bounded sets that are not relatively compact (see Exercise 14 in §1.7). We will also see later that the same is true for any infinite-dimensional p-normed space.

We close this introduction to boundedness with a result which will not be needed until the proof of Theorem 2.1.9. So you can either ignore it until then, or cover up the proof and treat the whole thing as an amusing exercise. For motivation observe that the only bounded linear subspace of a Hausdorff linear topological space is the trivial subspace $\{0\}$. Observe also that

$$\begin{gathered}\text{a circled subset of } \mathbb{C} \text{ endowed with the norm topology is bounded}\\ \text{precisely when it is not all of } \mathbb{C}.\end{gathered} \tag{1.6.488}$$

Lemma 1.6.45. *If a closed circled subset of $\mathbb{C}^n$ (the complex Euclidean space of dimension n) is unbounded, then it contains a nontrivial linear subspace.*

Proof. Suppose $A \subseteq \mathbb{C}^n$ is circled, closed, and unbounded. The goal is to show that A contains a nontrivial linear subspace. For starters, there exists a sequence $(x_j)_{j\in\mathbb{N}}$

of vectors in A with $\|x_j\| \to \infty$ as $j \to \infty$. By the Heine-Borel Theorem, upon replacing the original sequence with a subsequence if necessary, we may assume that the sequence of unit vectors $e_j := x_j/\|x_j\|$, indexed by $j \in \mathbb{N}$, converges to some unit vector $e \in \mathbb{C}^n$. We claim that the linear subspace spanned by the vector e lies entirely in A. To see this, fix $\lambda \in \mathbb{C}$. Because A is circled and $\lim_{j\to\infty} \|x_j\| = \infty$, the elements of the sequence $(\lambda e_j)_{j\in\mathbb{N}}$ are in A from a rank on, hence its limit λe belongs to A (since A is closed). Thus, span $\{e\} \subseteq A$, as claimed. □

As Exercise 11 in §1.7 shows, Lemma 1.6.45 is a purely finite-dimensional phenomenon.

1.6.7 Locally Bounded Spaces

The starting point is the following definition.

Definition 1.6.46. *A linear topological space X is said to be* `locally bounded` *if it has a bounded zero-neighborhood.*

Basic examples are listed in the next exercise.

Exercise 1.6.47. Any normed space is a Hausdorff, locally bounded, linear topological space. The same is true for every p-normed space with $0 < p \le 1$.

Hint: Show that, in all cases, the unit ball is a bounded zero-neighborhood.

What is not so clear, but nevertheless true, is that

the topology of every Hausdorff locally bounded linear topological space can be given by a p-norm. (1.6.489)

This remarkable result is due independently to Aoki (1942) and Rolewicz (1957) (see [23, Chapter 1] and the discussion in [33, § 1.1]). We shall return to this later, in §1.10. Here is the fact we *do* need about local boundedness.

Proposition 1.6.48. *If X is a locally bounded linear topological space, U is any bounded zero-neighborhood in X, and* $\{r_n\}_{n\in\mathbb{N}} \subseteq \mathbb{R}$ *is a sequence having* $0 \in X$ *as an accumulation point, then the family of neighborhoods* $\{r_n U : n \in \mathbb{N}\}$ *is a local base at the origin for X.*

Proof. Since multiplication by a non-zero scalar is a homeomorphism of X, $r_n U$ is indeed a zero-neighborhood, for each $n \in \mathbb{N}$. Thus we need only to show that given a zero-neighborhood W there exists $n \in \mathbb{N}$ such that $r_n U \subseteq W$. By Proposition 1.6.19, there exists a zero-neighborhood V satisfying

$$V \subseteq W \quad \text{and} \quad V = \text{circ}\,(V). \tag{1.6.490}$$

The boundedness of U insures that $tU \subseteq V$ for all sufficiently small positive numbers t, and hence $|r_{n_0}|U \subseteq V$ for some $n_0 \in \mathbb{N}$ large enough, since the sequence $\{r_n\}_{n\in\mathbb{N}}$ contains a convergent subsequence with limit zero. By the circledness of V, we may then write that $r_{n_0}U \subseteq V$, which completes the proof of the proposition due to the fact that $V \subseteq W$. □

Corollary 1.6.49. *The weak topology of an infinite-dimensional Hilbert space is not locally bounded.*

Proof. This result follows from Proposition 1.6.48 and Proposition 1.6.26. □

The space $C(\mathbb{D})$ introduced in Example 1.6.4 is an F-space (cf. Definition 1.6.6) which is *not* locally bounded. The same is true for its closed subspace $\mathscr{O}(\mathbb{D})$, consisting of all holomorphic functions on $\mathbb{D}$. The proofs of these statements are relegated to Exercise 15, in §1.7, where other examples of linear topological spaces that are not locally bounded may be found.

1.6.8 Distance to a Subspace

Let H be a Hilbert space. Consider Y a subspace of H, and its orthogonal $Y^\perp$. A straightforward application of the Pythagorean Theorem shows that every $x \in Y^\perp$ lies at distance $\|x\|$ from Y, i.e.,

$$\operatorname{dist}(x, Y) := \inf\{\|x - y\| : y \in Y\} = \|x\|. \tag{1.6.491}$$

Furthermore, it can be shown that the origin is the unique point of Y closest to x. Consequently, if in addition $x \neq 0$ and U is the closed ball in H with center at the origin and radius $\|x\|$, then $x \in U$, but $x \notin \frac{1}{2}U + Y$.

The next result, which is central to much of the subsequent discussion, emphasizes a similar phenomenon in every locally bounded linear topological space.

Lemma 1.6.50 (First Riesz). *Suppose X is a locally bounded linear topological space, and assume that $Y \subseteq X$, $Y \neq X$, is a closed subspace of X. Let U be a bounded zero-neighborhood in X. Then there exists $x \in U$ such that*

$$x \notin \tfrac{1}{2}U + Y. \tag{1.6.492}$$

Proof. We shall reason by contradiction. To this end, assume that there exists a bounded zero-neighborhood U such that

$$U \subseteq \tfrac{1}{2}U + Y. \tag{1.6.493}$$

Using the fact that $\frac{1}{2}Y = Y$, this immediately gives that $\frac{1}{2}U \subseteq \frac{1}{4}U + Y$. Substituting this last inclusion in (1.6.493) and using the fact that $Y + Y = Y$, allows us to conclude that

$$U \subseteq \left(\tfrac{1}{4}U + Y\right) + Y = \tfrac{1}{4}U + Y. \tag{1.6.494}$$

Iterating this procedure then yields

$$U \subseteq 2^{-n}U + Y \quad \text{for each} \quad n \in \mathbb{N}. \tag{1.6.495}$$

Moving on, Proposition 1.6.48 shows that the family of zero-neighborhoods $\{2^{-n}U : n \in \mathbb{N}\}$ is a local base at the origin for X. This, in concert with Corollary 1.6.32 and (1.6.495), shows that $U \subseteq \overline{Y}$. Thus, using the fact that Y is closed we obtain

$$U \subseteq Y. \tag{1.6.496}$$

Finally, since U is a zero-neighborhood of the linear topological space X, it absorbs every point of X (cf. part *(a)* in Theorem 1.6.18). This, (1.6.496), and the fact that Y is a subspace of X, implies $Y = X$, contradicting the hypothesis that Y is proper. This completes the proof of the lemma. □

Remark 1.6.51. The conclusion (1.6.492) of Lemma 1.6.50 can be rephrased as:

$$(x + Y) \cap \tfrac{1}{2}U = \varnothing. \tag{1.6.497}$$

This readily follows from the fact that Y is a subspace of X and, as such, $Y = -Y$.

If, in addition, U is taken to be symmetric then, thanks to Lemma 1.6.30, the conclusion (1.6.492) becomes equivalent to

$$(x + \tfrac{1}{2}U) \cap Y = \varnothing. \tag{1.6.498}$$

1.7 Additional Exercises: I

1. Let $X = \mathbb{C}^2$ (or $\mathbb{R}^2$ if you prefer). For $\epsilon > 0$ let

$$U_\epsilon = \{(z, 0) \in \mathbb{C}^2 : |z| < \epsilon\}. \tag{1.7.499}$$

   The collection of sets $\{U_\epsilon : \epsilon > 0\}$ satisfies all properties listed in Theorem 1.6.21 except one. Which one?
2. *The Indiscrete Topology*. Suppose X is any real or complex vector space. Show that the *indiscrete topology* $\tau := \{\varnothing, X\}$ is a vector topology on X.
3. Referring to the definition of p-norm, explain the restriction $p \leq 1$.
4. The notion of "circled" depends on the scalar field. Find a set that is circled in $\mathbb{R}^2$ over the reals, but not in $\mathbb{C}$ (with which $\mathbb{R}^2$ canonically identifies) over the complex numbers.
5. *A local base for* $C(\mathbb{D})$. Recall the F-space $C(\mathbb{D})$ introduced in Example 1.6.4. For each positive integer k and each $\epsilon > 0$ let

$$N(k, \epsilon) := \{f \in C(\mathbb{D}) : p_k(f) < \epsilon\}. \tag{1.7.500}$$

Show that the collection of all such sets forms a local base for the topology of uniform convergence on compact subsets of $\mathbb{D}$.

6. Let H be a finite-dimensional Hilbert space.

   (a) Prove that the weak topology coincides with the norm topology on H.
   (b) Deduce that the weak topology on H is metrizable.

7. *The weak-star topology of the dual of a Banach space.* Suppose X is a Banach space and X^* its dual space. For every finite subset F of X and every number $\epsilon > 0$ define

$$N(F, \epsilon) := \{\varphi \in X^* : |\varphi(f)| < \epsilon \text{ for every } f \in F\}. \quad (1.7.501)$$

   (a) Show that this collection of sets forms a local base for a Hausdorff vector topology on X^*. This topology is called the `weak-star topology` of X^*.
   (b) Show that if X is infinite dimensional then the weak-star topology of X^* is not metrizable.

8. Suppose X is a *separable* Banach space. Show that the *unit ball* of X^*, in the weak-star topology, is metrizable.
9. Show that for each finite subset F of an infinite-dimensional Banach space X there is a bounded linear functional φ_F on X such that $\varphi_F(f) = 0$ for each $f \in F$, yet φ_F has norm equal to the cardinality of F. Show that the net $\{\varphi_F\}$ converges to zero in the weak star topology of X^*. Why does this not contradict the Uniform Boundedness Principle?
10. Let $\mathcal{L}(X)$ denote the space of bounded linear operators on a Banach space X. For F a finite subset of X, Φ a finite subset of X^* and $\epsilon > 0$ define

$$N(F, \Phi, \epsilon) := \{T \in \mathcal{L}(X) : |\varphi(Tf)| < \epsilon \text{ for every } f \in F \text{ and } \varphi \in \Phi\}. \quad (1.7.502)$$

   Show that these sets form a local base at the origin for a Hausdorff vector topology on $\mathcal{L}(X)$, and show that if X is infinite dimensional then this topology is not metrizable.
11. Regarding Lemma 1.6.45, show that in an *infinite-dimensional* Hilbert space there are closed, circled, unbounded sets that do *not* contain nontrivial linear subspaces.
12. Show that a subset $\mathcal{F}$ of the linear topological space $C(\mathbb{D})$ introduced in Example 1.6.4 is bounded if and only if for every positive integer n:

$$\sup\{p_n(f) : f \in \mathcal{F}\} < +\infty. \quad (1.7.503)$$

   Hint: Use the result of Exercise 5 above.
13. Use Exercise 11 above, along with Montel's Theorem from complex analysis to show that the subspace $\mathscr{O}(\mathbb{D})$ of $C(\mathbb{D})$ consisting of all holomorphic functions on $\mathbb{D}$ has the same "Heine-Borel" property possessed by Hilbert space in its weak topology: *every bounded set is relatively compact.*

14. Use Exercise 11 above, along with the Arzela-Ascoli Theorem ("a subset of $C(\mathbb{D})$ is relatively compact if and only if, on every compact subset of $\mathbb{D}$, it is bounded and equicontinuous") to show that $C(\mathbb{D})$ has bounded subsets that are *not* relatively compact.
15. Show that these linear topological spaces are *not* locally bounded:
    (a) The F-space $C(\mathbb{D})$ introduced in Example 1.6.4.
    (b) The subspace $\mathscr{O}(\mathbb{D})$ of $C(\mathbb{D})$ that consists of all holomorphic functions on $\mathbb{D}$.
    (c) The dual of an infinite-dimensional Banach space, in its weak-star topology.
    (d) The collection of bounded operators on an infinite-dimensional Banach space, in its weak operator topology.
16. Call a net (x_n) in a Banach space X `weakly convergent` if there exists $x \in X$ such that for every continuous linear functional φ on X,

$$\lim_{n} \varphi(x_n) = \varphi(x). \tag{1.7.504}$$

Define a local base at the origin for a vector topology (the `weak topology`) whose convergent nets are precisely the weakly convergent ones. Model your definition on that of the weak-star topology given in Exercise 7 above. Show that this topology is Hausdorff, and that it is nonmetrizable whenever X is infinite dimensional.

To lighten the notation, we shall henceforth suppress the dependence of the dimension of a vector space X on the specific field of scalars $\mathbb{K}$ (over which X is considered), and simply write $\dim X$ in place of $\dim_{\mathbb{K}} X$. The same convention applies to the codimension of a subspace Y of X, i.e., we shall drop the dependence on field of scalars $\mathbb{K}$, writing $\operatorname{codim} Y$ in place of $\operatorname{codim}_{\mathbb{K}} Y$.

1.8 Finite-Dimensional Issues

Central to our impending investigations into operator theory are some fundamental results that deal with finite-dimensional phenomena in linear topological spaces. Here they are collected into one comprehensive theorem (Theorem 1.8.1 below) whose proof makes the subject of the rest of this section.

Theorem 1.8.1 (Finite Dimension). *Let X be a complex Hausdorff linear topological space. Then the following properties hold.*

(a) If $n := \dim X < +\infty$, then X is linearly homeomorphic to $\mathbb{C}^n$.
(b) Every finite-dimensional subspace of X is closed.
(c) X is finite dimensional if and only if it is locally compact (cf. Definition 1.1.27).
(d) Every closed subspace of finite codimension in X is topologically complemented in X (in the sense of Definition 1.8.12).

Moreover, similar results are valid in the case when X is a real Hausdorff linear topological space.

Remark 1.8.2. By part *(a)* of Theorem (1.8.1), every finite-dimensional subspace of a linear topological space has the Heine-Borel property (i.e., every closed and bounded subset is compact). Part *(c)* shows that, in fact, this property characterizes finite dimensionality.

The plan for the justification of the claims made in Theorem 1.8.1 is as follows: the one-dimensional version of part *(a)* is dealt with in Theorem 1.8.3; the one-dimensional version of part *(b)* is contained in Corollary 1.8.9; the one-dimensional version of part *(d)* is implied by Theorem 1.8.28; the proofs of parts *(a)* and *(b)* are completed in §1.8.5; the proof of part *(d)* is finished in §1.8.6; the proof of part *(c)* is done in §1.8.7.

The proof of Theorem 1.8.1 begins with the following, small but crucial step, the proof of part *(a)* for spaces of dimension one.

Theorem 1.8.3. *Suppose X is a complex Hausdorff linear topological space of dimension one. Select an arbitrary vector $e \in X \setminus \{0\}$. Then the map*

$$f : \mathbb{C} \longrightarrow X, \quad \textit{given by} \quad f(\lambda) := \lambda e \quad \textit{for all} \quad \lambda \in \mathbb{C}, \tag{1.8.505}$$

is a linear homeomorphism of $\mathbb{C}$ onto X.

Proof. It is clear from (1.8.505) that f is a linear one-to-one mapping of $\mathbb{C}$ onto X. Thus, invoking the continuity of scalar multiplication (here we use only continuity in the scalar variable), it follows that f is continuous. To prove that f is a homeomorphism it remains to show that f^{-1} is continuous, i.e., that f is open. By the linearity of f and the fact that translations are homeomorphisms, matters can be reduced to showing that the f-image of each zero-neighborhood in $\mathbb{C}$ contains a zero-neighborhood in X. In turn, every zero-neighborhood in $\mathbb{C}$ contains $\varDelta_\varepsilon$, the open disk centered at the origin with radius ε, for some $\varepsilon > 0$ sufficiently small. As such, to prove that f in (1.8.505) is open, it suffices to show that for each $\varepsilon > 0$ there exists

$$U \text{ zero-neighborhood in } X \text{ such that } \ U \subseteq f(\varDelta_\varepsilon). \tag{1.8.506}$$

Corollary 1.6.34 guarantees that the circled zero-neighborhoods form a local base at the origin for the topology of X. Since X is also Hausdorff and $e \neq 0$, this further implies that

$$\text{there exists } \ V \ \text{ zero-neighborhood in } X \text{ such that } \ \mathrm{circ}(V) = V \ \text{ and } \ e \notin V. \tag{1.8.507}$$

Next, we claim that

$$V \subseteq f(\varDelta_1). \tag{1.8.508}$$

To this end, let $v \in V$, and appeal to the surjectivity of f to write $v = \lambda e$ for some $\lambda \in \mathbb{C}$. To show that $v \in f(\varDelta_1)$ matters reduce to proving that $|\lambda| < 1$. Indeed, if $|\lambda|$ were ≥ 1 then because V is circled we would have $e = \lambda^{-1} v \in V$, contradicting the last assertion from (1.8.507). Finally, using (1.8.508) and the linearity of f, it follows that $U := \varepsilon V$ satisfies (1.8.506). This completes the proof of the theorem. □

Next on the agenda is the proof of part *(b)* in Theorem 1.8.1 for dimension one. Before starting this, however, we make a small detour through the issue of completeness in linear topological spaces.

1.8.1 Linear Topological Completeness

Central to the discussion here is the brand of completeness described in the definition below.

Definition 1.8.4. *A net $(x_\alpha)_{\alpha\in A}$ in a linear topological space X is said to be* Cauchy *if for every zero-neighborhood U there exists an index $\gamma = \gamma(U) \in A$ such that*

$$\alpha, \beta \in A \;\; \text{such that} \;\; \alpha, \beta \geq \gamma \Longrightarrow x_\alpha - x_\beta \in U. \tag{1.8.509}$$

If every Cauchy net in X converges, the linear topological space X is called complete in the linear topological sense.

This notion of linear topological completeness differs from the usual metric space concept in two fundamental ways. First there is the obvious "sequences to nets" generalization, but beyond this there is the fact that:

linear topological completeness depends only on the topology, while metric completeness depends on the metric. (1.8.510)

The following fundamental example illustrates how metric completeness depends on the metric:

Example 1.8.5. Consider the metric space $(\mathbb{R}, d)$ where $d : \mathbb{R} \times \mathbb{R} \longrightarrow [0, \infty)$ is the metric given by

$$d(x, y) := |\arctan x - \arctan y|, \qquad \forall\, x, y \in \mathbb{R}. \tag{1.8.511}$$

This metric (called arctangent metric) induces on $\mathbb{R}$ precisely the Euclidean topology, yet there are d-Cauchy sequences that are not d-convergent (for example, every sequence that converges to $+\infty$). Thus $\mathbb{R}$ is *not* metric complete for the metric d. However $\mathbb{R}$ clearly **is** metric complete for the canonical Euclidean metric.

We shall show next that familiar results like "Cauchy implies convergent" and "Euclidean space is complete" still hold for our new definition of completeness in the linear topological sense.

Proposition 1.8.6. *Any convergent net of a linear topological space is Cauchy.*

Proof. Suppose $(x_\alpha)_{\alpha\in A}$ is a convergent net in a linear topological space X. Thus there exists $x \in X$ such that

$$x_\alpha \to x \text{ in } X. \tag{1.8.512}$$

Fix next a zero-neighborhood $U \subseteq X$. In view of Theorem 1.6.18, there exists a circled zero-neighborhood V such that $V + V \subseteq U$. Using (1.8.512), one can find an index $\alpha(V) \in A$ such that $x - x_\alpha \in V$ whenever $\alpha \geq \alpha(V)$. Thus

$$\alpha, \beta \geq \alpha(V) \Longrightarrow x_\alpha - x_\beta = x_\alpha - x + x - x_\beta \in -V + V = V + V \subseteq U, \tag{1.8.513}$$

and, as such, the net $(x_\alpha)_{\alpha\in A}$ is Cauchy. □

Theorem 1.8.7. *If X is a linear topological space whose topology is induced by a complete, translation-invariant metric, then X is complete in the linear topological sense.*

Proof. Being metrizable, X has a countable local base $\{U_n : n \in \mathbb{N}\}$ consisting of closed zero-neighborhoods. Suppose $(x_\alpha)_{\alpha\in A}$ is a Cauchy net in X. Then for every $n \in \mathbb{N}$ there exists an index $\gamma(n) \in A$ such that

$$\alpha, \beta \in A \text{ and } \alpha, \beta \geq \gamma(n) \Longrightarrow x_\alpha - x_\beta \in U_n \tag{1.8.514}$$

and we can arrange for the function $\gamma : \mathbb{N} \to A$ to be increasing, i.e., $\gamma(n) \geq \gamma(m)$ for each $n, m \in \mathbb{N}$ such that $n \geq m$. Thus the *sequence* $(x_{\gamma(n)})_{n\in\mathbb{N}}$ is a Cauchy sequence for the metric of X (we use translation invariance for the first and only time here). Since X is complete as a metric space, this sequence converges to some $x \in X$.

We claim next that the full net $(x_\alpha)_{\alpha\in A}$ converges to x. For this we employ successively (1.8.514) above and the fact that each U_n, is closed to obtain for every $n \in \mathbb{N}$:

$$\begin{aligned} \alpha \in A \text{ and } \alpha \geq \gamma(n) &\Longrightarrow x_\alpha - x_{\gamma(k)} \in U_n \text{ for every integer } k \geq n \\ &\Longrightarrow x_\alpha - x = \lim_{k\to\infty} (x_\alpha - x_{\gamma(k)}) \in U_n. \end{aligned} \tag{1.8.515}$$

Thus $x_\alpha \to x$, as desired. □

Proposition 1.8.8. *Suppose X is a Hausdorff linear topological space and Y is a linear subspace whose relative topology is complete in the linear topological sense. Then Y is closed in X.*

Proof. Suppose x is a limit point of Y, i.e., there is a net $(y_\alpha)_{\alpha\in A}$ in Y that converges to x. By Proposition 1.8.6, the net $(y_\alpha)_{\alpha\in A}$ is Cauchy in Y, and the linear topological completeness of Y insures that this converges to some vector $y \in Y$. Since X is Hausdorff, limits of convergent nets are unique, and thus $x = y \in Y$, as desired. □

We can now prove part *(b)* of Theorem 1.8.1 for one dimensional linear topological spaces.

Corollary 1.8.9. *Every one dimensional subspace of a Hausdorff linear topological space is closed.*

Proof. Suppose X is a Hausdorff linear topological space and Y is a one dimensional subspace of X. Fix $0 \neq e \in Y$, then

$$Y = \text{span}\{e\}. \tag{1.8.516}$$

In Theorem 1.8.3 we proved that the linear isomorphism $\lambda \mapsto \lambda e$ is actually a homeomorphism of $\mathbb{C}$ onto Y. Therefore the Euclidean metric of $\mathbb{C}$ can be transported via this homeomorphism onto Y, yielding a metric

$$d(\lambda e, \nu e) := |\lambda - \nu| \tag{1.8.517}$$

for the relative topology of Y that is complete and translation invariant. By Theorem 1.8.7 the space Y is therefore complete in the linear topological sense, so by Proposition 1.8.8 it is closed in X. □

The Baire Category Theorem states that *in any complete metric space, or locally compact Hausdorff space, the intersection of every countable collection of dense open sets is itself dense*. In relation to this, we make the following comment.

Comment 1.8.10 *Although it appears that the notion of completeness in the linear topological sense is superior in every way to the metric one, there is one pitfall:* There are complete but nonmetrizable linear topological spaces for which the Baire Category Theorem fails. *For an example, see Exercise 13 in §1.9.*

1.8.2 Dissecting and Reassembling Linear Topological Spaces

The goal of this subsection is to generalize the one dimensional linear topological results proved so far to higher dimensions by induction. This will require dissecting linear topological spaces into subspaces, and then reassembling them from such component subspaces. *Continuous projections* will serve as tools for dissection, while *topological direct sums* will allow us to do the reassembly.

We start by recalling the following basic result from linear algebra.

Theorem 1.8.11 (Direct sums). *Let X be a vector space and assume that Y and Z are linear subspaces of X. Then X is the algebraic direct sum of Y and Z, that is, $X = Y \boxplus Z$ (see Definition 1.3.27), if and only if each $x \in X$ has a unique representation as $x = y + z$ with $y \in Y$ and $z \in Z$.*

Note that if Y and Z are algebraic complements of each other in a vector space X, then the `summation map` given by

$$S : Y \times Z \longrightarrow X, \quad S(y,z) := y + z, \quad \forall\, y \in Y \text{ and } \forall\, z \in Z, \tag{1.8.518}$$

is linear and bijective. If in addition X is a linear topological space and the algebraic complements Y and Z are equipped with the topologies inherited from X, then by the joint continuity of vector addition on X, it follows that S is a continuous map from $Y \times Z$ onto X.

Definition 1.8.12. *Let X be a linear topological space and Y and Z be linear subspaces of X such that $X = Y \boxplus Z$. According to Proposition 1.6.8, Y and Z become linear topological spaces themselves when equipped with the relative topology induced by X. Endow $Y \times Z$ with the product topology (cf. Definition 1.1.15), and consider $S^{-1} : X \longrightarrow Y \times Z$, the inverse of the summation map S from* (1.8.518).

If S^{-1} is continuous, then X is called the `topological direct sum of` *Y* `and` *Z, something denoted by writing*

$$X = Y \oplus Z. \tag{1.8.519}$$

In this situation Y and Z are said to be `topological complements of each other`, *and each is said to be* `topologically complemented in` *X.*

It follows from the above definition that

if X is a linear topological space and Y and Z are linear subspaces of X such that $X = Y \oplus Z$, then the summation map S from (1.8.518) is open; in particular, $V + W$ is open in X whenever V is open in Y and W is open in Z. (1.8.520)

The next result, which follows directly from Definition 1.8.12, takes this observation further by asserting that

if X is a linear topological space, Y and Z are linear subspaces of X such that $X = Y \boxplus Z$, then this algebraic direct sum is topological if and only if the topology of X "is" the product topology of $Y \times Z$. (1.8.521)

The reader is reminded that the notion of basis for a topology has been introduced in Definition 1.1.14.

Proposition 1.8.13. *Suppose a linear topological space X is the algebraic direct sum of linear subspaces Y and Z. Then $X = Y \oplus Z$ if and only if the sets of the form $V + W$, where V is open in Y and W is open in Z, form a basis for the topology of X.*

Proof. If $X = Y \oplus Z$ then, as noted earlier, $V + W$ is open in X whenever V is open in Y and W is open in Z. Let O be an arbitrary open set in X. Consider a point $x \in O$ and decompose $x = y + z$ with $y \in Y$ and $z \in Z$. Since the summation map (1.8.518) is continuous, there exist V_x open neighborhood of y in Y and W_x open neighborhood of z in Z such that $S(V_x \times W_x) \subseteq O$. Hence, $V_x \times W_x \subseteq O$ and, ultimately, $O = \bigcup_{x \in O}(V_x \times W_x)$. These considerations lead to the conclusion that $\{V + W : V \text{ open in } Y, W \text{ open in } Z\}$ is a basis for the topology of X (cf. Definition 1.1.14).

Conversely, suppose $V + W$ is open in X whenever V is open in Y and W is open in Z. Pick U open in $Y \times Z$. From Definition 1.1.15 we know that there exist V_i open in Y and W_i open in Z, for $i \in I$, such that $U = \bigcup_{i \in I}(V_i \times W_i)$. Then $S(U) = \bigcup_{i \in I} S(V_i \times W_i) = \bigcup_{i \in I}(V_i + W_i)$ is an open set in X, thanks to our working assumption. This proves that S is an open map, hence S^{-1} is continuous. According to Definition 1.8.12 we therefore have $X = Y \oplus Z$. □

Let us momentarily digress for the purpose of introducing relevant terminology in relation to generic topological spaces.

Definition 1.8.14. *Let $(\mathscr{X}, \tau_{\mathscr{X}})$ and $(\mathscr{Y}, \tau_{\mathscr{Y}})$ be two topological spaces. Call a given function $f : \mathscr{X} \to \mathscr{Y}$* `open` *provided f maps open subsets of $\mathscr{X}$ onto open subsets of $\mathscr{Y}$. Also, a function $f : \mathscr{X} \to \mathscr{Y}$ is said to be* `relatively open` *provided f maps open subsets of $\mathscr{X}$ onto relatively open subsets of $f(\mathscr{X})$, i.e., for each $O \in \tau_{\mathscr{X}}$ there exists $\mathcal{U} \in \tau_{\mathscr{Y}}$ such that $f(O) = f(\mathscr{X}) \cap \mathcal{U}$.*

Clearly, any function which is open is also relatively open. Also, for a surjective function, being open and being relatively open are identical qualities.

Proposition 1.8.15. *Let $(\mathscr{X}, \tau_{\mathscr{X}})$, $(\mathscr{Y}, \tau_{\mathscr{Y}})$, and $(\mathscr{Z}, \tau_{\mathscr{Z}})$ be topological spaces. Also, suppose $f : \mathscr{X} \to \mathscr{Y}$ is a relatively open function, and $g : \mathscr{Y} \to \mathscr{Z}$ is both relatively open and injective. Then the composition $g \circ f : \mathscr{X} \to \mathscr{Z}$ is relatively open.*

Proof. Let $O \in \tau_{\mathscr{X}}$. Since f is relatively open, there exists $\mathcal{U} \in \tau_{\mathscr{Y}}$ such that $f(O) = \mathcal{U} \cap f(\mathscr{X})$. Also, g is relatively open, there exists $\mathcal{V} \in \tau_{\mathscr{Z}}$ such that $g(\mathcal{U}) = \mathcal{V} \cap g(\mathscr{Y})$. Based on these we may then write

$$
\begin{aligned}
g(f(O)) &= g(\mathcal{U} \cap f(\mathscr{X})) = g(\mathcal{U}) \cap g(f(\mathscr{X})) \\
&= \mathcal{V} \cap g(\mathscr{Y}) \cap g(f(\mathscr{X})) = \mathcal{V} \cap g(f(\mathscr{X})),
\end{aligned}
\tag{1.8.522}
$$

where the second equality uses the fact that g is injective. In turn, (1.8.522) proves that the function $g \circ f : \mathscr{X} \to \mathscr{Z}$ is relatively open. □

Exercise 1.8.16. Let $(\mathscr{X}, \tau_{\mathscr{X}})$ and $(\mathscr{Z}, \tau_{\mathscr{Z}})$ be two topological spaces. Fix a set $\mathscr{Y} \subseteq \mathscr{Z}$ and consider it as a topological space, by equipping it with the relative topology induced by $\tau_{\mathscr{Z}}$ on $\mathscr{Y}$. Also, denote by ι the canonical inclusion of $\mathscr{Y}$ into $\mathscr{Z}$, and pick a function $f : \mathscr{X} \to \mathscr{Y}$ which is relatively open. Show that $\iota \circ f : \mathscr{X} \to \mathscr{Z}$ is relatively open.

Returning to the mainstream discussion we note a couple of useful corollaries.

Corollary 1.8.17. *Let X, Y be Hausdorff linear topological spaces and assume X_0, X_1 are linear subspaces of X such that $X = X_0 \oplus X_1$, and that Y_0, Y_1 are linear subspaces of Y. Finally, suppose $T : X \to Y$ is a linear mapping satisfying the following properties:*

$$
\begin{gathered}
T(X_0) \subseteq Y_0, \quad T(X_1) \subseteq Y_1, \quad T(X) = T(X_0) \oplus T(X_1), \\
T : X_0 \to Y_0 \text{ and } T : X_1 \to Y_1 \text{ are relatively open mappings.}
\end{gathered}
\tag{1.8.523}
$$

Then $T : X \to Y$ is a relatively open mapping.

Proof. Let O be open in X. The goal is to show that $T(O)$ is a relatively open set in $T(X)$. To this end, pick an arbitrary point $y \in T(O)$. Then there exists $x \in O$ such that $y = Tx$ and since $X = X_0 \oplus X_1$ we may decompose $x = x_0 + x_1$ with $x_0 \in X_0$, $x_1 \in X_1$. From Proposition 1.8.13 we also know that

$$\mathcal{B} := \{V_0 + V_1 : V_0 \text{ open in } X_0 \text{ and } V_1 \text{ open in } X_1\} \tag{1.8.524}$$

is a basis for the topology of X. Consequently, there exist V_0 open in X_0 along with V_1 open in X_1 such that $x \in (V_0 + V_1) \subseteq O$. Since $T : X_0 \to Y_0$ and $T : X_1 \to Y_1$ are relatively open mappings, there exist W_0 open set in Y_0 and W_1 open set in Y_1 such that $T(V_0) = T(X_0) \cap W_0$ and $T(V_1) = T(X_1) \cap W_1$. As such, on the one hand we have

$$T(X_0) \cap W_0 + T(X_1) \cap W_1 = T(V_0) + T(V_1) = T(V_0 + V_1) \subseteq T(O) \tag{1.8.525}$$

and

$$y = Tx \in T(V_0 + V_1) = T(V_0) + T(V_1) = T(X_0) \cap W_0 + T(X_1) \cap W_1. \tag{1.8.526}$$

On the other hand, we may invoke Proposition 1.8.13 in relation to the topological direct sum $T(X) = T(X_0) \oplus T(X_1)$ to conclude that $T(X_0) \cap W_0 + T(X_1) \cap W_1$ is open in $T(X)$. Together with (1.8.525)-(1.8.526) this shows that $T(O)$ is a neighborhood of y in the topology of $T(X)$. In view of the arbitrariness of the point $y \in T(O)$, this ultimately proves that $T(O)$ is open in the topology of $T(X)$. Hence, $T : X \to Y$ is a relatively open mapping. □

Here is the second corollary mentioned above.

Corollary 1.8.18. *If X is a Hausdorff linear topological space and Y and Z are linear subspaces of X such that $X = Y \oplus Z$, then Y and Z are closed in X.*

Proof. It is straightforward to check that Y and Z, equipped with the induced topologies from X, are Hausdorff linear topological spaces. Then, using Theorem 1.6.36 both $Y \times \{0\}$ and $\{0\} \times Z$ are closed subspaces of $Y \times Z$, and the homeomorphism S from (1.8.518) takes these subspaces respectively onto Y and Z, respectively. □

For an F-space (that is, a linear topological space given by a complete, translation-invariant metric), the preceding result has a converse:

Proposition 1.8.19. *If an F-space is the algebraic direct sum of two closed linear subspaces, then it is the topological direct sum of these subspaces.*

Proof. Suppose X is an F-space and $X = Y \boxplus Z$, where Y and Z are closed linear subspaces of X. We need to prove continuity for the inverse of the summation map $S : Y \times Z \to X$. Since Y and Z are closed in X, each is complete in the metric of X, hence the product $Y \times Z$ is also a complete metrizable linear topological space. Thus S, being a continuous linear map of $Y \times Z$ onto X, is, by the Open Mapping Theorem (see Theorem 1.6.7), an *open* map. Thus the inverse of S is continuous, and this establishes X as the topological direct sum of Y and Z. □

1.8.3 Projections

Projections are idempotent linear transformations on vector spaces.

Definition 1.8.20. *Given a vector space X, call a linear transformation $P : X \to X$ a* `projection` *if $P^2 := P \circ P$ satisfies*

$$P^2 = P. \tag{1.8.527}$$

It is immediate to check that if P is a projection on a vector space X and I is the identity map on X, then $I - P$ is also a projection on X, $\operatorname{ran} P = \ker(I - P)$, and $\ker P = \operatorname{ran}(I - P)$. Also,

$$Px = x \quad \text{for all} \quad x \in \operatorname{ran} P. \tag{1.8.528}$$

Remark 1.8.21. In relation to each algebraic direct sum decomposition $X = Y \boxplus Z$ of a given vector space X, there are `canonical projections` $P_Y : X \to X$ taking X onto Y and $P_Z : X \to X$ taking X onto Z, defined by $P_Y x := y$ and $P_Z x := z$, where $x = y + z$ is the unique decomposition of the vector $x \in X$ into components $y \in Y$ and $z \in Z$.

When more specificity is desired, we shall refer to $P_Y : X \to X$ as *the projection from X onto Y along Z*, with a similar piece of terminology attributed to P_Z. It turns out that the aforementioned canonical projections play a crucial role in deciding when an algebraic direct sum is topological:

Proposition 1.8.22. *Suppose a linear topological space X is the algebraic direct sum of linear subspaces Y and Z. If P_Y and P_Z denote the natural projections onto Y and Z, respectively, then the following conditions are equivalent:*

(1) $X = Y \oplus Z$, i.e., X is the topological direct sum of Y and Z;
(2) both P_Y and P_Z are continuous;
(3) either P_Y or P_Z is continuous;

Proof. The equivalence between *(1)* and *(2)* asserts that continuity of natural projections is equivalent to the continuity of S^{-1}, the inverse of the summation map S from (1.8.518). This, however, is immediate from the formula $S^{-1}(x) = (P_Y x, P_Z x)$, valid for all $x \in X$. Obviously, *(2)* implies *(3)*. Finally, since $P_Y + P_Z = I$, the identity operator on X, it clear that *(3)* implies *(2)*. □

Proposition 1.8.23 (Transitivity of Topological Complementation). *Let X be a Hausdorff linear topological space and suppose that Y and Z are linear subspaces of X such that $Y \subseteq Z$, Y is topologically complemented in Z and Z is topologically complemented in X. Then Y is topologically complemented in X.*

Proof. Since Y is topologically complemented in Z, there exists a closed linear subspace $W \subseteq Z$ such that

$$Z = Y \boxplus W, \tag{1.8.529}$$

and the associated canonical projections

$$P^Z_Y : Z \to Y \text{ and } P^Z_W : Z \to W \text{ are continuous.} \tag{1.8.530}$$

Likewise, since Z is topologically complemented in X, there exists a closed linear subspace $M \subseteq X$ such that

$$X = Z \boxplus M, \tag{1.8.531}$$

and the canonical projections

$$P^X_Z : X \to Z \text{ and } P^X_M : X \to M \text{ are continuous.} \tag{1.8.532}$$

Combining (1.8.529) and (1.8.531), we see that every $x \in X$ admits a unique representation of the form

$$x = z + m = (y + w) + m, \tag{1.8.533}$$

where $z \in Z$, $m \in M$, and $y \in Y$, $w \in W$. In particular,

$$X = Y \boxplus (W \boxplus M). \tag{1.8.534}$$

Set

$$N := W \boxplus M. \tag{1.8.535}$$

Then N is a linear subspace of X, and (1.8.534) shows that $X = Y \boxplus N$.

We next show that this direct sum is topological and yields Y as a topologically complemented subspace of X. To this end, consider the map

$$\Phi := P^Z_Y \circ P^X_Z : X \longrightarrow Y. \tag{1.8.536}$$

Since P^X_Z and P^Z_Y are continuous linear maps, their composition Φ is a continuous linear map from X into Y. We claim that Φ is the canonical projection from X onto Y along N.

Indeed, let $x \in X$ and write x as in (1.8.533), i.e.,

$$x = y + w + m, \qquad y \in Y,\ w \in W,\ m \in M. \tag{1.8.537}$$

Then $P^X_Z(x) = z = y + w$ and hence

$$\Phi(x) = P^Z_Y(P^X_Z(x)) = P^Z_Y(y + w) = y. \tag{1.8.538}$$

Thus Φ coincides with the algebraic projection from X onto Y with kernel equal to $N = W \boxplus M$. Since Φ is continuous and $Y = \Phi(X)$ is its range, we have exhibited a continuous linear projection from X onto Y. Based on Proposition 1.8.22, we may now conclude that Y is topologically complemented in X, as desired. □

Exercise 1.8.24. Let (X, τ) be a linear topological space, and suppose Y, Z are linear subspaces of X with the property that $X = Y \oplus Z$. Denote by $P_Y : X \to X$ the canonical projection of X onto Y, associated with this direct topological sum as in Remark 1.8.21.

Next, recall from Proposition 1.6.8 that $(Y, \tau|_Y)$ is itself a linear topological space, and suppose $Y = V \oplus W$ for some linear subspaces V, W of Y. Let $P_V : Y \to Y$ denote the canonical projection of Y onto V, associated with this direct topological sum as in Remark 1.8.21.

Show that the map

$$P : X \to X, \quad P(x) := P_V(P_Y(x)) \text{ for each } x \in X, \tag{1.8.539}$$

is a projection of X onto V, i.e., show that P is linear, continuous, $P^2 = P$, and $\operatorname{ran} P = V$.

The natural projections associated with a topological direct sum have an additional special property: *they are relatively open*, and this makes the subject of our next result.

Proposition 1.8.25. *Let X be a linear topological space and Y be a linear subspace of X which is topologically complemented in X. Then P_Y, the naturally associated projection of X onto Y, maps every open set in X onto a set that is open in the linear subspace $P_Y(X) = Y$ (where Y is equipped with the relative topology, inherited from X). Hence, P_Y is a relatively open mapping.*

Also, if $\iota : Y \to X$ is the canonical inclusion, then $\iota \circ P_Y$ is a relatively open mapping.

Proof. Let $U \neq \varnothing$ be an open set in X, and fix $y \in P_Y(U)$. The goal is to find a neighborhood V of y in Y contained in $P_Y(U)$. Since Y is topologically complemented in X there exists Z, a linear subspace of X, such that $X = Y \oplus Z$. Next, fix $z \in P_Z(U)$, and define $x := y + z$. In view of the fact that $y \in Y$ and $z \in Z$, it follows that $P_Y(x) = y$. Since U is non-void and open, and since by Proposition 1.8.13 the topology on X is the "product topology" from Y and Z, there exists an open subset V of Y and an open subset W of Z such that $x \in V + W \subseteq U$. In particular, this forces $y = P_Y(x) \in P_Y(V + W) = V$. Consequently, $y \in V = P_Y(V + W) \subseteq P_Y(U)$, so V is the desired neighborhood of y. This proves that P_Y is a relatively open mapping. The final claim in the statement is then a consequence of what we have just proved and Exercise 1.8.16. □

Proposition 1.8.26. *Let X, Y be two linear topological spaces. Suppose A, B are linear subspaces of X and C, D are linear subspaces of Y with the property that*

$$X = A \oplus B \text{ and } Y = C \oplus D. \tag{1.8.540}$$

Then

$$X \times Y = (A \times C) \oplus (B \times D). \tag{1.8.541}$$

Proof. Clearly, $A \times C$ and $B \times D$ are linear subspaces of $X \times Y$ satisfying

$$X \times Y = (A \times C) \boxplus (B \times D). \tag{1.8.542}$$

Denote by P_A the projection from X onto A along B, and denote by P_C the projection from Y onto C along D. Thanks to (1.8.540) and Proposition 1.8.22, these are all continuous functions. As such, Exercise 1.1.18 tells us that

$$X \times Y \ni (x, y) \mapsto P(x, y) := (P_A(x), P_C(y)) \in X \times Y \tag{1.8.543}$$

is itself a continuous function. Since P is actually the projection from $X \times Y$ onto $A \times C$ along $B \times D$, another appeal to Proposition 1.8.22 shows that the algebraic direct sum in (1.8.542) is actually topological. □

As we have seen so far, algebraic direct sums give rise to projections, and topological direct sums give rise to continuous projections. We can turn this around: if P is a projection on a vector space X then P acts as the identity operator on its range, and X decomposes into the algebraic direct sum $X = Y \boxplus Z$ where $Y = \operatorname{ran} P$ and $Z = \ker P$, with $P_Y = P$, and $P_Z = I - P$. If in addition X is a linear topological space and P a *continuous* projection on X, then by Proposition 1.8.22 this algebraic direct sum is a topological one, and P is a relatively open map.

We summarize this discussion in the theorem below.

Theorem 1.8.27. *If P is a continuous projection on a linear topological space X then the following properties hold:*

(a) $X = \operatorname{ran} P \oplus \ker P$.
(b) P is a relatively open map (in the sense of Definition 1.8.14).
(c) If, in addition, X is Hausdorff, then $\operatorname{ran} P$ *is a closed subspace of X.*

Proof. Parts *(a)* and *(b)* are discussed right before the statement of the theorem, whereas part *(c)* follows readily from part *(a)* and Corollary 1.8.18. □

The next order of business is to use the tools developed so far to treat the one-dimensional version of part *(d)* of Theorem 1.8.1.

1.8.4 Codimension One

Suppose Y is a linear subspace of a vector space X. Since every basis for Y can be extended to a basis for X, the linear span of the new basis vectors is a subspace complementary to Y. Moreover, since all bases for X have the same cardinality, all subspaces complementary to Y have the same dimension (possibly infinite); we call this cardinal number the `codimension` of Y (in X).

The next result implies the following particular case of part *(d)* of Theorem 1.8.1: *In a linear topological space every closed subspace of codimension one is topologically complemented.*

Theorem 1.8.28 (Codimension-One). *Suppose X is a linear topological space and Y a closed linear subspace of X such that X is the algebraic direct sum of Y and F, for some one-dimensional subspace F of X, i.e., $X = Y \boxplus F$. Then $X = Y \oplus F$.*

Proof. Fix $0 \neq e \in F$, so that $F = \{\lambda e : \lambda \in \mathbb{C}\}$. Now the natural projection of X onto F has the form $P_F x = \Lambda(x)e$, where Λ is a linear functional on X (that is, $\Lambda : X \to \mathbf{C}$ is a linear mapping). In particular, Λ maps the zero vector of X to the zero scalar.

Since $\operatorname{ran} P_F = F$ and $\ker P_F = Y$, by Theorem 1.8.27 we need only show that P_F is continuous, i.e., that Λ is continuous. Suppose this is not the case. Then, because translation in X is a homeomorphism, we know that Λ is not continuous at the origin, so there exists a positive number δ such that:

$$\begin{gathered}\text{for each zero-neighborhood } U \text{ of } X \\ \text{there exists } x_U \in U \text{ with } |\Lambda(x_U)| > \delta.\end{gathered} \tag{1.8.544}$$

Fix a circled zero-neighborhood V, and let $U := \delta V$, also a zero-neighborhood. Let $x' := \delta^{-1} x_U$. Then

$$x' \in V \quad \text{and} \quad |\Lambda(x')| > 1. \tag{1.8.545}$$

Since $X = Y \boxplus F$, one may write

$$x' = y + \Lambda(x')e, \quad \text{for some } y \in Y. \tag{1.8.546}$$

Multiply both sides of (1.8.546) by $\gamma := [\Lambda(x')]^{-1}$ (note that $0 < |\gamma| < 1$ by the second part of (1.8.545)), and rearrange the result to get

$$e = \gamma x' - \gamma y \in \gamma V - Y \subseteq V + Y, \tag{1.8.547}$$

where we have used the first part of (1.8.545), the fact that Y is a linear subspace, and that V is circled.

Accordingly, the assumption of non-continuity for Λ leads to the conclusion that $e \in V + Y$ for every circled zero-neighborhood V. Since circled zero-neighborhoods constitute a local base at the origin (by Proposition 1.6.19), this further implies that $e \in V + Y$ for *every* zero-neighborhood V. It follows from Corollary 1.6.32 that the vector e lies in $\overline{Y}$, the closure of Y in X. Since Y is closed to begin with, e belongs to Y, hence sits in $Y \cap F$, which is the trivial linear subspace. This contradicts our fundamental assumption that $e \neq 0$, and shows that Λ must be continuous, completing the proof of the theorem. □

The proof we have just given naturally permits us to recognize certain fundamental properties of linear functionals, collected in the next corollary.

Corollary 1.8.29. *Suppose X is a linear topological space, and Λ is a nontrivial linear functional on X. Then*

$$\ker \Lambda \textit{ is a linear subspace of codimension one in } X, \tag{1.8.548}$$

and the following properties of Λ are equivalent:

(a) Λ *is continuous on* X*;*
(b) $\ker \Lambda$ *is closed in* X*;*
(c) $\ker \Lambda$ *is not dense in* X*;*
(d) Λ *is bounded in some zero-neighborhood* (*i.e., the set* $\Lambda(U)$ *is bounded, in the field of scalars, for some zero-neighborhood* U *of* X).

Proof. First, $\ker \Lambda$ is a linear proper subspace of X since Λ is a nontrivial linear functional on X. In particular, it is possible to pick some $x_* \in X$ with $\Lambda(x_*) \neq 0$. Then $x_* \neq 0$, and if $\langle x_* \rangle$ denotes the linear span of x_*, it follows that $\langle x_* \rangle$ is a one-dimensional linear subspace of X with the property that $\langle x_* \rangle \cap \ker \Lambda = \{0\}$. Furthermore, for any given vector $x \in X$ if we define the scalar $\lambda := \Lambda(x)/\Lambda(x_*)$ then $x = \lambda x_* + (x - \lambda x_*)$ and $\Lambda(x - \lambda x_*) = 0$. This argument shows that

$$X = \langle x_* \rangle \oplus \ker \Lambda, \tag{1.8.549}$$

from which we conclude that (1.8.548) holds.

We next turn our attention to the equivalences claimed in the last part of the statement. If Λ is continuous, since the singleton $\{0\}$ is closed in $\mathbb{C}$, the null-space $\Lambda^{-1}(\{0\})$ of Λ is closed in X. This establishes the implication *(a)*$\Rightarrow$*(b)*. The implication *(b)*$\Rightarrow$*(c)* is immediate from the assumption that Λ is nontrivial, i.e., not identically zero on X. Moving on, we now prove the next link in the chain, namely *(c)*$\Rightarrow$*(d)*. If *(c)* holds then the complement of $\ker \Lambda$ has nonempty interior. Upon appealing to (1.6.306) and Proposition 1.6.19, it becomes apparent that this, in turn, implies that there exists $x_0 \in X$ and a circled zero-neighborhood N of X satisfying

$$(x_0 + N) \cap \ker \Lambda = \varnothing. \tag{1.8.550}$$

Then the linearity of Λ and the fact that N is circled imply that $\Lambda(N)$ is a circled subset of the complex field. Hence, by recalling (1.6.488), it suffices to check that the restriction of Λ to N is not surjective for *(d)* to hold true. To this end, suppose there exists $n \in N$ satisfying $\Lambda(x_0) = \Lambda(n)$. Hence $x_0 - n \in (x_0 + N) \cap \ker \Lambda$, contradicting (1.8.550). Consequently, $\Lambda(N) \subsetneq \mathbb{C}$, just as wanted.

Finally, we show that *(d)*$\Rightarrow$*(a)*. Seeking a contradiction, suppose that Λ is not continuous on X. Hence (1.8.544) holds for some $\delta > 0$. Since we are assuming *(d)*, there exists W a zero-neighborhood of X and a positive constant M with the property that $|\Lambda(\omega)| \leq M$ for each $\omega \in W$. Therefore, the zero-neighborhood $U := \frac{\delta}{M} W$ of X contradicts (1.8.544), finishing the proof of the corollary. □

Remark 1.8.30. (a) It follows from Corollary 1.8.29 that a subspace of codimension one need not always be closed. For example, Exercise 1 in §1.9 shows that if X is infinite dimensional and *locally bounded* then it has a discontinuous linear functional. By the Corollary 1.8.29, the nullspace of this functional is a non-closed (in fact dense) subspace of codimension one.

(b) In a more concrete vein, the classical Lebesgue space $L^p([0,1])$ has no nontrivial continuous linear functional when $0 < p < 1$ (see Exercise 7 in §1.9). Thus, when $0 < p < 1$ the space $L^p([0,1])$ has no proper closed subspaces of finite codimension. Indeed, the existence of such a subspace would bring with it a nontrivial topologically complemented finite-dimensional subspace (cf. Theorem 1.8.36), and this would imply the existence of nontrivial continuous linear functionals on L^p (see Exercise 14 in §1.9).

(c) If the continuous linear functionals on X separate points (in the sense of Definition 1.6.12), then the finite-dimensional subspaces of X are *all* topologically complemented (see Exercise 9 in §1.9). However if the space X is infinite dimensional and locally bounded, then according to remark *(a)* above, some closed subspaces of finite codimension are *not* topologically complemented. (Be sure you understand why the last two sentences do not contradict each other!)

1.8.5 Finite-Dimensional Denouement

The path is now clear for us to complete the proofs of parts *(a)* and *(b)* of Theorem 1.8.1.

Theorem 1.8.31. *Let X be a Hausdorff linear topological space. If* $\dim X = n < \infty$ *then X is linearly homeomorphic to $\mathbb{C}^n$.*

Proof. We proceed by induction on n. The case $n = 1$ is contained in Theorem 1.8.3. Suppose next that $n > 1$ and the result has been established for all spaces of dimension $n - 1$. Suppose X has dimension n. Then we have the algebraic decomposition $X = F_{n-1} \boxplus F_1$, where the subscript on the symbol "F" indicates the dimension of the subspace.

By the induction hypothesis, $F_{n-1} \approx \mathbb{C}^{n-1}$ (where "$\approx$" means in this context "is linearly homeomorphic to"), so as in the proof of Corollary 1.8.9 its topology is induced by a complete, translation-invariant metric. Thus by Theorem 1.8.7 and Proposition 1.8.8, F_{n-1} is closed in X. Theorem 1.8.28 gives that $X = F_{n-1} \oplus F_1$. Also, Theorem 1.8.3 asserts that $F_1 \approx \mathbb{C}$. Putting everything together, we may conclude that $X \approx \mathbb{C}^{n-1} \oplus \mathbb{C} \approx \mathbb{C}^n$. □

Corollary 1.8.32. *Every finite-dimensional Hausdorff linear topological space is locally compact.*

Proof. Straightforward from Theorem 1.8.31 and the fact that every Cartesian Product of finitely many copies of $\mathbb{C}$ is locally compact. □

Corollary 1.8.33. *Every finite-dimensional subspace of a Hausdorff linear topological space is complete in the linear topological sense. In particular, such a subspace is closed.*

Proof. This is immediate from Theorem 1.8.31, the fact that its topology is induced by a complete, translation-invariant metric (as in Corollary 1.8.9), Theorem 1.8.7, and Proposition 1.8.8. □

We pause to insert a simple corollary that plays an important role in all of our subsequent work on operator theory.

Corollary 1.8.34. *Suppose X is a finite-dimensional Hausdorff linear topological space and Y is any linear topological space. Then every linear transformation from X to Y is continuous.*

Proof. Suppose the dimension of X is n. We know from Theorem 1.8.31 that there is a linear homeomorphism that takes $\mathbb{C}^n$ onto X. Let $\{e_1, e_2, \dots, e_n\}$ be the image under this homeomorphism of the standard unit-vector basis for $\mathbb{C}^n$. Then for each vector x in X there exists a unique n-tuple of scalars $(\lambda_j(x) : 1 \leq j \leq n)$ such that

$$x = \sum_{j=1}^{n} \lambda_j(x) e_j, \tag{1.8.551}$$

where each of the linear functionals λ_j inherits the continuity of the corresponding coordinate functionals on $\mathbb{C}^n$.

Suppose $T : X \to Y$ is a linear transformation. As such, $T(0) = 0$. To establish its continuity we need only show it is continuous at the origin. To this end, let V be a zero-neighborhood in Y. Our goal is to find a zero-neighborhood U in X such that $T(U) \subseteq V$. To see this, in a first stage we invoke (1.6.412) to select zero-neighborhood V_* in Y whose n-fold sum with itself is contained in V. From Proposition 1.6.19 we know that the circled zero-neighborhoods form a local base $\mathcal{U}$ at the origin in Y. In addition, Theorem 1.6.21 tells that every element of $\mathcal{U}$ is absorbing. In view of these and Definition 1.1.19, we may therefore choose a zero-neighborhood W in Y which is circled, absorbing, and contained in V_*. In particular, the n-fold sum of W with itself is contained in V. Because W is circled and absorbing, there exists some $\epsilon_W \in (0, \infty)$ such that

$$a \cdot T(e_j) \in W \quad \text{for each } a \in \mathbb{C} \text{ with } |a| < \epsilon_W, \text{ and each } j \in \{1, \dots, n\}. \tag{1.8.552}$$

(Parenthetically, we note that this may also be seen directly from (1.6.404).) Use this number to define

$$U := \{x \in X : |\lambda_j(x)| < \epsilon_W \text{ for } 1 \leq j \leq n\} = \bigcap_{j=1}^{n} \lambda_j^{-1}(\varDelta_W) \tag{1.8.553}$$

where $\varDelta_W$ is the open disk in $\mathbb{C}$, centered at the origin and of radius ϵ_W. Since each functional $\lambda_j : X \to \mathbb{C}$ is continuous and maps the zero vector of X to the zero scalar, the last equality establishes U as the finite intersection of zero-neighborhoods in X, so it is itself a zero-neighborhood in X. By linearity,

$$T(U) = \left\{ \sum_{j=1}^{n} \lambda_j(x) T(e_j) : x \in U \right\}. \tag{1.8.554}$$

By the definition of U we have $|\lambda_j(x)| < \epsilon_W$ for every $x \in U$ and $1 \leq j \leq n$, so thanks to (1.8.552), the set W contains each of the summands $\lambda_j(x) T(e_j)$ with $j \in \{1, \dots, n\}$, as x ranges through U. Thus $T(U)$ lies in the n-fold sum of W with itself, which in turn sits in V. □

1.8.6 Finite-Dimensional Extensions

Our next goal is to complete the proof of part *(d)* of Theorem 1.8.1, i.e., to extend Theorem 1.8.28 to higher codimensions. For this we need to study what happens when a closed subspace is "extended" by a finite number of dimensions.

If a subspace has finite codimension in a larger vector space, call the larger space a `finite-dimensional extension` of the smaller one. The next result asserts that every finite-dimensional extension of a closed subspace of X is closed (recall from item *(a)* in Remark 1.8.30 that subspaces of finite codimension need not automatically be closed).

Theorem 1.8.35 (Finite-Dimensional Extension). *Let X be a Hausdorff linear topological space. If Y is a closed linear subspace of X and F is a finite-dimensional linear subspace of X, then $Y + F$ is closed in X.*

Proof. We want to show that $Z := Y + F$ is closed. Upon replacing F, if necessary, by a smaller subspace we may assume that $Y \cap F = \{0\}$, i.e., that $Z = Y \boxplus F$. We proceed by induction on $n := \dim F < \infty$.

The case $n = 1$: It is a simple matter to check that Y is still closed in the smaller linear topological space Z (which has the relative topology inherited from X). Since Y is a closed subspace of Z having codimension one, Theorem 1.8.28 guarantees that $Z = Y \oplus F$. In particular P_F, the natural projection from Z onto F, is continuous. To show Z is closed in X let $(z_\alpha)_{\alpha\in A}$ be a net in Z that converges in X to some vector x. There remains to prove that $x \in Z$.

For each index α we have the unique decomposition $z_\alpha = y_\alpha + f_\alpha$, where $y_\alpha \in Y$ and $f_\alpha \in F$. Since P_F is continuous, the image net $(f_\alpha)_{\alpha\in A}$ converges to $P_F(x)$ in F. This yields that

$$y_\alpha = z_\alpha - f_\alpha \to x - P_F(x). \tag{1.8.555}$$

Since Y is closed in X, this shows that $x - P_F(x) \in Y$. Hence, $x = (x - P_F(x)) + P_F(x)$ belongs to $Y + F$. Therefore the desired conclusion, $x \in Z$, follows.

The induction step: Suppose $n > 1$ and assume the result holds for all linear subspaces of X of dimension $n - 1$. Suppose $\dim F = n$. Write $F = F_1 + F_{n-1}$ where on the right the subscript is the dimension of the subspace. Thus,

$$Z = (Y + F_1) + F_{n-1}. \tag{1.8.556}$$

By the one dimensional case, $Y + F_1$ is closed in X, therefore Z is a closed subspace of X thanks to the induction hypothesis. □

We can now prove a slightly more precise version of part *(d)* of Theorem 1.8.1.

Theorem 1.8.36 (Finite Codimension). *Let X be a Hausdorff linear topological space. If Y is a closed linear subspace of X of finite codimension and F is any algebraic complement of Y, then actually $X = Y \oplus F$.*

Proof. We proceed by induction on the codimension n of Y. The case $n = 1$ is covered by Theorem 1.8.28. Suppose next that $n > 1$ and that the result we seek to establish is true for all closed subspaces of codimension $n - 1$ of X. Assume that Y is a subspace of codimension n in X. Let F be an algebraic complement for Y, so $\dim F = n$ and $X = Y \boxplus F$.

Now F is linearly homeomorphic to $\mathbb{C}^n$ by Theorem 1.8.31, so we can write $F = F_{n-1} \oplus F_1$, where again "subscript = dimension." Let $Z := Y \boxplus F_{n-1}$. This is an algebraic direct sum decomposition of the linear topological space Z, in which Y is still closed, so it is a topological direct sum: $Z = Y \oplus F_{n-1}$ by the induction hypothesis.

However, $X = Z \boxplus F_1$, where Z, being a finite-dimensional extension of the closed subspace Y, is also closed in X (cf. Theorem 1.8.35). So by Theorem 1.8.28, this algebraic direct sum decomposition of X is topological. Thus, we have

$$X = Z \oplus F_1 = (Y \oplus F_{n-1}) \oplus F_1 = Y \oplus (F_{n-1} \oplus F_1), \tag{1.8.557}$$

where we leave it to the reader to verify the "associative law" for topological direct sums that is operating in the last equality. Thus it follows from (1.8.557) that, as wanted, $X = Y \oplus F$. □

Here is a versatile version of the theorem we just proved.

Corollary 1.8.37. *Assume X is a Hausdorff linear topological space, and Y is a closed linear subspace of X of finite codimension in X. Then there exists a finite dimensional linear subspace F of X such that $X = Y \oplus F$.*

Proof. This is a direct consequence of Theorem 1.8.36, Theorem 1.3.29, and Exercise 1.3.41. □

We conclude by presenting a very desirable result, essentially asserting that being a closed subspace of finite codimension is a hereditary property.

Theorem 1.8.38. *Let X be a Hausdorff linear topological space, and suppose Z is a closed subspace of finite codimension in X. Then any linear subspace Y of X with the property that $Z \subseteq Y$ is necessarily closed and of finite codimension in X.*

Proof. Let M be a linear subspace of Y such that $Z \boxplus M = Y$, and let N be a linear subspace of X such that $Y \boxplus N = X$. Then $M \cap N \subseteq Y \cap N = \{0\}$, so $M \cap N = \{0\}$. As such, it makes sense to define $W := M \boxplus N$. This is a linear subspace of X with the property that $Z \cap W = \{0\}$. Indeed, if $z \in Z$ is written as $z = m + n$ for some $m \in M$ and $n \in N$ then $Y \ni z - m = n \in N$ hence $n = 0$. In turn, this implies $Z \ni z = m \in N$ which also forces $m = 0$. Ultimately, $z = 0$, proving that $Z \cap W = \{0\}$. Since we also have $Z + W = Z + (M + N) = (Z + M) + N = Y + N = X$, we conclude that $Z \boxplus W = X$. As a consequence, $\dim W = \operatorname{codim}_{\mathbb{K}} Z < +\infty$, hence $\dim N \leq \dim W < +\infty$ and

$\dim M \leq \dim W < +\infty$. Since Z is a closed linear subspace of X, and M is a finite-dimensional linear subspace of X, Theorem 1.8.35 guarantees that $Y = Z + M$ is a closed linear subspace of X. In addition,

$$\operatorname{codim}_{\mathbb{K}} Y = \dim N \leq \dim W = \operatorname{codim}_{\mathbb{K}} Z < +\infty, \tag{1.8.558}$$

so Y also has finite codimension in X. □

1.8.7 Heine-Borel Converse

There remains to prove part *(c)* of Theorem 1.8.1, namely the "Heine-Borel" characterization of finite dimensionality. Like many arguments to come, this one hinges on the "first Riesz lemma," recorded in Lemma 1.6.50.

Recall from Definition 1.1.27 that a topological space is *locally compact* if every point has a compact neighborhood.

Exercise 1.8.39. Let (X, τ) be a topological vector space. Then (X, τ) is locally compact (as a topological space) if and only if there exists a compact set $V \in \mathcal{N}(0; \tau)$ if and only if there exists a relatively compact set $V \in \mathcal{N}(0; \tau)$.

Theorem 1.8.40. *Let X be a Hausdorff linear topological space. If X is locally compact, then it is finite dimensional.*

Proof. Suppose X is infinite dimensional. We will prove that no zero-neighborhood in X is relatively compact. If X is not locally bounded this is automatic (since relatively compact sets are subsets of compact sets, therefore, by Theorem 1.6.40 and Proposition 1.6.39, are bounded), so we may assume X is locally bounded. Thus, this and Proposition 1.6.19 imply that X has a bounded, circled zero-neighborhood U, and since the positive multiples of U form a local base for the zero-neighborhoods of X (see Proposition 1.6.48) we need only show that U is not relatively compact.

Knowing that X is nontrivial, U cannot be the singleton $\{0\}$, since this is not absorbing (and this violates item *(a)* in Theorem 1.6.18). Thus, it is possible to choose a vector non-zero vector $e_1 \in U$. Use this to define the one-dimensional subspace $F_1 := \operatorname{span}\{e_1\}$ of X. By Corollary 1.8.33, this is closed in X, and it is certainly not equal to the infinite-dimensional space X. As such, Lemma 1.6.50 guarantees that there exists $e_2 \in U$ with $e_2 \notin \frac{1}{2}U + F_1$. Thus $e_2 \notin \frac{1}{2}U + e_1$, and by the circledness of U, the same is true with e_1 and e_2 reversed.

Let $F_2 := \operatorname{span}\{e_1, e_2\}$, a two dimensional (closed) subspace of X, and apply Lemma 1.6.50 in order to produce $e_3 \in U$ with $e_3 \notin \frac{1}{2}U + F_2$. Thus $e_3 \notin \frac{1}{2}U + e_j$, $j = 1, 2$, so again by circledness it is also true that $e_j \notin \frac{1}{2}U + e_3$, for $j = 1, 2$.

Continuing in this manner we produce an infinite sequence $\{e_j\}_{j \in \mathbb{N}}$ of vectors in U such that for $j \neq k$ we have $e_j \notin \frac{1}{2}U + e_k$. To prove that U is not relatively compact, we need only show that the infinite set $\{e_j\}_{j \in \mathbb{N}}$ has no limit point (cf. Theorem 1.1.39). To this end, suppose, by way of contradiction, that some $x \in X$ is a limit point of this set. Choose a circled zero-neighborhood V in X with $V+V \subseteq \frac{1}{2}U$. Then the neighborhood $x + V$ of x contains infinitely many vectors e_j. Suppose e_j and e_k are two of these ($j \neq k$). Then $e_j - e_k \in V - V = V + V \subseteq \frac{1}{2}U$, which contradicts the fact that $e_j \notin e_k + \frac{1}{2}U$. □

We will use the last part of this argument often enough in the sequel to warrant giving its conclusion official status.

Proposition 1.8.41. *Suppose A is a subset of a Hausdorff linear topological space X and assume $\{x_j\}_{j\in\mathbb{N}}$ is an infinite sequence in A with this property: there exists a zero-neighborhood U in X such that $x_j \notin x_k + U$ whenever $j, k \in \mathbb{N}$ satisfy $j \neq k$. Then A is not relatively compact in X.*

1.9 Additional Exercises: II

1. Show that *every infinite-dimensional locally bounded topological vector space space has a discontinuous linear functional*. Thus every such space has a one dimensional subspace that is not topologically complemented, or equivalently, a subspace of codimension one that is dense.
   Outline: Start with a bounded zero-neighborhood U. Show that U contains a countable linearly independent set of vectors $\{e_n\}$. If necessary, extend this set to a basis for X. Define $\Lambda(e_n) := n$ and take Λ to be 0 on the remaining basis vectors (if there are any). Extend Λ linearly to the whole space. Show that $e_n/\sqrt{n} \to 0$, whereas $|\Lambda(e_n/\sqrt{n})| \to \infty$. Thus Λ is not continuous.
2. If $0 < p \leq 1$ and X is any separable p-Banach space X, then there is a continuous linear map taking ℓ^p onto X.

   Outline of the proof:

   (i) We know that the separable p-Banach space X has a countable, dense subset. Show that its closed unit ball $X_1 := \{x \in X : \|x\|_p \leq 1\}$ also has a countable dense subset.
   (ii) Let $D := \{x_n;\ n \in \mathbb{N}\}$ be a countable dense subset of X_1 guaranteed by (i). Show that for each $f \in \ell^p$ the series $\sum_n f(n)x_n$ converges in the p-norm of X to a vector $Tf \in X$, and that the mapping $T : \ell^p \to X$ so defined is linear and continuous.
   (iii) For the map T defined in (ii) show that $T(\ell^p) = X$.
   Hint: Fix $x \in X$. We desire to find $f \in \ell^p$ such that $Tf = x$. To do so, assume (without loss of generality) that $x \in X_1$, and choose $x_{n_1} \in D$ such that $x - x_1 \in \frac{1}{2}X_1$. Now show that there exists $x_{n_2} \in D$ with $x - x_{n_1} - \frac{1}{2}x_{n_2} \in \frac{1}{4}X_1$. Continue inductively to obtain a subsequence $(x_{n_k})_{k\in\mathbb{N}}$ of D such that for each positive integer k we have

$$x - \left(x_{n_1} + \frac{1}{2}x_{n_2} + \cdots + \frac{1}{2^{k-1}}x_{n_k}\right) \in \frac{1}{2^k}X_1. \tag{1.9.559}$$

Prove that the series $\sum_{j=1}^{\infty} \frac{1}{2^j} x_{n_j}$ converges in X to x, and that there is an element $f \in \ell^p$, with $Tf = x$.

3. Even if $0 < p < 1$, the space ℓ^p still has enough continuous linear functionals to separate points (cf. Definition 1.6.12).
Hint: For each positive integer n the "evaluation functional" $e_n : f \mapsto f(n)$ is linear and continuous on ℓ^p.
4. From Exercise 2 above, we know that there exists some continuous linear map T taking ℓ^p onto the space $L^p([0, 1])$. For a given exponent $0 < p < 1$, let M be the null space of this surjective map $T : \ell^p \to L^p([0, 1])$ and suppose Λ is a continuous linear functional on ℓ^p. Show that if $\Lambda = 0$ on the closed subspace M of ℓ^p, then $\Lambda = 0$ on all of ℓ^p.
5. When $0 < p < 1$, the Hahn-Banach Theorem fails in ℓ^p.
Note: Exercise 3 shows that this failure does not happen "trivially." We know that the classical Hahn-Banach Theorem guarantees that every Banach space has enough continuous linear functionals to separate points (cf. Definition 1.6.12). More generally, by Theorem 1.12.2, the same is true in every Hausdorff locally convex topological vector space.
Hint: From Exercise 4 it follows that for $0 < p < 1$ the space ℓ^p does not have enough continuous linear functionals to separate M from points not in M (in fact, the dual space of ℓ^p can't separate any $f \in \ell^p \setminus M$ from M). As a consequence, for $0 < p < 1$ the classical Hahn-Banach Theorem fails in ℓ^p.
6. ℓ^p is not isomorphic to a Banach space when $0 < p < 1$ (cf. [41]).
7. (see [10])

   (a) Show that if $0 < p < 1$ then the convex hull of the unit ball of $L^p([0, 1])$ is *the whole space*.
   Suggestion: For $f \in L^p([0, 1])$ show that you can divide the unit interval into n subintervals over each of which f^p has the same integral. Using this decomposition write f as an average of n functions, each of which, if n is large enough, lies in the unit ball.
   (b) Show that if $0 < p < 1$ then the only continuous linear functional on $L^p([0, 1])$ is the zero-functional.

8. Suppose X has a nontrivial closed subspace of finite codimension. Show that X has nontrivial continuous linear functionals.
9. Suppose X has enough continuous linear functionals to separate points (cf. Definition 1.6.12). Show that every finite-dimensional subspace of X is topologically complemented (for more on this see also Corollary 1.12.3).
Hint: Begin with dimension one, say $Y := \text{span}\{v\}$ for some $0 \neq v \in X$. If X^* separates points in X, then there exists $f \in X^*$ with $f(v) \neq 0$. Take $Z := \ker f$ and observe that each $x \in X$ may be decomposed as

$$x = \Big(x - \frac{f(x)}{f(v)} v\Big) + \Big(\frac{f(x)}{f(v)} v\Big) \tag{1.9.560}$$

with $x - \frac{f(x)}{f(v)}v$ in Z and $\frac{f(x)}{f(v)}v$ in Y. Moreover, since f is continuous it follows that the assignments $x \mapsto x - \frac{f(x)}{f(v)}v$ and $x \mapsto \frac{f(x)}{f(v)}v$ are continuous. Finally, the fact that $f(v) \neq 0$ ensures that $Z \cap Y = \{0\}$. Thus, Z is a topologic complement of Y in X. This takes care of the case when $\dim Y = 1$. Arbitrary finite-dimensional spaces are then treated inductively, iterating this method.

10. *An infinite-dimensional linear topological space on which every linear functional is continuous.* By contrast with the situation described in Exercise 1, there are infinite-dimensional linear topological spaces for which *every* linear functional is continuous.
To construct one, let X be any infinite-dimensional complex (or real) vector space, and let X' denote the space of all linear functionals on X. Endow X with the topology of pointwise convergence it gets as a collection of linear functionals on X'. More precisely, for each subset F of X' and each positive number ε let

$$N(F,\varepsilon) := \{x \in X : |\Lambda(x)| < \varepsilon \text{ for every } \Lambda \in F\}. \tag{1.9.561}$$

(a) Show that the collection of all such sets forms a local base at the origin for a Hausdorff vector topology $\mathcal{T}$ on X.
(b) Show that a net $\{x_\alpha\}$ converges to 0 in X if and only if $\Lambda(x_\alpha) \to 0$ for every $\Lambda \in X'$.
(c) Show that every linear functional on X is $\mathcal{T}$-continuous.

11. *An infinite-dimensional linear topological space, every subspace of which is complemented.* Show that the linear topological space $(X,\mathcal{T})$ of the last problem has these properties:

(a) It is complete.
(b) Every linear transformation on it is continuous.
(c) Every subspace of it is topologically complemented.

12. Show that the weak topology on an infinite-dimensional Hilbert space is not complete. Can you extend this result to the dual of an infinite-dimensional Banach space?
13. *The strict topology.* Consider the following topology on the Hilbert space ℓ^2. If $g \in c_0$ (the space of complex sequences that converge to zero), define a seminorm p_g on ℓ^2 by:

$$p_g(f) := \|fg\|, \quad f \in \ell^2. \tag{1.9.562}$$

(a) Show that the collection of all "seminorm balls" of the form

$$N(g,\varepsilon) := \{f \in \ell^2 : p_g(f) < \varepsilon\} \tag{1.9.563}$$

is a local base at the origin for a Hausdorff vector topology β on ℓ^2 that lies between the weak topology and the norm topology, and is strictly weaker than the norm topology.
(b) Show that the linear topological space (ℓ^2,β) is not metrizable.

(c) Show that (ℓ^2,β) is complete, so in view of Exercise 12 above, β is different from the weak topology.
(d) Show that part (c) notwithstanding, β coincides with the weak topology on *bounded* subsets of ℓ^2.
(e) Show that even though (ℓ^2,β) is complete, it does not "support" the Baire Category Theorem. (**Suggestion:** use the fact that the unit ball of the space ℓ^2 is β-compact).

14. Suppose X is a Hausdorff linear topological space, and Y is a finite-dimensional proper subspace of X which is topologically complemented in X. Use this to construct a nontrivial continuous linear functional Λ on X.
Hint: With n denoting the dimension of Y, choose a basis $\{y_1, \dots, y_n\}$ of Y and define $\pi : Y \to \mathbb{C}$ by setting $\pi(c_1 y_1 + \cdots + c_n y_n) := c_1$ for each choice of scalars $c_1, \dots, c_n$. Note that π is linear and continuous. With Z denoting a topologically complement of Y in X, bring in the projection $P_Y : X \to Y$ (which is continuous, thanks to Proposition 1.8.22), and finally define $\Lambda := \pi \circ P_Y$.

1.10 The Aoki-Rolewicz Theorem in Topological Vector Spaces

A well-known result in the theory of topological vector spaces is as follows (see, e.g., the discussion in [23, p. 5] and [33, § 1.1]):

Theorem 1.10.1. *Let (X, τ) be a Hausdorff topological vector space with a countable base of neighborhoods at the origin. Then τ is metrizable (cf. Definition 1.2.6) and, in fact, may be given by a translation-invariant metric.*

Corollary 1.10.2. *If X is a locally bounded Hausdorff topological vector space, then the topology on X is metrizable (cf. Definition 1.2.6).*

Proof. Since X is locally bounded, there exists a bounded neighborhood of the origin $B \subseteq X$. Proposition 1.6.48 then guarantees that $\mathcal{U} := \{\frac{1}{n}B : n \in \mathbb{N}\}$ is a countable base of zero-neighborhoods in X. Keeping in mind that the topology on X is Hausdorff, Theorem 1.10.1 applies and gives that the topology on X is indeed metrizable. □

Let X be a vector space (over the field $\mathbb{K}$), and assume that B is an absorbent subset of X. Then the `Minkowski functional` (or `gauge`) associated with B is defined as

$$\|x\|_B := \inf\{\lambda > 0 : \lambda^{-1}x \in B\}, \qquad \forall\, x \in X. \tag{1.10.564}$$

Note that the quality of B being absorbent ensures that the above infimum is taken over a nonempty set. Hence, $\|x\|_B$ is a well-defined nonnegative real number for every $x \in X$. From second part of item *(8)* in Proposition 1.6.16, we know that δB continues to be absorbent for each $\delta > 0$. Bearing this in mind, one can easily check from definitions that

$$\|\cdot\|_{\delta B} = \delta^{-1}\|\cdot\|_B \quad \text{for each} \quad \delta > 0. \tag{1.10.565}$$

Moreover,

$$B \subseteq \{x \in X : \|x\|_B \leq 1\}. \tag{1.10.566}$$

If, in addition, B is balanced (cf. item *(a)* in Definition 1.6.14) then

$$\{x \in X : \|x\|_B < 1\} \subseteq B, \tag{1.10.567}$$

and we claim that we also have the homogeneity property

$$\|\alpha x\|_B = |\alpha| \, \|x\|_B, \quad \forall\, \alpha \in \mathbb{K}, \ \forall\, x \in X. \tag{1.10.568}$$

Indeed, $x \in X$ is such that $\|x\|_B < 1$ then there exists $\lambda \in (0, 1)$ with the property that $\lambda^{-1}x \in B$. Hence, $x = \lambda(\lambda^{-1}x) \in B$, given that B is balanced. This establishes (1.10.567). As far as (1.10.568) is concerned, since the set B is balanced, for every $\lambda \in (0, \infty)$, $\alpha \in \mathbb{K}$, and $x \in X$ we have $\lambda^{-1}\alpha x \in B$ if and only if $\lambda^{-1}|\alpha|x \in B$. Bearing this in mind, for each $\alpha \in \mathbb{K}$ and $x \in X$ we may write

$$\begin{aligned} \|\alpha x\|_B &= \inf\{\lambda > 0 : \lambda^{-1}\alpha x \in B\} \\ &= \inf\{\lambda > 0 : (\lambda|\alpha|^{-1})^{-1}x \in B\} \\ &= |\alpha| \, \|x\|_B. \end{aligned} \tag{1.10.569}$$

To continue, assume that B is a bounded neighborhood of the origin in a Hausdorff topological vector space X. Recall from item *(a)* of Theorem 1.6.18 that, automatically, B is is absorbent; in particular it makes sense to consider the Minkowski functional associated with B as in (1.10.564). These considerations and the fact that the topological vector space X is Hausdorff then entail that, for every $x \in X$,

$$\|x\|_B = 0 \iff x = 0. \tag{1.10.570}$$

The left-pointing implication is clear, so we focus on the right-pointing implication in (1.10.570). Assuming $\|x\|_B = 0$, it follows that there exists a sequence $\{\lambda_j\}_{j\in\mathbb{N}}$ of positive real numbers converging to zero such that $x \in \bigcap_{j\in\mathbb{N}} \lambda_j B$. If $x \neq 0$, since the topology on X is Hausdorff, there exists zero-neighborhood of X with $x \notin V$. However, B is bounded so there exists $\delta > 0$ such that $tB \subseteq V$ for each $t \in (0, \delta)$. Since $\lambda_j \to 0$ as $j \to \infty$, this shows that there exists $j \in \mathbb{N}$ with the property that $\lambda_j B \subseteq V$. In particular, x must be in V, leading to a contradiction. This concludes the proof of (1.10.570).

Next, work under the assumption that B is a bounded and balanced neighborhood of the origin of a Hausdorff topological vector space X (as before, B is therefore absorbing). Observe that $B + B$ is also bounded and, hence,

$$\text{there exists a constant } c \in (0, \infty) \text{ such that } B + B \subseteq cB. \tag{1.10.571}$$

We claim that, for $c \in (0, \infty)$ as in (1.10.571),

$$\|x+y\|_B \le c\cdot\max\{\|x\|_B,\|y\|_B\},\quad \forall\, x,y\in X. \tag{1.10.572}$$

To justify (1.10.572), fix $x,y\in X$ and select a real number $\lambda_* > \max\{\|x\|_B,\|y\|_B\}$. From (1.10.564), this implies the existence of positive numbers $\lambda_1,\lambda_2 \in (0,\lambda_*)$ satisfying $\lambda_1^{-1}x, \lambda_2^{-1}y\in B$. Then

$$\lambda_*^{-1}x = \frac{\lambda_1}{\lambda_*}(\lambda_1^{-1}x)\in B, \tag{1.10.573}$$

with the membership implied by the fact that B is balanced and $\lambda_1/\lambda_*\in(0,1)$. In a similar fashion, we also obtain $\lambda_*^{-1}y\in B$. Consequently,

$$\lambda_*^{-1}x+\lambda_*^{-1}y\in B+B\subseteq cB, \tag{1.10.574}$$

hence $(c\lambda_*)^{-1}(x+y)\in B$. In view of (1.10.564), this forces $\|x+y\|_B\le c\lambda_*$. Sending λ_* monotonically to $\max\{\|x\|_B,\|y\|_B\}$ ultimately produces (1.10.572).

In particular, from (1.10.568), (1.10.570), (1.10.572), and (1.10.566)-(1.10.567), we conclude that

> if B is a bounded and balanced neighborhood of the origin in a Hausdorff topological vector space X, then (B is absorbent and) the Minkowski functional $\|\cdot\|_B$, associated with B as in (1.10.564), is a quasi-norm on X. (1.10.575)

For further use, let us also note that (1.10.572) forces c to be ≥ 2, thus

$$\text{if } c\in(0,\infty) \text{ is as in (1.10.571) then necessarily } c\ge 2. \tag{1.10.576}$$

We now proceed to review some basic notions of convexity. Recall from (1.3.81) that if X is a given vector space, then $V\subseteq X$ is said to be a *convex set* provided $\alpha V+(1-\alpha)V\subseteq V$ whenever $\alpha\in[0,1]$. Also, recall from (1.6.349) that a topological vector space (X,τ) is called *locally convex* if it has a local base at the origin consisting of convex zero-neighborhoods. More generally, given any $p\in(0,1]$, call a topological vector space (X,τ) `locally` p-`convex` if it has a local base at the origin consisting of p-convex zero-neighborhoods. For each $p\in(0,1]$, the notion of p-convexity is defined as follows: a subset V of a vector space X is said to be p-`convex` if

$$\alpha_1,\alpha_2\in[0,1] \text{ with } \alpha_1^p+\alpha_2^p=1 \implies \alpha_1V+\alpha_2V\subseteq V. \tag{1.10.577}$$

Parenthetically, we wish to note that 1-convex sets are precisely the convex sets mentioned earlier. A related, stronger concept is as follows: given $p\in(0,1]$, a subset V of a vector space X is said to be `absolutely` p-`convex` if

$$\alpha_1,\alpha_2\in\mathbb{K} \text{ with } |\alpha_1|^p+|\alpha_2|^p\le 1 \implies \alpha_1V+\alpha_2V\subseteq V. \tag{1.10.578}$$

Lemma 1.10.3. *Pick $p\in(0,1]$. If X is a vector space and if $B\subseteq X$ is an absorbent, balanced, p-convex set, then the Minkowski functional associated with B*

as in (1.10.564) *satisfies*

$$\|x + y\|_B^p \leq \|x\|_B^p + \|y\|_B^p, \qquad \forall\, x, y \in X. \tag{1.10.579}$$

Proof. Fix $x, y \in X$ and select two real numbers, $a > \|x\|_B$ and $b > \|y\|_B$. Reasoning as in the proof of (1.10.573) we presently obtain

$$a^{-1}x \in B \ \text{ and } \ b^{-1}x \in B. \tag{1.10.580}$$

By virtue of (1.10.580) and the p-convexity of B, we have

$$\begin{aligned} x + y &= (a^p + b^p)^{1/p} \cdot \Big[\frac{a}{(a^p + b^p)^{1/p}}(a^{-1}x) + \frac{b}{(a^p + b^p)^{1/p}}(b^{-1}y)\Big] \\ &\in (a^p + b^p)^{1/p} \cdot \Big[\frac{a}{(a^p + b^p)^{1/p}}B + \frac{b}{(a^p + b^p)^{1/p}}B\Big] \\ &\in (a^p + b^p)^{1/p} \cdot B, \end{aligned} \tag{1.10.581}$$

since $\Big(\frac{a}{(a^p+b^p)^{1/p}}\Big)^p + \Big(\frac{b}{(a^p+b^p)^{1/p}}\Big)^p = 1$. We therefore have $(a^p + b^p)^{-1/p}(x + y) \in B$, which goes to show that $\|x + y\|_B \leq (a^p + b^p)^{1/p}$ by (1.10.564). Sending $a \searrow \|x\|_B$ and $b \searrow \|y\|_B$ then readily gives (1.10.579). □

Lemma 1.10.3 plus the discussion in relation to (1.10.568) and (1.10.570) imply the following result.

Proposition 1.10.4. *If X is a Hausdorff locally bounded topological vector space and if $B \subseteq X$ is an absorbent, bounded, balanced, p-convex set for some $p \in (0, 1]$, then the Minkowski functional associated with B as in* (1.10.564) *has the property that $\|\cdot\|_B^p$ is a p-norm on X (cf. Definition 1.4.1).*

It is important to point out that in general, a Hausdorff locally bounded topological vector space is not necessarily locally convex. In this light, the next result does the next best thing, namely it guarantees that the space under discussion is locally p-convex for some $p \in (0, 1]$.

Theorem 1.10.5 (Aoki-Rolewicz). *Let (X, τ) be a Hausdorff topological vector space such that there exists a bounded and balanced neighborhood B of the origin in X. In particular, $B + B$ is bounded*[4] *and, since B is a zero-neighborhood in X, there exists a positive constant c (which is actually ≥ 2; cf.* (1.10.576)) *such that $B + B \subseteq cB$. Use this constant to define*

$$p := \frac{1}{\log_2 c} \in (0, 1] \tag{1.10.582}$$

and, for each $x \in X$, set

[4] see item *(a)* in Theorem 1.6.40

$$\|x\|_* := \inf\Big\{\sum_{i=1}^{N} \|x_i\|_B^p : N \in \mathbb{N},\ x_1, \dots, x_N \in X \text{ are such that } \sum_{i=1}^{N} x_i = x\Big\}, \tag{1.10.583}$$

where $\|\cdot\|_B$ *is the Minkowski functional associated with* B *as in* (1.10.564) (*recall from part (a) in Theorem 1.6.18 that* B *is absorbent*).

Then $\|\cdot\|_*$ *is a* p*-norm on* X*, with the property that* $\|\cdot\|_*^{1/p}$ *and* $\|\cdot\|_B$ *are equivalent quasi-norms on* X (*in the sense of Definition 1.4.15*).

Furthermore, both the p*-norm topology on* $(X, \|\cdot\|_*)$ (*cf. Definition 1.4.6*) *and* $\tau_{\|\cdot\|_B}$ (*cf.* (1.10.575) *and Exercise 1.4.18*) *coincide with the original topology* τ *on* X*. In particular, the topological vector space* (X, τ) *is locally* p*-convex, and the original topology* τ *on* X *is metrizable via a translation-invariant distance.*

Note that Theorem 1.10.5 is applicable to any *locally bounded Hausdorff topological vector space* (X, τ). Indeed, in such a scenario, there exists a bounded zero-neighborhood B in (X, τ) which, thanks to item *(2)* in Proposition 1.6.19 and Proposition 1.6.39, may also be assumed to be balanced (aka circled). Thus, at long last, we are in a position to justify the claim made in (1.6.489).

We also wish to remark that

$$B + B \subseteq 2B \text{ whenever } B \text{ is a convex set,} \tag{1.10.584}$$

since for all $x, y \in B$ it follows that $x + y = 2(\frac{1}{2}x + \frac{1}{2}y) \in 2B$. Consequently, in such a scenario, we can take $c = 2$ in (1.10.582) which forces $p = 1$. Thus, in the case in which (X, τ) is a Hausdorff topological vector space such that there exists a convex, bounded, and balanced neighborhood of the origin in X, it follows that the topology τ is normable (i.e., it coincides with the topology induced by a genuine norm on X). See also Exercise 1.10.7 in this regard.

Regarding Theorem 1.10.5, the original references are [2], [38], and more timely expositions may be found in [23] and [33]. Theorem 1.10.5 is essentially a particular version of [33, Theorem 3.27, pp. 108-109] recalled below, itself a special case of a more general metrization result on groupoids from [33].

Theorem 1.10.6. *Let* $(X, \|\cdot\|)$ *be a (nontrivial) quasi-normed space. Then*

$$\kappa := \sup_{\substack{x,y\in X \\ \text{not both zero}}} \left(\frac{\|x+y\|}{\max\{\|x\|, \|y\|\}} \right) \in [2, \infty) \tag{1.10.585}$$

and if, with

$$p := \frac{1}{\log_2 \kappa} \in (0, 1], \tag{1.10.586}$$

for each $x \in X$ *one defines*

$$\|x\|_\# := \inf\left\{ \sum_{i=1}^{N} \|x_i\|^p : N \in \mathbb{N},\ x_1, \dots, x_N \in X \text{ with } \sum_{i=1}^{N} x_i = x \right\}, \tag{1.10.587}$$

it follows that $\|\cdot\|_{\#}$ *is a p-norm on X satisfying*

$$\kappa^{-2}\|x\| \leq \|x\|_{\#}^{1/p} \leq \|x\| \quad \textit{for each} \quad x \in X. \tag{1.10.588}$$

In particular, the p-norm $\|\cdot\|_{\#}$ *induces the same topology on X as the original quasi-norm* $\|\cdot\|$.

Here are the details in the proof of Theorem 1.10.5:

Proof of Theorem 1.10.5. Keeping in mind (1.10.575), we may invoke Theorem 1.10.6 for the choice $\|\cdot\| := |\!|\!|\cdot|\!|\!|_B$, and this shows that $\|\cdot\|_*$ defined in (1.10.583) is a p-norm on X with the property that

$$c^{-2}|\!|\!|x|\!|\!|_B \leq \|x\|_*^{1/p} \leq |\!|\!|x|\!|\!|_B \quad \text{for each} \quad x \in X. \tag{1.10.589}$$

In particular, $\|\cdot\|_*^{1/p}$ and $|\!|\!|\cdot|\!|\!|_B$ are equivalent quasi-norms on X. Also, thanks to the last property in the statement of Theorem 1.10.6, the p-norm topology on $(X, \|\cdot\|_*)$ coincides with $\tau_{|\!|\!|\cdot|\!|\!|_B}$, the topology induced on X by the quasi-norm $|\!|\!|\cdot|\!|\!|_B$.

Since B is a bounded subset of (X, τ), given any $V \in \mathcal{N}(0; \tau)$ it follows that there exists $\varepsilon > 0$ such that $\varepsilon B \subseteq V$. In view of (1.10.567), we therefore conclude that

$$\{x \in X : |\!|\!|x|\!|\!|_B < \varepsilon\} \subseteq \varepsilon B \subseteq V. \tag{1.10.590}$$

On the other hand, (1.10.589) implies

$$\{x \in X : \|x\|_* < c^{-2p}\varepsilon^p\} \subseteq \{x \in X : |\!|\!|x|\!|\!|_B < \varepsilon\} \tag{1.10.591}$$

which, together with (1.10.590), shows that

$$\{x \in X : \|x\|_* < c^{-2p}\varepsilon^p\} \subseteq V. \tag{1.10.592}$$

Given that $\|\cdot\|_*$ is a p-norm, the ball $\{x \in X : \|x\|_* < c^{-2p}\varepsilon^p\}$ is open in τ_*, the p-norm topology on $(X, \|\cdot\|_*)$ (cf. Definition 1.4.6 and Proposition 1.4.7). Based on this and (1.10.592) we deduce that $V \in \mathcal{N}(0; \tau_*)$. The argument so far shows that $\mathcal{N}(0; \tau) \subseteq \mathcal{N}(0; \tau_*)$, hence $\tau \subseteq \tau_*$.

To prove the opposite inclusion, start with an arbitrary $W \in \mathcal{N}(0; \tau_*)$. Then there exists $r > 0$ with the property that

$$\{x \in X : \|x\|_* < r\} \subseteq W. \tag{1.10.593}$$

But, as seen from (1.10.566) and (1.10.589), for each $\varepsilon \in (0, r^{1/p})$ we have

$$\varepsilon B \subseteq \{x \in X : |\!|\!|x|\!|\!|_B < r^{1/p}\} \subseteq \{x \in X : \|x\|_* < r\}. \tag{1.10.594}$$

In concert with (1.10.593) this proves that

$$\varepsilon B \subseteq W \quad \text{for each} \quad \varepsilon \in (0, r^{1/p}). \tag{1.10.595}$$

Since $\varepsilon B \in \mathcal{N}(0;\tau)$ for every $\varepsilon > 0$, we further deduce that $W \in \mathcal{N}(0;\tau)$. Thus, $\mathcal{N}(0;\tau_*) \subseteq \mathcal{N}(0;\tau)$, so $\tau_* \subseteq \tau$.

All together, at this stage we have shown that τ and τ_* are identical and, as noted earlier, τ_* also coincides with $\tau_{|\cdot|_B}$. There remains to observe that, since $\|\cdot\|_*$ is a p-norm on X, the family of balls $B_j := \{x \in X : \|x\|_* < 1/j\}$ indexed by $j \in \mathbb{N}$ is a (countable) base of neighborhoods at the origin in $(X,\tau_*) = (X,\tau)$, consisting of p-convex sets in the sense of (1.10.577). This renders the topological vector space (X,τ) locally p-convex. Thus, all claims in the final paragraph in the statement of Theorem 1.10.5 have been justified, so the proof of Theorem 1.10.5 is complete. □

Exercise 1.10.7. If $(X, \|\cdot\|)$ is a quasi-normed space with the property that

$$\|x+y\| \leq 2\max\{\|x\|, \|y\|\}, \quad \forall\, x, y \in X, \tag{1.10.596}$$

show that $\tau_{\|\cdot\|}$ coincides with the topology induced by a norm on X.

Hint: Use the fact that (1.10.596) forces κ in (1.10.585) to be 2, hence p in (1.10.586) is presently equal to 1.

1.11 Quasi-Banach Spaces Revisited

The following result may be regarded as a version of Theorem 1.10.5 in the class of quasi-normed spaces.

Theorem 1.11.1 (Aoki-Rolewicz). *Let $(X, \|\cdot\|)$ be a quasi-normed space. Then there exists some p-norm $\|\cdot\|_{\#}$ on X for some $p \in (0,1]$ with the property that the quasi-norm[5] $\|\cdot\|_{\#}^{1/p}$ is equivalent (in the sense of Definition 1.4.15) to the original quasi-norm $\|\cdot\|$ on X.*

In particular, the topology $\tau_{\|\cdot\|}$ induced by $\|\cdot\|$ on X (as in Exercise 1.4.18) is metrizable, since it coincides with the topology τ_d induced by the (translation-invariant) distance on X

$$d(x,y) := \|x-y\|_{\#} \ \text{for all} \ \ x,y \in X. \tag{1.11.597}$$

Furthermore, $\tau_{\|\cdot\|}$ coincides with the p-norm topology on $(X, \|\cdot\|_{\#})$, discussed in Definition 1.4.6. Also,

$$\begin{aligned}&(X,\tau_{\|\cdot\|}) \text{ is a Hausdorff, locally bounded, and}\\ &\text{locally } p\text{-convex, topological vector space,}\end{aligned} \tag{1.11.598}$$

and each ball

$$B_{\|\cdot\|}(0,r) := \{x \in X : \|x\| < r\}, \quad r \in (0,\infty), \tag{1.11.599}$$

[5] see Exercise 1.4.14

is an absorbent, bounded, and balanced neighborhood of the origin in $(X, \tau_{\|\cdot\|})$.

As a corollary, every quasi-Banach space is a p-Banach space for some index $p \in (0, 1]$, *so*

$$\text{the collection of all quasi-Banach spaces coincides with the collection of all } p\text{-Banach spaces with } p \in (0,1] \text{ arbitrary.} \tag{1.11.600}$$

Consequently, in view of this and (1.6.343),

$$\text{every quasi-Banach space is an } F\text{-space.} \tag{1.11.601}$$

Proof. If we define $\|\cdot\|_{\#}$ as in (1.10.585)-(1.10.587), then Theorem 1.10.6 ensures that $\|\cdot\|_{\#}$ is a p-norm on X for some $p \in (0, 1]$, with the property that $\|\cdot\|_{\#}^{1/p}$ is equivalent to $\|\cdot\|$, and the topology $\tau_{\|\cdot\|}$ coincides with the topology τ_d if the distance d is as (1.11.597). The latter is readily seen to further coincide with the p-norm topology on $(X, \|\cdot\|_{\#})$. Thus, since $\tau_{\|\cdot\|}$ coincides with the p-norm topology on $(X, \|\cdot\|_{\#})$, we may invoke Proposition 1.6.2 to conclude that

$$(X, \tau_{\|\cdot\|}) \text{ is a Hausdorff linear topological space.} \tag{1.11.602}$$

Also, since (1.10.588) implies that

$$B_{\|\cdot\|}(0, r) \subseteq B_d(0, r^p) \ \text{ for all } \ r \in (0, \infty), \tag{1.11.603}$$

and, as seen from part *(2)* of Proposition 1.2.5,

$$B_d(0, r) \in \mathcal{N}(x; \tau_d) = \mathcal{N}(x; \tau_{\|\cdot\|}) \ \text{ for all } \ r \in (0, \infty), \tag{1.11.604}$$

we deduce that

$$B_{\|\cdot\|}(0, r) \in \mathcal{N}(0; \tau_{\|\cdot\|}) \ \text{ for each } \ r \in (0, \infty). \tag{1.11.605}$$

With this in hand, we then readily deduce from definitions that, for each radius $r \in (0, \infty)$, the ball $B_{\|\cdot\|}(0, r)$ is an absorbent, bounded, and balanced neighborhood of the origin in the Hausdorff linear topological space $(X, \tau_{\|\cdot\|})$.

Finally, since $\|\cdot\|_{\#}$ is a p-norm on X, the family of balls $B_j := \{x \in X : \|x\|_{\#} < 1/j\}$ indexed by $j \in \mathbb{N}$ is a (countable) base consisting of p-convex neighborhoods at the origin in the p-norm topology on $(X, \|\cdot\|_{\#})$. Given that the latter topology coincides with $\tau_{\|\cdot\|}$, we ultimately see that the topological vector space $(X, \tau_{\|\cdot\|})$ is indeed locally p-convex. □

Regarding the converse direction in (1.11.605), we have the following result:

Exercise 1.11.2. If $(X, \|\cdot\|)$ is a quasi-normed space, show that

$$\begin{gathered}\text{for any } x \in X \text{ and any } V \in \mathcal{N}(x; \tau_{\|\cdot\|}) \text{ there exists } r \in (0, \infty) \\ \text{such that } B_{\|\cdot\|}(x, r) := \{y \in X : \|x - y\| < r\} \subseteq V.\end{gathered} \tag{1.11.606}$$

As a consequence, for each $x \in X$, the collection of balls $\{B_{\|\cdot\|}(x, r)\}_{r>0}$ is a local base at x in the topological space $(X, \tau_{\|\cdot\|})$ (cf. Definition 1.1.19).

Hint: Given any $x \in X$ and $V \in \mathcal{N}(x; \tau_{\|\cdot\|})$, there exists $\mathcal{U} \in \tau_{\|\cdot\|}$ with the property that $x \in \mathcal{U} \subseteq V$. The definition of $\tau_{\|\cdot\|}$ in Exercise 1.4.18 then guarantees that there exists $r \in (0, \infty)$ for which $B_{\|\cdot\|}(x, r) \subseteq \mathcal{U}$.

The reader is reminded that the quasi-norm of a linear operator acting between two quasi-normed spaces has been introduced in Exercise 1.4.23.

Proposition 1.11.3. *Consider a linear operator T mapping a quasi-normed space $(X, \|\cdot\|_X)$ into a quasi-normed space $(Y, \|\cdot\|_Y)$. Then, with the topologies $\tau_{\|\cdot\|_X}$ on X and $\tau_{\|\cdot\|_Y}$ on Y considered as in Exercise 1.4.18, it follows that*

$$\begin{gathered} T : (X, \tau_{\|\cdot\|_X}) \to (Y, \tau_{\|\cdot\|_Y}) \text{ is a continuous function (in the sense of Definition 1.1.7)} \\ \text{if and only if} \\ T : X \to Y \text{ is bounded in the sense of Exercise 1.4.23 (i.e., } \|T\|_{X\to Y} < \infty). \end{gathered} \tag{1.11.607}$$

Proof. The left-pointing implication in (1.11.607) is a direct consequence of the definitions of $\tau_{\|\cdot\|_X}$, $\tau_{\|\cdot\|_Y}$ in Exercise 1.4.18 and the fact that, as seen from (1.4.241),

$$\|Tx - Ty\|_Y \leq \|T\|_{X\to Y}\|x - y\|_X \quad \text{for all} \quad x, y \in X. \tag{1.11.608}$$

Next, to justify the right-pointing implication in (1.11.607), start from the realization that we have $B_{\|\cdot\|_Y}(0, 1) \in \mathcal{N}(0; \tau_{\|\cdot\|_Y})$ (cf. (1.11.599)). Since $T0 = 0$, Definition 1.1.7 guarantees the existence of some $V \in \mathcal{N}(0; \tau_{\|\cdot\|_X})$ such that $T(V) \subseteq B_{\|\cdot\|_Y}(0, 1)$. But, according to (1.11.606), one can find $r \in (0, \infty)$ such that $B_{\|\cdot\|_X}(0, r) \subseteq V$, thus we obtain

$$T(B_{\|\cdot\|_X}(0, r)) \subseteq B_{\|\cdot\|_Y}(0, 1). \tag{1.11.609}$$

Now, given any $\lambda \in (0, r)$ and any $x \in X \setminus \{0\}$, it follows that $\lambda \frac{x}{\|x\|_X}$ belongs to $B_{\|\cdot\|_X}(0, r)$, so this vector gets mapped by T into $B_{\|\cdot\|_Y}(0, 1)$, by (1.11.609). Consequently,

$$\left\| T\left(\lambda \frac{x}{\|x\|_X}\right) \right\|_Y < 1 \tag{1.11.610}$$

from which we conclude that $\|Tx\|_Y < \lambda^{-1}\|x\|_X$. Based on this and (1.4.241) we ultimately see that

$$\|T\|_{X\to Y} = \sup_{x\in X\setminus\{0\}} \frac{\|Tx\|_Y}{\|x\|_X} \leq r^{-1}, \tag{1.11.611}$$

so $\|T\|_{X\to Y} < \infty$, as wanted. □

Exercise 1.11.4. Let $(X, \|\cdot\|)$ be a quasi-normed space and $Y \subseteq X$ be a finite-dimensional subspace of X. Show that Y is closed in the topology induced on X by the quasi-norm $\|\cdot\|$.

Hint: Use Theorem 1.11.1 and item *(b)* of Theorem 1.8.1.

Combining Theorem 1.6.7 with (1.11.601) yields the following important result:

Theorem 1.11.5 (Open Mapping Theorem for Quasi-Banach spaces). *Suppose $(X, \|\cdot\|_X)$ and $(Y, \|\cdot\|_Y)$ are two quasi-Banach spaces. Then every continuous surjective linear operator $T : X \to Y$ is an open map.*

See also [23] and [33] for related material.

Moving on, let $(X, \|\cdot\|_X)$ be a quasi-normed space. Given a linear subspace Y of X, consider the quotient space X/Y and define

$$\big\|[x]_{X/Y}\big\|_{X/Y} = \|x + Y\|_{X/Y} := \inf_{y\in Y} \|x + y\|_X \text{ for each } x \in X. \tag{1.11.612}$$

If Y is closed, then (1.11.612) becomes a quasi-norm on the quotient space X/Y, and the associated modulus of concavity (cf. Definition 1.4.12) satisfies

$$\rho(X/Y) = \rho(X). \tag{1.11.613}$$

Moreover, the canonical projection

$$\begin{aligned}&\pi : (X, \|\cdot\|_X) \longrightarrow (X/Y, \|\cdot\|_{X/Y}) \text{ defined by}\\ &\pi(x) := [x]_{X/Y} = x + Y \in X/Y \text{ for each } x \in X,\\ &\text{is a well-defined, linear, and continuous operator.}\end{aligned} \tag{1.11.614}$$

One can check without difficulty that replacing $\|\cdot\|_X$ in (1.11.612) by an equivalent quasi-norm on X produces an equivalent quasi-norm on the quotient space X/Y. Finally, we remark that if $\|\cdot\|_X$ is actually a p-norm for some $p \in (0, 1]$, then $\|\cdot\|_{X/Y}$ is also a p-norm, and the canonical projection, defined as in (1.11.614), is still a well-defined, linear, and continuous operator.

Lemma 1.11.6. *Let $(X, \|\cdot\|_X)$ be a p-Banach space for some $p \in (0, 1]$ and let Y be a closed linear subspace of X. Then $(X/Y, \|\cdot\|_{X/Y})$ is also a p-Banach space. Also, if X_o is a closed linear subspace of X which contains Y, then X_o/Y is a closed linear subspace of X/Y.*

As a corollary of this and Theorem 1.11.1, if $(X, \|\cdot\|_X)$ is a quasi-Banach space and Y is a closed linear subspace of X, then $(X/Y, \|\cdot\|_{X/Y})$ is also a quasi-Banach space, and for each closed linear subspace X_o of X which contains Y it follows that X_o/Y is a closed linear subspace of X/Y.

Proof. From the discussion preceding the statement of the current lemma, we already know that $\|\cdot\|_{X/Y}$ is a p-norm on X/Y. To prove completeness, suppose $\{\xi_n\}_{n\in\mathbb{N}}$ is a Cauchy sequence in $(X/Y, \|\cdot\|_{X/Y})$. We wish to prove this sequence converges, and for this it is enough to produce a convergent subsequence. Begin by inductively choosing a subsequence $\{\xi_{n_k}\}_{k\in\mathbb{N}}$ such that

$$\sum_{k=1}^{\infty} \|\xi_{n_k} - \xi_{n_{k+1}}\|_{X/Y} < \infty. \tag{1.11.615}$$

This may be accomplished as follows. Since $\{\xi_n\}_{n\in\mathbb{N}} \subseteq X/Y$ is Cauchy, there exists $n_1 \in \mathbb{N}$ with the property that $\|\xi_n - \xi_m\|_{X/Y} < 2^{-1}$ whenever $n, m \geq n_1$. Next, there exists $n_2 > n_1$ with the property that $\|\xi_n - \xi_m\|_{X/Y} < 2^{-2}$ whenever $n, m \geq n_2$. Repeat this process, to produce a strictly increasing sequence $\{n_k\}_{k\in\mathbb{N}}$ of integers such that $\|\xi_{n_k} - \xi_{n_{k+1}}\|_{X/Y} < 2^{-k}$ for all $k \in \mathbb{N}$. From this, (1.11.615) follows.

Going further, pick $x_1 \in \xi_{n_1}$ and use the definition in (1.11.612) to conclude that for each $k \in \mathbb{N}$ there exists a vector

$$x_{k+1} \in \xi_{n_{k+1}} - \xi_{n_k} \quad \text{with} \quad \|x_{k+1}\|_X \leq \|\xi_{n_{k+1}} - \xi_{n_k}\|_{X/Y} + 2^{-k}. \tag{1.11.616}$$

Together, (1.11.615) and (1.11.616) ensure that $\sum_{k=1}^{\infty} \|x_k\|_X < \infty$. From this, the fact that

$$\Big\| \sum_{k=M}^{N} x_k \Big\|_X \leq \sum_{k=M}^{N} \|x_k\|_X \quad \text{for all } M, N \in \mathbb{N} \text{ with } N > M \tag{1.11.617}$$

since $\|\cdot\|_X$ is a p-norm, and the completeness of X, we then conclude that the series $\sum_{k=1}^{\infty} x_k$ converges in $(X, \|\cdot\|_X)$. By continuity of the quotient map (1.11.614), the series

$$\xi_{n_1} + \sum_{k=1}^{\infty} (\xi_{n_{k+1}} - \xi_{n_k}) = \sum_{k=0}^{\infty} \pi(x_{k+1}) \tag{1.11.618}$$

converges in X/Y to $\pi\Big(\sum_{k=0}^{\infty} x_{k+1} \Big)$. Given that

$$\xi_{n_{M+1}} = \xi_{n_1} + \sum_{k=1}^{M} (\xi_{n_{k+1}} - \xi_{n_k}) \quad \text{for each} \quad M \in \mathbb{N}, \tag{1.11.619}$$

this implies that the subsequence $\{\xi_{n_{M+1}}\}_{M\in\mathbb{N}}$ is convergent in the quotient p-norm, as desired. Ultimately, this establishes that $(X/Y, \|\cdot\|_{X/Y})$ is indeed a p-Banach space.

Finally, consider a closed linear subspace X_o of X which contains Y. We wish to show that X_o/Y is closed in $(X/Y, \|\cdot\|_{X/Y})$. With this in mind, pick a sequence $\{\xi_n\}_{n\in\mathbb{N}} \subseteq X_o/Y$ which converges in $(X/Y, \|\cdot\|_{X/Y})$ to some $\xi \in X/Y$, with the goal of proving that $\xi \in X_o/Y$. Since $\{\xi_n\}_{n\in\mathbb{N}}$ is Cauchy in $(X/Y, \|\cdot\|_{X/Y})$, we may proceed as in the first part of the proof and produce vectors $x_1, x_2, \dots, x_k, \dots \in X_o$ satisfying (1.11.616). As before, this ensures that the series $\sum_{k=1}^{\infty} x_k$ converges in $(X, \|\cdot\|_X)$, and since X_o is closed, it follows that the sum $x := \sum_{k=0}^{\infty} x_{k+1}$ belongs to X_o. Ultimately,

$$\xi = \xi_{n_1} + \sum_{k=1}^{\infty} (\xi_{n_{k+1}} - \xi_{n_k}) = \pi\Big(\sum_{k=0}^{\infty} x_{k+1} \Big) = \pi(x) \in X_o/Y, \tag{1.11.620}$$

as wanted. □

We close with a proposition which elaborates on the nature of operators acting between quasi-Banach spaces which are invertible from the right.

Proposition 1.11.7. *Let X, Y be two quasi-Banach spaces and let $T : X \to Y$ be a linear and continuous operator. Then the following three statements are equivalent:*

(i) T has an inverse from the right (i.e., there exists a linear and continuous operator $S : Y \to X$ with the property that $T \circ S = I_Y$, the identity on Y);
(ii) T is surjective and $\ker T$ *has a closed algebraic complement in X;*
(iii) T is surjective and $\ker T$ *is topologically complemented in X.*

Proof. Assume first that T has an inverse from the right, $S : Y \to X$. We claim that

$$\operatorname{ran} S \text{ is closed in } X \text{ and } \operatorname{ran} S \cap \ker T = \{0\}. \tag{1.11.621}$$

To justify this claim, assume that $\{y_j\}_{j\in\mathbb{N}}$ is a sequence in Y with the property that $\{Sy_j\}_{j\in\mathbb{N}}$ converges in X to some vector $x \in X$. Given that T is continuous and S is an inverse from the right for T, it follows that $\{y_j\}_{j\in\mathbb{N}} = \{TSy_j\}_{j\in\mathbb{N}}$ converges in Y to Tx. Hence, since S is continuous, $\{Sy_j\}_{j\in\mathbb{N}}$ converges in X to $S(Tx)$. Consequently, $x = S(Tx) \in \operatorname{ran} S$ since the topology on X is Hausdorff. Given that the topology on X is also first-countable, this reasoning shows that $\operatorname{ran} S$ is indeed closed in X. Assume next that $x \in \operatorname{ran} S \cap \ker T$. Then there exists $y \in Y$ such that $x = Sy$ which further implies that $0 = Tx = T(Sy) = y$. Hence $y = 0$ which ultimately forces $x = Sy$ to be $0 \in X$. This concludes the proof of (1.11.621). Granted this, in order to conclude that $\ker T$ has a closed algebraic complement in X it suffices to show that

$$\operatorname{ran} S \boxplus \ker T = X. \tag{1.11.622}$$

To see that this is the case, observe that each $x \in X$ may be decomposed as

$$x = (x - S(Tx)) + S(Tx) \text{ with } (x - S(Tx)) \in \ker T \text{ and } S(Tx) \in \operatorname{ran} S. \tag{1.11.623}$$

Thus, (1.11.622) holds. Given that the existence of a right-inverse implies ontoness, this finishes the proof of the implication *(i)* $\Rightarrow$ *(ii)*.

Next, since T is continuous, it follows that $\ker T$ is closed in X. Bearing this in mind, the fact that *(ii)* $\Rightarrow$ *(iii)* is seen from Proposition 1.8.19 and (1.11.601).

As regards the implication *(iii)* $\Rightarrow$ *(i)*, assume that T is surjective and that there exists a linear subspace Z of X with the property that

$$Z \oplus \ker T = X. \tag{1.11.624}$$

In particular, Z is closed (cf. Corollary 1.8.18). This makes Z a quasi-Banach space and forces $T\big|_Z$, the restriction of T to Z, to be bijective. Of course, $T\big|_Z : Z \to Y$ remains continuous. As a consequence of the Open Mapping Theorem (cf. Theorem 1.11.5),

$$(T\big|_Z)^{-1} : Y \longrightarrow Z \text{ is continuous.} \tag{1.11.625}$$

Denote by ι the embedding of Z into X and set

$$S := \iota \circ (T\big|_Z)^{-1} : Y \longrightarrow X. \tag{1.11.626}$$

Then S is linear and since $\iota : Z \to X$ is linear and continuous, it follows from (1.11.625)-(1.11.626) that $S : Y \to X$ is also continuous. We now claim that

$$T \circ S = I_Y, \quad \text{the identity on } Y. \tag{1.11.627}$$

To prove this, assume that $y \in Y$ is arbitrary. Since $T : X \to Y$ is onto, there exists $x \in X$ such that $Tx = y$. Use now (1.11.624) in order to decompose $x = x_1 + x_2$ with $x_1 \in Z$ and $x_2 \in \ker T$. In particular, $y = Tx = T(x_1 + x_2) = Tx_1$, thus $x_1 = (T\big|_Z)^{-1}y$. In turn, this and (1.11.626) imply that $Sy = x_1$ hence, ultimately, $T(Sy) = Tx_1 = y$. Since $y \in Y$ has been arbitrarily chosen, this establishes (1.11.627), thus finishing the proof of the implication *(iii)* ⇒ *(i)*. □

1.12 A Foray into Locally Convex Topological Vector Spaces

The main goal in this section is to substantiate the claim that *local convexity* (which we have encountered sporadically in the past; cf., e.g., (1.6.349)) is the key condition that allows most of functional analysis to work smoothly. The great facilitator in this endeavor is the Hahn–Banach Extension Theorem which gives a concrete mechanism for constructing global linear functionals from partially defined ones, as long as they are dominated by a "seminorm." The local convexity is used, via the template of Minkowski functionals, to produce lots of such "seminorms," ultimately ensuring the existence of an abundance of continuous linear functionals on the topological vector space in question.

In essence, this is the route we take to establish the main result in this section, recorded in the form of Theorem 1.12.2. The availability of topological duals with good properties further opens the door for considering weak and weak-$*$ topologies, though we shall not pursue this here. We begin by making a couple of basic definitions.

We begin by formally recording the version of the Hahn-Banach Extension Theorem needed here (for a proof see, e.g., [39, Theorem 3.2, p. 57]).

Theorem 1.12.1 (Hahn-Banach Extension Theorem). *Assume X is a real vector space, and suppose $p : X \to \mathbb{R}$ is a function which is subadditive and positive homogeneous of degree one, i.e.,*

$$\begin{gathered} p(x+y) \leq p(x) + p(y) \ \textit{ as well as } \ p(\lambda x) = \lambda p(x) \\ \textit{for all } \ x, y \in X \ \textit{ and } \ \lambda \in [0, \infty). \end{gathered} \tag{1.12.628}$$

Also, let Y be a linear subspace of X, and let $f : Y \to \mathbb{R}$ be a linear map satisfying

$$f(y) \leq p(y) \ \textit{ for all } \ y \in Y. \tag{1.12.629}$$

Then there exists a linear map $\Lambda : X \to \mathbb{R}$ such that

$$\Lambda|_Y = f \ \textit{ and } \ -p(-x) \leq \Lambda(x) \leq p(x) \ \textit{ for all } \ x \in X. \tag{1.12.630}$$

Here is the separation theorem alluded to earlier.

Theorem 1.12.2. *Let (X,τ) be a Hausdorff locally convex topological vector space over $\mathbb{K}$ (where $\mathbb{K}$ is either $\mathbb{R}$ or $\mathbb{C}$). Then for every nonzero vector $x_0 \in X$ there exists a continuous linear functional $f \in X^*$ such that $f(x_0) \neq 0$. In particular, the topological dual X^* separates points of X so, in the parlance of Definition 1.6.12,*

$$\textit{every Hausdorff locally convex topological vector space is dual-rich.} \tag{1.12.631}$$

Proof. It is enough to prove the main claim in the statement when $\mathbb{K} = \mathbb{R}$. To indicate how the scenario when $\mathbb{K} = \mathbb{C}$ then follows from it, use the case of real scalars (bearing in mind that every vector space over the complex numbers is automatically a vector space over the real numbers) to construct a function $f : X \to \mathbb{R}$ which is continuous, additive, homogeneous with respect to multiplication by real numbers, and such that $f(x_0) \neq 0$. Then

$$\widetilde{f} : X \to \mathbb{C} \text{ defined by } \widetilde{f}(x) := f(x) - i f(ix) \text{ for each } x \in X \tag{1.12.632}$$

is additive, homogeneous of degree one with respect to multiplication by complex numbers, and satisfies

$$\operatorname{Re} \widetilde{f} = f. \tag{1.12.633}$$

In particular, having $f(x_0) \neq 0$ implies that $\widetilde{f}(x_0) \neq 0$. Moreover, it is visible from (1.12.632) that if $f : (X,\tau) \to \mathbb{R}$ is continuous, then so is $\widetilde{f} : (X,\tau) \to \mathbb{C}$. In view of this reduction step, there remains to treat the case when X is a real vector space, which we shall henceforth assume.

Since X is locally convex and Hausdorff, there exists $U \in \mathcal{N}(0;\tau)$ which is a convex set not containing x_0. Recall from item *(a)* in Theorem 1.6.18 that every U is absorbing. As in (1.10.564), define the Minkowski functional of U:

$$p_U(x) := \inf \{\lambda > 0 : \lambda^{-1} x \in U\}, \qquad \forall\, x \in X. \tag{1.12.634}$$

Since U is absorbing, the set in (1.12.634) over which the infimum is taken is nonempty, so $0 \leq p_U(x) < \infty$ for all $x \in X$. Thus,

$$p_U : X \to [0,\infty) \text{ is a well-defined function.} \tag{1.12.635}$$

It is immediate from its definition that (compare with (1.10.568))

$$p_U(\lambda x) = \lambda\, p_U(x) \text{ for each } x \in X \text{ and } \lambda \in [0,\infty). \tag{1.12.636}$$

We also claim that

$$x \in X \text{ and } a \in \big(p_U(x), \infty\big) \implies a^{-1} x \in U. \tag{1.12.637}$$

To justify this, suppose $x \in X$ and $a \in (0,\infty)$ are such that $p_U(x) < a$. Then (1.12.634) implies that there exists some $\lambda \in (0,a)$ such that $\lambda^{-1} x \in U$. Given that $0 \in U$ and U is convex, we may write

$$a^{-1}x = \frac{\lambda}{a} \cdot (\lambda^{-1}x) + \frac{a-\lambda}{a} \cdot 0 \in U, \tag{1.12.638}$$

from which (1.12.637) now follows.

Continuing on, reasoning as in the proof of Lemma 1.10.3, with $p := 1$ and with (1.12.637) in place of (1.10.580), we obtain

$$p_U(x+y) \leq p_U(x) + p_U(y) \quad \text{for all} \quad x, y \in X. \tag{1.12.639}$$

As is apparent from (1.12.634),

$$p_U(x) \in (0, 1] \quad \text{for each } x \in U. \tag{1.12.640}$$

Let us also observe that, since $x_0 \notin U$, from (1.12.637) (with $x := x_0$ and $a := 1$) we conclude that

$$p_U(x_0) \geq 1. \tag{1.12.641}$$

To proceed, consider the one-dimensional subspace

$$Y := \text{span}_{\mathbb{R}}\{x_0\} = \{\lambda x_0 : \lambda \in \mathbb{R}\}. \tag{1.12.642}$$

Define a linear function $f : Y \to \mathbb{R}$ by setting

$$f(\lambda x_0) := \lambda \quad \text{for each} \quad \lambda \in \mathbb{R}. \tag{1.12.643}$$

For any $\lambda \in [0, \infty)$ we have $y = \lambda x_0 \in Y$ and

$$f(\lambda x_0) = \lambda \leq \lambda p_U(x_0) = p_U(\lambda x_0), \tag{1.12.644}$$

thanks to (1.12.643), (1.12.641), and (1.12.636). If $\lambda \in (-\infty, 0)$ then

$$f(\lambda x_0) = \lambda < 0 \leq p_U(\lambda x_0), \tag{1.12.645}$$

by (1.12.635). All together, this proves that

$$f(y) \leq p_U(y) \quad \text{for each} \quad y \in Y. \tag{1.12.646}$$

By Theorem 1.12.1 then there exists a linear map $\Lambda : X \to \mathbb{R}$ such that

$$\Lambda|_Y = f \quad \text{and} \quad -p_U(-x) \leq \Lambda(x) \leq p_U(x) \quad \text{for all} \quad x \in X. \tag{1.12.647}$$

In particular, from the double inequality above and (1.12.640) we see that

$$|\Lambda(x)| \leq 1 \quad \text{for each} \quad x \in V := U \cap (-U) \in \mathcal{N}(0; \tau). \tag{1.12.648}$$

In concert with the positive homogeneity property of Λ, the above property further implies that for every $\varepsilon > 0$ we have

$$|\Lambda(x)| \leq \varepsilon \quad \text{for each} \quad x \in \varepsilon V \in \mathcal{N}(0; \tau). \tag{1.12.649}$$

Having established this, we now conclude (see item *(1)* in Definition 1.1.7) that $\Lambda : (X, \tau) \to \mathbb{R}$ is continuous. Finally, since Λ extends f, we have

$$\Lambda(x_0) = f(x_0) = 1, \tag{1.12.650}$$

by (1.12.643), so $\Lambda(x_0) \neq 0$. Thus for every nonzero $x_0 \in X$ there exists a continuous linear functional which does not vanish at x_0. The proof of Theorem 1.12.2 is therefore complete. □

One of the main interests in the separation property established in Theorem 1.12.2 is the complementation property described in the corollary below.

Corollary 1.12.3. *Let X be a Hausdorff locally convex topological vector space over $\mathbb{K}$ (where $\mathbb{K}$ is either $\mathbb{R}$ or $\mathbb{C}$). Then every finite-dimensional subspace of X is topologically complemented.*

Proof. This is a direct consequence of Theorem 1.12.2 and Exercise 9 in §1.9. □

We conclude this section by returning to the result recorded in (1.6.356), which brings into focus the issue of *normability*. In this regard, we wish to mention (cf., e.g., [39, Theorem 1.39, p. 30]) that:

a topological vector space X is normable if and only if its origin has a convex bounded neighborhood. (1.12.651)

In concert with (1.6.356) (itself a standard consequence of the classical Hahn-Banach theorem), and Exercise 9 in §1.9, this shows that:

a topological vector space X which is both locally bounded and locally convex (or, equivalently, whose origin has a convex and bounded neighborhood) is necessarily dual-rich; in particular, any finite-dimensional subspace F of X is topologically complemented in X.
(1.12.652)

Chapter 2
Riesz Theory in Topological Vector Spaces

Abstract The *Riesz Theory* of compact operators comprises the study of eigenvalues, eigensubspaces, and generalized eigensubspaces of compact operators, culminating in a "Jordan decomposition" of sorts. Although the theory is well known in the context of Banach spaces, not many people know that it goes through undiminished in general linear topological spaces. Elucidating this aspect is the main goal in the present chapter. Its material is divided into three sections.

First, §2.1 addresses elementary Riesz theory. It begins with §2.1.1, where we introduce and examine the class of compact linear operators on linear topological spaces. In §2.1.2, the focus shifts to eigenvalues, followed by §2.1.3, which provides a closer analysis of bounded operators. This sequence culminates in §2.1.4, where the preceding ideas converge in the study of compact perturbations of the identity.

Second, §2.2 offers an algebraic interlude focused on range and null-space sequences associated with linear operators. Here, topological considerations are set aside to highlight the purely algebraic aspects of the theory.

Third, in §2.3, we advance to higher-level Riesz theory. The central result demonstrates that, on linear topological spaces, compact perturbations of the identity exhibit behavior analogous to that of operators on finite-dimensional vector spaces: specifically, their range and null-space sequences eventually stabilize. After establishing this striking fact, we explore its implications for the spectral properties of compact operators. As a concluding highlight, we prove that for $0 < p < 1$, the only compact operator supported by the p-Banach space $L^p([0, 1])$ is the zero operator – a property even stronger than the space's triviality of its dual.

D. Mitrea et al., *Fredholm Theory in Topological Vector Spaces*, Springer Monographs in Mathematics, https://doi.org/10.1007/978-3-032-24738-4_2

2.1 Elementary Riesz Theory

Here we study compact operators on topological vector spaces, with special emphasis on their eigenvalues and eigensubspaces. For the reader's convenience, below we review some basic terms (compare with notions introduced earlier in Definition 1.3.16):

Terminology. The terms `operator`, `linear operator`, `linear map`, and `linear transformation`, will all be used interchangeably, and they all mean the same thing. If T is any linear transformation of a vector space X into itself we write:

1. $\ker T$ for the `null-space` (or `kernel`) of T, i.e., the set of vectors in X taken to zero by T;
2. $\operatorname{ran} T$ for $T(X)$, the `range` (or `image`) of T.
3. $\lambda \in \mathbb{C}$ is called an `eigenvalue` of the operator T if there exists some vector $x \in X \setminus \{0\}$, itself called an `eigenvector`, with the property that $Tx = \lambda x$.

Both $\ker T$ and $\operatorname{ran} T$ are linear subspaces of X. If $\operatorname{ran} T$ is finite dimensional we say T is a `finite-rank` operator. The dimension of $\operatorname{ran} T$ is called the `rank` of T. A number $\lambda \in \mathbb{C}$ is an eigenvalue of T if and only if $\lambda I - T$ is not injective, i.e., $\ker(\lambda I - T)$ is non-trivial, where I is the identity operator on X. In such a case, the dimension of $\ker(\lambda I - T)$ is referred to as the `multiplicity` of the eigenvalue λ.

An `invariant subspace` of a given linear operator T acting from a vector space X into itself is a linear subspace Y of X that is taken into (not necessarily onto) itself by said operator, i.e.,

$$T(Y) \subseteq Y. \tag{2.1.1}$$

In such a case, by the restriction of T to Y we shall mean the operator

$$T|_Y : Y \longrightarrow Y, \quad (T|_Y)x := Tx \text{ for each } x \in Y. \tag{2.1.2}$$

We denote the collection of `continuous` linear operators on a linear topological space X by

$$\mathcal{L}(X) := \{T : X \to X \,:\, T \text{ is linear and continuous}\}. \tag{2.1.3}$$

Whenever one wishes to emphasize the dependence on the linear topology τ on X, write $\mathcal{L}(X, \tau)$ in place of just $\mathcal{L}(X)$.

Under the usual operations of scalar multiplication, pointwise addition, and operator multiplication (composition) $\mathcal{L}(X)$ is a (typically non-commutative) algebra over the complex numbers. Until later on, when it's really needed, we make no effort to topologize $\mathcal{L}(X)$.

If X is a Hausdorff linear topological space and $T \in \mathcal{L}(X)$ then $\ker T$ is closed in X, but this may not be the case for $\operatorname{ran} T$ (see Exercises 2.1.6 below). Furthermore, if X is infinite dimensional, then there may be finite-rank operators that are *not* continuous on X; see item *(f)* in Theorem 2.1.3 for details.

2.1.1 Compact Operators

We begin by introducing the class of compact linear operators on a generic linear topological space.

Definition 2.1.1. *A linear operator K on a linear topological space (X,τ) is said to be* `compact` *if there is a zero-neighborhood V in X such that $K(V)\subseteq X$ is relatively compact in τ (cf. Definition 1.1.33). Denote the set of all compact operators on X by $\mathcal{K}(X)$; that is,*

$$\mathcal{K}(X) := \{K : X \to X \ :\ K \text{ is linear and compact}\}. \tag{2.1.4}$$

Also, write $\mathcal{K}(X,\tau)$ in place of just $\mathcal{K}(X)$, when one wishes to emphasize the dependence on the linear topology τ on X.

It is reassuring to note, as we do in the proposition below, that this definition of compact linear operators on linear topological spaces reduces to its more familiar version when the linear topology is induced by a quasi-norm (as in Exercise 1.4.18).

Proposition 2.1.2. *Let $(X,\|\cdot\|)$ be a quasi-normed space, and denote by $\tau_{\|\cdot\|}$ the topology induced by the quasi-norm on X (as in Exercise 1.4.18); in particular, $(X,\tau_{\|\cdot\|})$ is a Hausdorff topological vector space (cf. Theorem 1.11.1). Also, consider a linear map $K : X\to X$. Then $K\in\mathcal{K}(X,\tau_{\|\cdot\|})$ if and only if*

$$\text{for each sequence } \{x_j\}_{j\in\mathbb{N}}\subseteq X \text{ with } \|x_j\|\leq 1 \text{ for each } j\in\mathbb{N} \text{ there exists a subsequence } \{x_{j_k}\}_{k\in\mathbb{N}} \text{ with the property that } \{Kx_{j_k}\}_{k\in\mathbb{N}} \text{ converges in } (X,\tau_{\|\cdot\|}). \tag{2.1.5}$$

Proof. First assume that $K\in\mathcal{K}(X,\tau_{\|\cdot\|})$. Then, according to Definition 2.1.1, there exists some $V\in\mathcal{N}(0;\tau_{\|\cdot\|})$ such that $K(V)$ is a relatively compact set in $(X,\tau_{\|\cdot\|})$. In view of Exercise 1.11.2, there exists $r\in(0,\infty)$ such that $B_{\|\cdot\|}(0,r)\subseteq V$. Consider now a sequence $\{x_j\}_{j\in\mathbb{N}}\subseteq X$ with the property that $\|x_j\|\leq 1$ for each $j\in\mathbb{N}$, and pick some $\lambda\in(0,r)$. Then $\{\lambda x_j\}_{j\in\mathbb{N}}\subseteq B_{\|\cdot\|}(0,r)\subseteq V$, so $\{K(\lambda x_j)\}_{j\in\mathbb{N}}\subseteq K(V)$. Recall from Theorem 1.11.1 that

$$(X,\tau_{\|\cdot\|}) = (X,\tau_d) \text{ where } d \text{ is the distance defined in (1.11.597).} \tag{2.1.6}$$

Bearing this in mind and relying on Corollary 1.2.12, it is then possible to select a subsequence $\{K(\lambda x_{j_k})\}_{k\in\mathbb{N}}$ which converges in $(X,\tau_{\|\cdot\|})$. Since $K(\lambda x_{j_k}) = \lambda Kx_{j_k}$ for each $k\in\mathbb{N}$, and since multiplication by λ is a homeomorphism, we conclude that $\{Kx_{j_k}\}_{k\in\mathbb{N}}$ converges in $(X,\tau_{\|\cdot\|})$, as wanted.

In the converse direction, suppose the property formulated in (2.1.5) holds. From Theorem 1.11.1 we know that $B_{\|\cdot\|}(0,1)\in\mathcal{N}(0;\tau_{\|\cdot\|})$. Let $\{y_j\}_{j\in\mathbb{N}}$ be a sequence in $K(B_{\|\cdot\|}(0,1))$. Then for each $j\in\mathbb{N}$ there exists $x_j\in B_{\|\cdot\|}(0,1)$ such that $y_j = Kx_j$. This gives rise to a sequence $\{x_j\}_{j\in\mathbb{N}}\subseteq X$ with $\|x_j\|<1$ for each $j\in\mathbb{N}$. Thanks to our current working hypothesis, formulated in (2.1.5), there exists a subsequence $\{x_{j_k}\}_{k\in\mathbb{N}}$ with the property that $\{Kx_{j_k}\}_{k\in\mathbb{N}}$ converges in $(X,\tau_{\|\cdot\|})$. Since $Kx_{j_k} = y_{j_k}$ for each $k \in \mathbb{N}$, it follows that $\{y_{j_k}\}_{k\in\mathbb{N}}$ converges in $(X, \tau_{\|\cdot\|})$. Granted this, and once more recalling (2.1.6) (cf. also Remark 1.2.13), we may now appeal to Corollary 1.2.12 to conclude that $K(B_{\|\cdot\|}(0, 1))$ is a relatively compact subset of $(X, \tau_{\|\cdot\|})$. In view of Definition 2.1.1, this places K in $\mathcal{K}(X, \tau_{\|\cdot\|})$. □

Our next theorem collects a number of key properties, basically amounting to a basic toolbox for handling jobs involving compact operators on topological vector spaces.

Theorem 2.1.3. *Suppose (X, τ) is a linear topological space. Then the following properties hold:*

(a) Every compact operator on X is continuous; in particular $\mathcal{K}(X) \subseteq \mathcal{L}(X)$. If, in addition, X is Hausdorff, then $\mathcal{K}(X)$ becomes a linear subspace of $\mathcal{L}(X)$.
(b) $\mathcal{K}(X)$ is a right ideal in $\mathcal{L}(X)$; i.e.,

$$K \circ L \in \mathcal{K}(X) \ \textit{for all} \ K \in \mathcal{K}(X) \ \textit{and} \ L \in \mathcal{L}(X). \tag{2.1.7}$$

If, in addition, X is assumed to be Hausdorff, $\mathcal{K}(X)$ becomes a two-sided ideal in $\mathcal{L}(X)$; that is,

$$J \circ K \circ L \in \mathcal{K}(X) \ \textit{for all} \ K \in \mathcal{K}(X) \ \textit{and} \ J, L \in \mathcal{L}(X). \tag{2.1.8}$$

(c) If X is Hausdorff, then every continuous finite-rank linear operator on X is compact.
(d) The restriction of a compact linear operator on X to a closed invariant subspace (cf. (2.1.1)-(2.1.2)) is again a compact operator.
(e) There exists no compact operator on an infinite-dimensional Hausdorff linear topological space with the property that it is an open mapping (in particular, a linear homeomorphism).
(f) Every compact projection on a Hausdorff linear topological space has finite rank. However, in any infinite-dimensional locally bounded topological vector space there exist finite rank projections which are not continuous (hence also not compact).
(g) Every compact operator on X maps bounded sets into relatively compact sets.

Proof. As regards the first claim in part *(a)*, consider a compact operator K on X. With the goal of showing that K is continuous, let W be a zero-neighborhood in X. Since the operator K is compact, there is a zero-neighborhood V in X such that $K(V)$ is relatively compact in X. In particular, $K(V)$ is bounded, by Exercise 1.6.41. As such, it is absorbed by W, hence there exists $\varepsilon > 0$ with $K(\varepsilon V) = \varepsilon \cdot K(V) \subseteq W$. Since εV is itself a zero-neighborhood in X, we conclude from this and (1.1.4) that K is continuous at zero. By linearity, K is therefore continuous on X.

For the second claim in part *(a)*, assume X is also Hausdorff. Since multiplication by non-zero scalars is a homeomorphism on X, it is clear that $\mathcal{K}(X)$ is stable under multiplication by scalars. As such, it remains to show that $\mathcal{K}(X)$ is also stable under addition. To this end, assume $K_1, K_2 \in \mathcal{K}(X)$. Then one can find $V_j \in \mathcal{N}(0; \tau)$ such that $K_j(V_j)$ is relatively compact in X, for $j = 1, 2$. Define $V := V_1 \cap V_2 \in \mathcal{N}(0; \tau)$, and note that

$$(K_1 + K_2)V \subseteq K_1(V) + K_2(V) \subseteq K_1(V_1) + K_2(V_2) \subseteq \overline{K_1(V_1)} + \overline{K_2(V_2)}. \tag{2.1.9}$$

Since $\overline{K_j(V_j)}$ is compact in X for $j = 1, 2$, Exercise 1.6.9 applies and gives that the last set in (2.1.9) is compact. Given that we are currently assuming X to be Hausdorff, it follows from (2.1.9) and Proposition 1.1.34 that $(K_1 + K_2)V$ is relatively compact in X. Thus, $K_1 + K_2 \in \mathcal{K}(X)$, as wanted.

As regards the first claim in *(b)*, assume $K \in \mathcal{K}(X)$ and $L \in \mathcal{L}(X)$. Then there exists some $V \in \mathcal{N}(0; \tau)$ such that $K(V)$ is relatively compact in (X, τ). Also, since $L : (X, \tau) \to (X, \tau)$ is a continuous function with $L0 = 0$, from (1.1.4) we see that there exists some $W \in \mathcal{N}(0; \tau)$ with the property that $L(W) \subseteq V$. Consequently, $(K \circ L)(W) = K(L(W)) \subseteq K(V)$, making $(K \circ L)(W)$ relatively compact in (X, τ), by the first part in Proposition 1.1.35. Since $K \circ L$ is linear, Definition 2.1.1 labels it compact, hence $K \circ L \in \mathcal{K}(X)$.

Concerning the second claim in *(b)*, make the additional assumption that (X, τ) is Hausdorff, and pick $K \in \mathcal{K}(X)$ and $J \in \mathcal{L}(X)$. In light of what we have proved already, (2.1.8) follows as soon as we show that $J \circ K \in \mathcal{K}(X)$. To this end, first recall from Definition 2.1.1 that there exists $V \in \mathcal{N}(0; \tau)$ such that $K(V)$ is relatively compact in (X, τ). Since $J : (X, \tau) \to (X, \tau)$ is a continuous function, Exercise 1.1.36 permits us to conclude that $(J \circ K)(V) = J(K(V))$ is relatively compact in (X, τ). Given that $J \circ K$ is also linear, it follows from Definition 2.1.1 that $J \circ K$ belongs to $\mathcal{K}(X)$, as promised.

Consider now the claim made in part *(c)*. Once again make the additional assumption that (X, τ) is Hausdorff, and suppose $P \in \mathcal{L}(X)$ has finite-rank. In particular, $Y := P(X)$ is a finite-dimensional linear subspace of X. From Proposition 1.6.8 we see that $(Y, \tau|_Y)$ becomes a Hausdorff topological vector space itself. Keeping this in mind, Theorem 1.8.31 ensures that there exists a linear homeomorphism $\Phi : Y \to \mathbb{C}^n$, where $n := \dim Y$. Denote by $B(0, 1)$ the standard open unit ball in $\mathbb{C}^n$, centered at the origin, and set $W := \Phi^{-1}(B(0, 1))$. Since $B(0, 1)$ is a relatively compact neighborhood of the origin in $\mathbb{C}^n$ and Φ^{-1} is a linear homeomorphism, it follows that W is a relatively compact zero-neighborhood in $(Y, \tau|_Y)$. From item *(b)* in Theorem 1.8.1 we know that Y is a closed subspace of (X, τ), so we may invoke the last part in Proposition 1.1.35 to conclude that W is relatively compact in (X, τ). Bring in the linear operator

$$\begin{aligned} &R : (X, \tau) \longrightarrow \left(Y, \tau\big|_Y\right) \text{ defined by} \\ &Rx := Px \text{ for each vector } x \in X. \end{aligned} \tag{2.1.10}$$

In view of the continuity of P from (X, τ) to itself, Exercise 1.1.11 tells us that R is also continuous in the above context. Together with (1.1.4), this guarantees the existence of a zero-neighborhood V in (X, τ) such that $P(V) = R(V) \subseteq W$. Hence, $P(V)$ is relatively compact in (X, τ), thanks to the first part in Proposition 1.1.35 and the fact that W is relatively compact in (X, τ). This argument shows that P

is a compact operator on X, as desired. To deal with part *(d)*, assume $K \in \mathcal{K}(X)$ and Y is a closed linear subspace of X which is invariant under K. First, $(Y, \tau|_Y)$ is itself a topological vector space, thanks to Proposition 1.6.8. We know that there exists a zero-neighborhood V in X such that $K(V)$ is relatively compact in X. Then $W := V \cap Y$ is a zero-neighborhood in $(Y, \tau|_Y)$, by Exercise 1.1.2. If $K|_Y$ denotes the restriction of K to Y (cf. (2.1.1)-(2.1.2)), we have $(K|_Y)W = K(V \cap Y)$, a subset of $K(V)$. Then the first part in Proposition 1.1.35 tells us that $(K|_Y)W$ is a relatively compact set in (X, τ). Since this set is included in Y, itself a closed subset of (X, τ), the second part in Proposition 1.1.35 then implies that $(K|_Y)W$ is relatively compact in $(Y, \tau|_Y)$. Hence, $K|_Y \in \mathcal{K}(Y, \tau|_Y)$, as wanted.

Moving on, we shall prove *(e)* reasoning by contradiction. To set the stage, assume (X, τ) is an infinite-dimensional Hausdorff linear topological space for which there exists an open mapping $K \in \mathcal{K}(X)$. Then there exists some $V \in \mathcal{N}(0; \tau)$ with the property that $K(V)$ is relatively compact in (X, τ). Since K an open mapping, $K(V)$ is also a zero-neighborhood in X. Thus, X possesses a relatively compact zero-neighborhood, which renders X a locally compact Hausdorff linear topological space. According to Theorem 1.8.40, we necessarily have $\dim X < +\infty$, contradicting the present working assumption.

As far as the first claim in item *(f)* is concerned, assume (X, τ) is a Hausdorff linear topological space, and suppose $P \in \mathcal{K}(X)$ satisfies $P^2 = P$. The goal is to show that P has finite rank. Given that P is linear, it follows that $Y := P(X)$ is a linear subspace of X. Moreover, since P is continuous and X is Hausdorff, item *(c)* in Theorem 1.8.27 ensures that Y is closed in (X, τ). Since P is compact, there is a zero-neighborhood V in X such that $P(V)$ is relatively compact in (X, τ). The second part in Proposition 1.1.35 then guarantees that $P(V)$ is relatively compact in $(Y, \tau|_Y)$. In view of the fact that $V \cap Y \subseteq P(V)$ and that $V \cap Y$ is a zero-neighborhood in $(Y, \tau|_Y)$, we conclude that the bigger set $P(V)$ is itself a zero-neighborhood in $(Y, \tau|_Y)$. As this was shown earlier to be relatively compact in $(Y, \tau|_Y)$, it follows that $(Y, \tau|_Y)$ is a locally compact Hausdorff linear topological space. Having established this, Theorem 1.8.40 then shows that $\dim Y < +\infty$, ultimately proving that P has finite rank.

To treat the second claim in item *(f)*, suppose (X, τ) is an infinite-dimensional locally bounded topological vector space. From Exercise 1 in §1.9 we know that there exists a one-dimensional linear subspace Y of X that is not topologically complemented. Nonetheless, Theorem 1.3.29 guarantees that Y has an algebraic complement in X. That is, there exists a linear subspace Z of X that $Y \boxplus Z = X$. Define $P : X \to X$ by setting

$$\begin{array}{l} Px := y \text{ for each } x \in X, \text{ where } y \text{ is the unique vector} \\ \text{in } Y \text{ for which there exists } z \in Z \text{ such that } x = y + z. \end{array} \tag{2.1.11}$$

Then P is linear, $P^2 = P$, and $P(X) = Y$, hence P is a projection onto Y. If P were continuous, then Proposition 1.8.22 would imply that $X = Y \oplus Z$, which would make Z a topological complement of Y. This conflicts with the fact that Y does not possess any topological complements. The contradiction we just arrived at shows that P is not continuous, so we have constructed a finite rank projection which is not continuous on X.

Finally, to justify the claim in item *(g)* pick some $K \in \mathcal{K}(X)$ and assume B is a bounded set (X, τ). The fact that K is compact ensures the existence of some $V \in \mathcal{N}(0; \tau)$ such that $K(V)$ is relatively compact in (X, τ). Given that V absorbs B, there exists some $\varepsilon > 0$ with the property that $\varepsilon B \subseteq V$. This inclusion further implies $\varepsilon K(B) = K(\varepsilon B) \subseteq K(V)$ so, by the first part in Proposition 1.1.35, the set $\varepsilon K(B)$ is relatively compact in (X, τ). Since multiplication by non-zero scalars is a homeomorphism on X, we ultimately conclude that $K(B)$ is relatively compact in (X, τ), as wanted. □

The next proposition further helps put matters into perspective.

Proposition 2.1.4. *Let X be a Hausdorff linear topological space. Then each of the following five statements implies the other four:*

(i) X is finite dimensional.
(ii) X is locally compact.
(iii) There exists $K \in \mathcal{K}(X)$ which is open.
(iv) The identity map on X is a compact operator.
(v) $\mathcal{K}(X) = \mathcal{L}(X)$.

Proof. The equivalence *(i)*⇔*(ii)* comes from item *(c)* in Theorem 1.8.1. Collectively, items *(a)* and *(c)* in Theorem 2.1.3 show that *(i)*⇒*(v)*. Also, trivially, *(v)*⇒*(iv)* and *(iv)*⇒*(iii)*. Finally, the fact that *(iii)*⇒*(i)* is implied by *(e)* in Theorem 2.1.3. □

Proposition 2.1.5. *Let $(X, \|\cdot\|)$ be a p-Banach space for some $0 < p \leq 1$, and let $\{K_n\}_{n\in\mathbb{N}} \subseteq \mathcal{K}(X)$ be a sequence with the property that $\|K_n - K\| \to 0$ as $n \to \infty$ (cf. Definition 1.4.26), for some $K \in \mathcal{L}(X)$. Then K belongs to $\mathcal{K}(X)$.*

Proof. Set $B(x, r) := \{y \in X : \|y - x\| < r\}$ for each $x \in X$ and $r > 0$. Also, abbreviate $B := B(0, 1)$. Fix $\varepsilon > 0$ and choose $n \in \mathbb{N}$ such that

$$\|K - K_n\| < \frac{\varepsilon}{2}. \tag{2.1.12}$$

Because K_n is compact, the set $K_n(B)$ is relatively compact. Hence there exist $y_1, \dots, y_M \in X$ such that

$$K_n(B) \subseteq \bigcup_{i=1}^{M} B\Big(y_i, \frac{\varepsilon}{2}\Big). \tag{2.1.13}$$

We claim that the same centers $y_1, \dots, y_M$ provide an ε-cover of $K(B)$. Indeed, for any $x \in B$, pick $i \in \{1, \dots, M\}$ with $\|K_n x - y_i\| < \varepsilon/2$, which is possible by (2.1.13). Then, by (1.4.249) and (2.1.12),

$$\|Kx - y_i\| \leq \|Kx - K_n x\| + \|K_n x - y_i\| < \frac{\varepsilon}{2} + \frac{\varepsilon}{2} = \varepsilon. \tag{2.1.14}$$

Consequently,

$$K(B) \subseteq \bigcup_{i=1}^{M} B(y_i, \varepsilon). \tag{2.1.15}$$

Since $\varepsilon > 0$ was arbitrary, it follows that $K(B)$ is totally bounded with respect to the distance (1.4.212) (cf. Definition 1.2.14). Keeping in mind that when equipped with this distance X is complete, Theorem 1.2.15 then allows us to conclude that $K(B)$ is relatively compact in X. Therefore, K is compact. □

Proposition 2.1.6 (Diagonal Operators). *Fix a sequence $(\lambda_n)_{n\in\mathbb{N}}$ of complex numbers, and let T be the associated* `diagonal operator`, *defined on complex sequences as follows*

$$Tx := (\lambda_n x_n)_{n\in\mathbb{N}} \text{ for each complex sequence } x = (x_n)_{n\in\mathbb{N}}. \tag{2.1.16}$$

Show that:

(a) $Tx \in \ell^2$ for each $x \in \ell^2$ if and only if $(\lambda_n)_{n\in\mathbb{N}}$ is a bounded sequence.
(b) If $(\lambda_n)_{n\in\mathbb{N}}$ is a bounded sequence of complex numbers, then T is continuous on ℓ^2 and $\|T\| = \sup\{|\lambda_n| : n \in \mathbb{N}\}$.
(c) T is compact on $\ell^2 \Leftrightarrow \lambda_n \to 0$, as $n \to \infty$.
(d) If $(\lambda_n)_{n\in\mathbb{N}}$ is a bounded sequence, then ran T *is closed in $\ell^2 \Leftrightarrow$ zero is not a limit point (cf. Definition 1.1.37) of the set $\{\lambda_n : n \in \mathbb{N}\}$.*

A comment pertaining to terminology is in order. If we consider the infinite diagonal matrix with entries $\lambda_1, \lambda_2, \lambda_3, \ldots$ along the diagonal and zeros elsewhere, i.e.,

$$D = \begin{pmatrix} \lambda_1 & 0 & 0 & 0 & \cdots \\ 0 & \lambda_2 & 0 & 0 & \cdots \\ 0 & 0 & \lambda_3 & 0 & \cdots \\ 0 & 0 & 0 & \lambda_4 & \cdots \\ \vdots & \vdots & \vdots & \vdots & \ddots \end{pmatrix} \tag{2.1.17}$$

then the action of the operator T in (2.1.16) on any complex sequence $x = (x_n)_{n\in\mathbb{N}}$ may be regarded as the multiplication (from the left) of x regarded as an infinite vertical column with the diagonal matrix D. Hence the name "diagonal operator."

We now turn to the task of proving Proposition 2.1.6.

Proof of Proposition 2.1.6. Let us deal with the right-pointing implication in *(a)*. Seeking a contradiction, assume $Tx \in \ell^2$ for each $x \in \ell^2$ and yet the sequence $(\lambda_n)_{n\in\mathbb{N}}$ is unbounded. Then for each $j \in \mathbb{N}$ there exists $n_j \in \mathbb{N}$ such that $|\lambda_{n_j}| > j$. Arrange matters so that $n_{j+1} > n_j$ for each $j \in \mathbb{N}$ and, for each $n \in \mathbb{N}$, define

$$x_n := \begin{cases} 1/j & \text{if } n = n_j \text{ for some } j \in \mathbb{N}, \\ 0 & \text{if } n \notin \{n_j : j \in \mathbb{N}\}. \end{cases} \tag{2.1.18}$$

Since $x := (x_n)_{n\in\mathbb{N}} \in \ell^2$, the assumption on T implies that $(\lambda_n x_n)_{n\in\mathbb{N}} \in \ell^2$, hence

$$\infty > \sum_{n\in\mathbb{N}} |\lambda_n x_n|^2 = \sum_{j\in\mathbb{N}} |\lambda_{n_j} x_{n_j}|^2 \geq \sum_{j\in\mathbb{N}} |j/j|^2 = \infty, \tag{2.1.19}$$

and this contradiction proves the right-pointing implication in *(a)*. For the left-pointing implication in *(a)*, suppose $(\lambda_n)_{n\in\mathbb{N}}$ is a bounded sequence, say

$$M := \sup\{|\lambda_n| : n \in \mathbb{N}\} < \infty. \tag{2.1.20}$$

Then for each $x = (x_n)_{n\in\mathbb{N}} \in \ell^2$ it is clear from definitions that

$$\|Tx\|_{\ell^2} \leq M\|x\|_{\ell^2}. \tag{2.1.21}$$

Since the right-hand side above is finite, we conclude that $Tx \in \ell^2$, which takes care of *(a)*.

To treat *(b)*, define the following unit vectors in ℓ^2

$$e_n := (\delta_{nj})_{j\in\mathbb{N}} \ \text{ for each } \ n \in \mathbb{N}, \tag{2.1.22}$$

where δ_{nj} is the usual Kronecker symbol, and consider their individual linear spans

$$\langle e_n \rangle := \{\alpha e_n : \alpha \in \mathbb{C}\} \ \text{ for each } \ n \in \mathbb{N}. \tag{2.1.23}$$

To proceed, we distinguish two cases. First, if the supremum in (2.1.20) is attained, say $M = \lambda_{n_*}$ for some $n_* \in \mathbb{N}$, then $Te_{n_*} = \lambda_{n_*} e_{n_*}$ from which we deduce that $\|Te_{n_*}\|_{\ell^2} = |\lambda_{n_*}| = M$. In concert with (2.1.21) and Proposition 1.4.28, all conclusions in *(b)* follow. Second, if the supremum in (2.1.20) is not attained, then there exists an increasing sequence $\{n_j\}_{j\in\mathbb{N}}$ of positive integers with the property that $|\lambda_{n_j}| \to M$ as $j \to \infty$. Note that $Te_{n_j} = \lambda_{n_j} e_{n_j}$ for every $j \in \mathbb{N}$, so we have $\|Te_{n_j}\|_{\ell^2} = |\lambda_{n_j}| \to M$ as $j \to \infty$. Together with (2.1.21) and Proposition 1.4.28, all conclusions in *(b)* once again follow.

Turning attention to *(c)*, let $\lambda = (\lambda_n)_{n\in\mathbb{N}}$ be an arbitrary sequence. First, under the assumption that λ does not converge to zero, we shall prove that T cannot be a compact operator on ℓ^2. The current working assumption implies that there exists $\varepsilon > 0$ and a subsequence $\{\lambda_{n_j}\}_{j\in\mathbb{N}}$ of λ with the property that $|\lambda_{n_j}| \geq \varepsilon$ for each $j \in \mathbb{N}$. Since for any two distinct integers $k, m \in \mathbb{N}$ we have

$$\|Te_{n_k} - Te_{n_m}\|_{\ell^2} = \|\lambda_{n_k} e_{n_k} - \lambda_{n_m} e_{n_m}\|_{\ell^2} = (|\lambda_{n_k}|^2 + |\lambda_{n_m}|^2)^{1/2} \geq \sqrt{2}\varepsilon, \tag{2.1.24}$$

it follows that $\{Te_{n_j}\}_{j\in\mathbb{N}}$ does not contain any subsequence which is Cauchy in ℓ^2, hence has no subsequence convergent in ℓ^2. Since all e_n's are unit vectors in ℓ^2, this proves that T cannot be a compact operator on ℓ^2.

Second, assume now the sequence $\lambda = (\lambda_n)_{n\in\mathbb{N}}$ converges to zero; in particular, λ is bounded. For each $k \in \mathbb{N}$, define $\lambda^{(k)}$ to be the sequence which has the same first k terms as λ and all subsequent terms are zero. Associate with the sequence $\lambda^{(k)}$ the diagonal operator T_k in the same manner T has been associated with λ in (2.1.16). This and item *(b)* then imply that T_k is a continuous finite rank operator on ℓ^2. In particular, T_k is compact on ℓ^2 by item *(c)* in Theorem 2.1.3. Another appeal to *(b)* shows that

$$\|T - T_k\| = \sup_{n>k} |\lambda_n| \to 0 \quad \text{as} \quad k \to \infty, \tag{2.1.25}$$

with the convergence implied by the current working assumption. Given that $\mathcal{K}(\ell^2)$ is closed under the operator norm convergence (cf. Proposition 2.1.5), this ultimately places T in $\mathcal{K}(\ell^2)$, finishing the proof of *(c)*.

Pressing on, as a preamble to *(d)*, fix a sequence $\lambda = (\lambda_n)_{n\in\mathbb{N}}$ of complex numbers and define its support as

$$\operatorname{supp} \lambda := \{n \in \mathbb{N} : \lambda_n \neq 0\}. \tag{2.1.26}$$

To proceed, make the assumption that λ is bounded and associate with it the linear operator T as in (2.1.16). Then item *(b)* guarantees that $T : \ell^2 \to \ell^2$ is a bounded operator. In relation to this, observe that

$$\operatorname{ran} T \subseteq Y := \{y \in \ell^2 : \operatorname{supp} y \subseteq \operatorname{supp} \lambda\} = \bigcap_{n\in\mathbb{N}\setminus\operatorname{supp}\lambda} \langle e_n \rangle^{\perp}. \tag{2.1.27}$$

Indeed, the first inclusion above is clear from definitions, and so is the last equality. In particular, this last equality in (2.1.27) readily implies that Y is a closed subspace of ℓ^2. Next, with ℓ^2_c denoting the subspace of ℓ^2 consisting of sequences with finite support,

$$\ell^2_c := \big\{y = (y_1, y_2, \dots) \in \ell^2 : \text{only finitely many terms are nonzero}\big\}, \tag{2.1.28}$$

define

$$Y_c := \{y \in \ell^2_c : \operatorname{supp} y \subseteq \operatorname{supp} \lambda\} \tag{2.1.29}$$

which, by design, is a linear subspace of Y. Naturally truncating sequences (by making all terms past a certain rank be zero) shows that, in fact, Y_c is dense in Y (in the ℓ^2 norm). Moreover, it is apparent from the first inclusion in (2.1.27) and definitions that

$$Y_c \subseteq \operatorname{ran} T \subseteq Y. \tag{2.1.30}$$

We are now ready to begin in earnest the proof of *(d)*. Throughout, make the assumption that $\lambda = (\lambda_n)_{n\in\mathbb{N}}$ is a bounded sequence of complex numbers. In one direction, if $\operatorname{ran} T$ is closed, from (2.1.30), and the fact that Y_c is dense in the closed space Y, we deduce that

$$\operatorname{ran} T = Y. \tag{2.1.31}$$

As such, for each $y = (y_n)_{n\in\mathbb{N}} \in Y$ arbitrary, there exists $x = (x_n)_{n\in\mathbb{N}} \in \ell^2$ with the property that $Tx = y$. Since, in light of (2.1.16), this forces $x_n = y_n/\lambda_n$ for each $n \in \operatorname{supp} \lambda$, we conclude from the membership of x in ℓ^2 that

$$\sum_{n\in\operatorname{supp}\lambda} \left|\frac{y_n}{\lambda_n}\right|^2 < +\infty. \tag{2.1.32}$$

In view of the arbitrariness of the sequence y in Y, we conclude that (2.1.32) happens whenever $(y_n)_{n\in\text{supp}\,\lambda}$ is a complex sequence satisfying $\sum_{n\in\text{supp}\,\lambda}|y_n|^2 < +\infty$. (Indeed, any such sequence indexed by supp λ may be extended to a sequence indexed by $\mathbb{N}$, simply including additional zero terms, and this new sequence now lies in Y, thanks to its definition in (2.1.27).) Invoking what we have proved in item *(a)*, therefore permits us to conclude that $\{\frac{1}{\lambda_n} : n \in \text{supp}\,\lambda\}$ is bounded. In turn, the latter property prevents the number zero from being a limit point for $\{\lambda_n : n \in \mathbb{N}\}$. This establishes the right-pointing implication in *(d)*.

For the left-pointing implication in *(d)*, continue to assume that $\lambda = (\lambda_n)_{n\in\mathbb{N}}$ is a bounded complex sequence, and make the assumption that zero is not a limit point for $\{\lambda_n : n \in \mathbb{N}\}$. The latter ensures the existence of some $\varepsilon > 0$ with the property that

$$|\lambda_n| \geq \varepsilon \quad \text{for each} \quad n \in \text{supp}\,\lambda. \tag{2.1.33}$$

Upon recalling that the space Y introduced in (2.1.27) is closed in ℓ^2, the conclusion we presently seek follows as soon as we show that ran $T = Y$. Recall from (2.1.30) that the left-to-right inclusion always holds. To prove the right-to-left inclusion, pick $y = (y_n)_{n\in\mathbb{N}} \in Y$ and define

$$x_n := \begin{cases} y_n/\lambda_n & \text{if } n \in \text{supp}\,\lambda, \\ 0 & \text{if } n \notin \text{supp}\,\lambda. \end{cases} \tag{2.1.34}$$

Thanks to (2.1.33), it follows that $x = (x_n)_{n\in\mathbb{N}} \in \ell^2$, while from (2.1.34) and (2.1.16) we see that $Tx = y$. This finishes the proof of ran $T = Y$ which, in turn, gives that ran T is closed in ℓ^2. The proof of Proposition 2.1.6 is therefore complete. □

Exercise 2.1.7. Take X to be the linear subspace of ℓ^2 consisting of sequences with finite support (i.e., $X := \ell^2_c$ used in the proof of Proposition 2.1.6), still topologized by the ℓ^2-norm. Show that it is not possible for a compact diagonal operator to map X onto itself. Can a surjective compact diagonal operator exist on ℓ^2?

Hint: For the first claim, if $T : \ell^2_c \to \ell^2_c$ is a surjective compact diagonal operator, associated with a complex sequence $\lambda = (\lambda_n)_{n\in\mathbb{N}}$ as in (2.1.16), use the fact that $Te_n = \lambda_n$ for each n to conclude, much as in the proof of item *(c)* in Proposition 2.1.6, that $\lambda_n \to 0$ as $n \to \infty$. Bring in the natural extension of T to the entire ℓ^2, by setting $\widetilde{T}x := (\lambda_n x_n)_{n\in\mathbb{N}}$ for each $x = (x_n)_{n\in\mathbb{N}} \in \ell^2$. Use item *(b)* in Proposition 2.1.6 to conclude that $\widetilde{T} : \ell^2 \to \ell^2$ is a continuous linear operator. Given an arbitrary sequence $x = (x_n)_{n\in\mathbb{N}} \in \ell^2$, for each $m, n \in \mathbb{N}$ set $x_n^{(m)} := x_n$ if $n \leq m$ and $x_n^{(m)} := 0$ if $n > m$. For each $m \in \mathbb{N}$ this gives rise to a sequence $x^{(m)} = (x_n^{(m)})_{n\in\mathbb{N}} \in \ell^2_c$ and $x^{(m)} \to x$ in ℓ^2 as $m \to \infty$. Conclude that $\widetilde{T}x^{(m)} \to \widetilde{T}x$ in ℓ^2 as $m \to \infty$ and, further, that $Tx^{(m)} \to \widetilde{T}x$ in ℓ^2 as $m \to \infty$. Use the compactness of $T : \ell^2_c \to \ell^2_c$ to obtain that $\{Tx^{(m)}\}_{m\in\mathbb{N}}$ has a convergent subsequence to an element in ℓ^2_c, and ultimately show that $\widetilde{T}x$ is the limit of this subsequence. Conclude that $\widetilde{T}x \in \ell^2_c$ for each $x \in \ell^2$, and use this to prove that sequence λ has finite support. Finally, show that this prevents $T : \ell^2_c \to \ell^2_c$ from being surjective.

For the last question, use items *(c)* and *(d)* in Proposition 2.1.6.

2.1.2 Eigenvalues

For the diagonal operators of Proposition 2.1.6 it is easy to check that the eigenvalues are precisely the elements of the diagonal sequence $\{\lambda_n\}_{n\in\mathbb{N}}$, with a particular element having multiplicity equal to the number of times it appears in the sequence. Thus for *compact* diagonal operators on ℓ^2, since the diagonal sequence converges to zero, the eigenvalues all have finite multiplicity, and they can only cluster at zero. The next couple of results show that this phenomenon is a general feature of compact operators.

Theorem 2.1.8 (Finite-Multiplicity). *If X is a Hausdorff linear topological space and if K is a compact operator on X, then every non-zero eigenvalue of K has finite multiplicity.*

Proof. Suppose λ is a non-zero eigenvalue of K. We want to show that $\ker(\lambda I - K)$ is finite dimensional. Since

$$\ker(\lambda I - K) = \ker(\lambda(I - \lambda^{-1}K)) = \ker(I - \lambda^{-1}K), \tag{2.1.35}$$

and since scalar multiples of compact operators are again compact, we may, without loss of generality, suppose that 1 is the eigenvalue under consideration. Our goal is to show that the eigensubspace $N := \ker(I - K)$ has finite dimension. Note that is closed in X since K is continuous, and

$$N = \{x \in X : Kx = x\} \tag{2.1.36}$$

from which we see that N is K-invariant. Item *(d)* in Theorem 2.1.3 therefore implies that K is a compact operator on N (which is a Hausdorff linear topological space when equipped with the relative topology from X). But K is also the identity on N, so Proposition 2.1.4 guarantees that $\dim N < +\infty$, which is the claim we seek. □

Theorem 2.1.9 (Global Eigenvalue Theorem). *Suppose K is a compact operator on a Hausdorff linear topological space X. Then the non-zero eigenvalues of K form either a finite (possibly empty) set, or a sequence converging to zero.*

Proof. We prove the contrapositive form of the theorem: Suppose $T \in \mathcal{L}(X)$ has a countable sequence of distinct non-zero eigenvalues $\{\lambda_n\}_{n\in\mathbb{N}}$ that is bounded away from zero:

$$\delta := \inf_{n\in\mathbb{N}} |\lambda_n| > 0. \tag{2.1.37}$$

We will show that T is not compact, i.e., for every zero-neighborhood V, the image $T(V)$ is not relatively compact in X. With this in mind, fix a zero-neighborhood V, which in light of the conclusion we seek may, without loss of generality, be assumed to be closed and circled. Indeed, this may be arranged based on Corollary 1.6.34 and the first part in Proposition 1.1.35.

For each positive integer n there is an eigenvector e_n for λ_n:

$$e_n \in X \setminus \{0\} \text{ and } Te_n = \lambda_n e_n. \tag{2.1.38}$$

We claim that $\{e_n\}_{n\in\mathbb{N}}$ is a linearly independent set. It suffices to show that for each $N \in \mathbb{N}$ the set $E_N := \{e_n\}_{1\leq n\leq N}$ is linearly independent, which can be done by induction on N. This is obviously true for $N = 1$. Assuming E_N is a linearly independent set for some $N \in \mathbb{N}$, consider the situation when $\sum_{j=1}^{N+1} c_j e_j = 0$ for a family of complex numbers $c_1, \dots, c_{N+1}$. After applying T to this equality and using (2.1.38) we arrive at $\sum_{j=1}^{N+1} \lambda_j c_j e_j = 0$. Subtracting from the previous equality $\lambda_{N+1} \sum_{j=1}^{N+1} c_j e_j = 0$ yields

$$\sum_{j=1}^{N} c_j(\lambda_j - \lambda_{N+1})e_j = 0. \tag{2.1.39}$$

In view of the fact that E_N is linearly independent and the eigenvalues are distinct, we obtain that $c_j = 0$ for $1 \leq j \leq N$. But $\sum_{j=1}^{N+1} c_j e_j = 0$, thus $c_{N+1} = 0$ as well, and the desired conclusion follows.

In view of the linear independence of $\{e_n\}_{n\in\mathbb{N}}$, the subspace

$$F_n := \operatorname{span}\{e_1, e_2, \dots, e_n\} \tag{2.1.40}$$

has dimension n, for each $n \in \mathbb{N}$. Thus $\{F_n\}_{n\in\mathbb{N}}$ is a strictly increasing sequence of finite-dimensional, T-invariant subspaces.

Recall that by the Finite-Dimension Theorem (stated as Theorem 1.8.31), each F_n is linearly homeomorphic to the Euclidean space $\mathbb{C}^n$. Note further that because F_n has a basis of eigenvectors of T that correspond to non-zero eigenvalues, T is one-to-one on F_n.

There are two possibilities, which we study next, starting with:

`Case I`: *V contains a nontrivial subspace of some F_n.*

Then because T is one-to-one on F_n, the image $T(V)$ contains a nontrivial subspace, and is therefore unbounded (see Proposition 1.6.39 and the right-pointing implication in (1.6.485)), so not relatively compact (according to item *(c)* in Theorem 1.6.40).

`Case II`: *V contains no nontrivial subspace of any F_n.*

In this case, fix $n \in \mathbb{N}$. Since $V \cap F_n$ is closed and circled (recall that both V and F_n are so), Lemma 1.6.45 insures that $V \cap F_n$ is bounded. Moreover, as $V \cap F_n$ is a zero-neighborhood in F_n, Lemma 1.6.50 along with Remark 1.6.51 may be invoked (with $X := F_n$, $Y := F_{n-1}$, $U := V \cap F_n$) and guarantees the existence of a vector $x_n \in V \cap F_n$ such that

$$x_n + \frac{1}{2}(V \cap F_n) \text{ is disjoint from } F_{n-1}. \tag{2.1.41}$$

We claim, that in fact,

$$x_n + \frac{1}{2}V \text{ is disjoint from } F_{n-1}. \tag{2.1.42}$$

Indeed, if there exist $v \in V$ and $f_{n-1} \in F_{n-1}$ such that $x_n + \frac{1}{2}v = f_{n-1}$, then

$$v = 2f_{n-1} - 2x_n \in F_{n-1} + F_n \subseteq F_n. \tag{2.1.43}$$

Thus, $v \in V \cap F_n$ hence $f_{n-1} \in F_{n-1} \cap (x_n + \frac{1}{2}(V \cap F_n))$, contradicting (2.1.41). This establishes (2.1.42).

Going further, upon representing x_n as a linear combination of the eigenvectors $\{e_1, e_2, \dots, e_n\}$ we see that $Tx_n - \lambda_n x_n \in F_{n-1}$. Now fix $n > m$ and observe that $Tx_m \in F_m \subseteq F_{n-1}$, so

$$\begin{aligned} Tx_n - Tx_m &= (Tx_n - \lambda_n x_n) - Tx_m + \lambda_n x_n \\ &\in F_{n-1} - F_{n-1} + \lambda_n x_n \\ &= \lambda_n (x_n + F_{n-1}). \end{aligned} \tag{2.1.44}$$

Since (2.1.42) implies that $x_n + F_{n-1}$ is disjoint from $\frac{1}{2}V$, we see that

$$Tx_n - Tx_m \notin \frac{1}{2}\lambda_n V \supseteq \frac{\delta}{2}V \tag{2.1.45}$$

where the last containment follows from the circledness of V and (2.1.37). This circledness, together with (2.1.45), also implies that

$$Tx_n - Tx_m \notin \frac{\delta}{2}V \quad \text{for all } n, m \in \mathbb{N} \text{ with } n \neq m. \tag{2.1.46}$$

Thus, we have shown that for every closed and circled zero-neighborhood V and every distinct pair m, n of positive integers,

$$Tx_n \notin Tx_m + \frac{\delta}{2}V. \tag{2.1.47}$$

From this it follows that the infinite subset $\{Tx_n\}_{n \in \mathbb{N}}$ of $T(V)$ has no limit point in X. According to Theorem 1.1.40, this shows that $T(V)$ is not relatively compact in X. Ultimately, this argument proves that T is not a compact operator. □

2.1.3 Bounded Operators

As a consequence of the Global Eigenvalue Theorem we know that the eigenvalues of a compact operator form a bounded subset of $\mathbb{C}$. On the other hand, for some classes of spaces this is true for *every* continuous linear operator. To give a precise statement, recall the piece of notation introduced in (1.4.246).

Proposition 2.1.10. *If X is a p-normed space for some $0 < p \leq 1$ and T a continuous linear operator on X, then the eigenvalues of T all lie in the closed disk of radius $\|T\|^{1/p}$ centered at the origin.*

Proof. Suppose $Tx = \lambda x$ for some scalar λ, where x is a non-zero vector in X. Upon replacing x by $x/\|x\|^{1/p}$ we may assume x is a unit vector, since

$$\left\| \frac{x}{\|x\|^{1/p}} \right\| = \left(\frac{1}{\|x\|^{1/p}} \right)^p \|x\| = 1. \tag{2.1.48}$$

Then, based on assumptions and (1.4.249), we may estimate

$$|\lambda|^p = \|\lambda x\| = \|Tx\| \leq \|T\|, \tag{2.1.49}$$

as promised. □

The following exercise shows that the property mentioned at the beginning of this section does not, however, hold in general linear topological spaces.

Exercise 2.1.11. Use the symbol $\mathbb{C}^{\mathbb{N}}$ to denote the space of complex-valued functions on the natural numbers $\mathbb{N}$, endowed with the topology of pointwise convergence, that is, the topology on $\mathbb{C}^{\mathbb{N}}$ whose zero-neighborhoods are all sets of the form

$$N(F, \varepsilon) := \{f \in \mathbb{C}^{\mathbb{N}} : |f(n)| < \varepsilon \text{ for all } n \in F\} \tag{2.1.50}$$

where F runs through all finite subsets of $\mathbb{N}$ and ε through all positive real numbers. Show that

$$d(f, g) := \sum_{n=1}^{\infty} \frac{2^{-n}|f(n) - g(n)|}{1 + |f(n) - g(n)|}, \qquad \forall f, g \in \mathbb{C}^{\mathbb{N}}. \tag{2.1.51}$$

is a metric on $\mathbb{C}^{\mathbb{N}}$ which induces the same topology on $\mathbb{C}^{\mathbb{N}}$ as the one introduced above. You can also think of $\mathbb{C}^{\mathbb{N}}$ as the topological direct sum (or topological product) of countably many copies of the complex scalars.

Consider the following class of operators on $\mathbb{C}^{\mathbb{N}}$. For each fixed function $\omega \in \mathbb{C}^{\mathbb{N}}$ define the multiplication operator M_ω on $\mathbb{C}^{\mathbb{N}}$ by $M_\omega f := \omega f$ for $f \in \mathbb{C}^{\mathbb{N}}$. If you prefer to think of $\mathbb{C}^{\mathbb{N}}$ as the space of complex sequences, then this is just the diagonal operator induced by the sequence $\omega = \{\omega(n)\}_{n \in \mathbb{N}}$ as in Proposition 2.1.6.

Show that M_ω is a linear continuous operator on $\mathbb{C}^{\mathbb{N}}$, and that its eigenvalues are precisely the values of ω. Conclude that M_ω can have an unbounded set of eigenvalues and, in fact, its eigenvalues can even be dense in $\mathbb{C}$.

Hint: For the first claim use Exercise 1.2.3.

Exercise 2.1.12. Find an example of a linear topological space X and a continuous linear operator on X for which the set of eigenvalues is the entire complex plane.

Hint: Take X to be $\mathcal{O}(\mathbb{C})$, the space of holomorphic functions in the complex plane. This becomes a linear topological space when equipped with the (unique) topology τ for which the sets

$$N(K, \varepsilon) := \Big\{f \in \mathcal{O}(\mathbb{C}) : \sup_{z \in K} |f(z)| < \varepsilon\Big\}, \tag{2.1.52}$$

where K runs through all compact subsets of $\mathbb{C}$ and ε through all positive real numbers, form a local base at the origin. Check that the complex differentiation operator $T : X \to X$ defined as $Tf := f'$ for each $f \in \mathcal{O}(\mathbb{C})$ is linear and continuous on X (here we may want to use interior estimates for holomorphic functions), and that every $\lambda \in \mathbb{C}$ is an eigenvalue for T (think of a suitable exponential function).

Given a linear topological space X, there exists a class of operators intermediate between $\mathcal{K}(X)$ and $\mathcal{L}(X)$ for which the eigenvalues always form a bounded set.

Definition 2.1.13. *A linear operator T defined on a linear topological space X is said to be* bounded *if there exists a zero-neighborhood U of X such that $T(U)$ is a bounded subset of X. Denote the set of bounded operators on X by $\mathcal{B}(X)$; i.e.,*

$$\mathcal{B}(X) := \{T : X \longrightarrow X \ : \ T \ \text{ linear and bounded}\}. \tag{2.1.53}$$

If one wishes to emphasize the dependence on the linear topology τ on X, write $\mathcal{B}(X, \tau)$ for the collection of all bounded operators on (X, τ).

We have already encountered the notion of "bounded" linear operator, in the context of quasi-normed and p-normed spaces (cf. Exercise 1.4.23 and Definition 1.4.26), so it is reassuring to learn that the newly minted label of boundedness in Definition 2.1.13 is compatible with these earlier versions. This (and more) makes the subject of our next theorem.

Theorem 2.1.14. *Given a linear topological space X, recall $\mathcal{K}(X)$ from* (2.1.4), *$\mathcal{L}(X)$ from* (2.1.3), *and $\mathcal{B}(X)$ from* (2.1.53). *Then*

$$\mathcal{K}(X) \subseteq \mathcal{B}(X) \subseteq \mathcal{L}(X), \tag{2.1.54}$$

and

$$\mathcal{B}(X) = \mathcal{L}(X) \ \text{ if } X \text{ is locally bounded.} \tag{2.1.55}$$

Simply put, every compact operator is bounded, every bounded operator is continuous, and every continuous linear operator on a locally bounded topological vector space is bounded.

Moreover, if $(X, \|\cdot\|)$ is a quasi-normed space, or a p-normed space for some $0 < p \leq 1$, and if $T : X \to X$ is a linear map, then T is bounded in the sense of Exercise 1.4.23 or Definition 1.4.26, respectively, if and only if $T \in \mathcal{B}(X, \tau)$, where τ is the vector space topology $\tau_{\|\cdot\|}$ associated with $\|\cdot\|$ as in Exercise 1.4.18, or the p-norm topology induced by $\|\cdot\|$ on X as in Definition 1.4.6, respectively.

In particular, if X is a quasi-normed space, or a p-normed space, then $\mathcal{B}(X)$ and $\mathcal{L}(X)$ are identical.

Proof. The first inclusion in (2.1.54) is a consequence of definitions and item *(c)* in Theorem 1.6.40. To justify the second inclusion in (2.1.54), fix $T \in \mathcal{B}(X)$ and pick a zero-neighborhood W of X. Since T is bounded, there exists a zero-neighborhood U of X such that $T(U)$ is a bounded subset of X. In particular, $T(U)$ is absorbed by W, so there exists $\varepsilon > 0$ with the property that $\varepsilon T(U) \subseteq W$. Thus, $T(\varepsilon U) \subseteq W$, and since

εU remains a zero-neighborhood in X we conclude from item *(1)* in Definition 1.1.7 that T is continuous at zero. Consequently, $T \in \mathcal{L}(X)$, as wanted.

For the next portion of the proof, work under the additional assumption that the space X is locally bounded. Thus, there exists a zero-neighborhood V of X which is bounded. Pick $T \in \mathcal{L}(X)$. Since T is continuous at $0 \in X$, it follows from item *(1)* in Definition 1.1.7 that there exists a zero-neighborhood U of X with the property that $T(U) \subseteq V$. Since Proposition 1.6.39 now implies that $T(U)$ is bounded in X, we see from Definition 2.1.13 that $T \in \mathcal{B}(X)$. Thus, in the present scenario, we also have $\mathcal{L}(X) \subseteq \mathcal{B}(X)$. In concert with (2.1.54), this proves (2.1.55).

Finally, all remaining claims in the statement are consequences of (2.1.55), Proposition 1.4.28, and Proposition 1.11.3 (also keeping in mind Exercise 1.6.47 and (1.11.598)). □

Theorem 2.1.14 should be contrasted with the result below which shows that continuous operators need not, in general, be bounded.

Theorem 2.1.15. *Recall the topological vector space $\mathbb{C}^{\mathbb{N}}$, introduced earlier in Exercise 2.1.11, and suppose T is a linear operator mapping the vector space $\mathbb{C}^{\mathbb{N}}$ into itself. Then the following properties are equivalent:*

(a) T is bounded;
(b) T is compact;
(c) T is a finite-rank continuous operator.

Proof. We already know that *(c)* ⇒ *(b)* ⇒ *(a)*, where the first implication uses part *(c)* of Theorem 2.1.3 and the second implication is apparent from (2.1.54). It remains to show that *(a)* ⇒ *(c)*. Since, as seen from (2.1.54), every bounded operator on $\mathbb{C}^{\mathbb{N}}$ is continuous, what is left to prove is that every bounded operator on $\mathbb{C}^{\mathbb{N}}$ has a finite-dimensional range. The key is that zero-neighborhoods in $\mathbb{C}^{\mathbb{N}}$ are very large. To this end, suppose T is a bounded operator on $\mathbb{C}^{\mathbb{N}}$. Then T takes some zero-neighborhood into a bounded set. In particular, with notation introduced in (2.1.50), there exists a finite subset F of $\mathbb{N}$ and an $\varepsilon > 0$ such that $T(N(F,\varepsilon))$ is a bounded subset of $\mathbb{C}^{\mathbb{N}}$.

Now $N(F,\varepsilon)$ contains the linear subspace

$$F^o := \{f \in \mathbb{C}^{\mathbb{N}} : f(j) = 0 \text{ for every } j \in F\} \tag{2.1.56}$$

so $T(N(F,\varepsilon))$ contains $T(F^o)$. But since the topology of $\mathbb{C}^{\mathbb{N}}$ is Hausdorff (in fact, metrizable) and $T(N(F,\varepsilon))$ is bounded, $T(N(F,\varepsilon))$ cannot contain a nontrivial subspace (see Proposition 1.6.39 and the right-pointing implication in (1.6.485)). As such, we must have $T(F^o) = \{0\}$. Now

$$\mathbb{C}^{\mathbb{N}} = F^o \boxplus N \tag{2.1.57}$$

where N is the linear span of the standard unit vectors $\{e_j : j \in F\}$ (that is, $e_j(k) = 1$ if $j = k$, and zero otherwise). Hence, $T(\mathbb{C}^{\mathbb{N}}) = T(N)$, and since N is finite dimensional, so is the range of the linear operator T. □

It is of interest to further augment Theorem 2.1.14 with the following result.

Proposition 2.1.16. *Let X be a linear topological space, and consider a linear mapping $T : X \to X$. Then*

$$\text{if } T \in \mathcal{B}(X) \text{ then } T \text{ maps bounded subsets of } X \text{ into bounded subsets of } X, \tag{2.1.58}$$

and

$$\begin{aligned}&\text{under the additional assumption that } X \text{ is locally bounded, if } T \text{ maps}\\ &\text{bounded subsets of } X \text{ into bounded subsets of } X \text{ then } T \in \mathcal{B}(X).\end{aligned} \tag{2.1.59}$$

As a corollary, given any locally bounded linear topological space X and any linear mapping $T : X \to X$, then $T \in \mathcal{B}(X)$ if and only if T maps bounded subsets of X into bounded subsets of X.

In particular, if $(X, \|\cdot\|)$ is a quasi-normed space, or a p-normed space, then a linear mapping $T : X \to X$ belongs to $\mathcal{B}(X)$ if and only if T maps $\|\cdot\|$-bounded subsets of X into $\|\cdot\|$-bounded subsets of X.

Proof. Fix an operator $T \in \mathcal{B}(X)$ and pick a bounded subset B of X. On the one hand, since T is bounded, there exists a zero-neighborhood U of X such that $T(U)$ is a bounded subset of X. On the other hand, since B is bounded, B is absorbed by U, therefore there exists some $\varepsilon > 0$ with the property that $\varepsilon B \subseteq U$. The latter forces $T(\varepsilon B) \subseteq T(U)$, and since $T(U)$ is a bounded we conclude that $T(\varepsilon B)$ is bounded, based on Proposition 1.6.39. Since $T(\varepsilon B) = \varepsilon T(B)$, another application of Proposition 1.6.39 shows that $T(B)$ is bounded, which is what we set to prove. This establishes (2.1.58).

Next, make the additional assumption that X is locally bounded. Hence, there exists a zero-neighborhood V of X which is bounded. Also, suppose the linear operator T maps bounded subsets of X into bounded subsets of X. Then $T(V)$ is bounded, placing T in $\mathcal{B}(X)$. This takes care of (2.1.59). All other claims in the statement follow from what we have proved already. □

Theorem 2.1.17 (The weak topology revisited). *Let $(H, \langle\cdot,\cdot\rangle)$ be a Hilbert space, and denote by $\|\cdot\|$ the norm associated with the inner product (as in Proposition 1.5.3). Recall the vector space topology $\tau_{\|\cdot\|}$ induced by this norm as in Definition 1.4.6. Also recall the weak topology τ_w on H introduced in Proposition 1.6.22. Then, for every linear transformation T on H, it follows that T is continuous on (H, τ_w) if and only if T is continuous on $(H, \tau_{\|\cdot\|})$. Succinctly put,*

$$\mathcal{L}(H, \tau_w) = \mathcal{L}(H, \tau_{\|\cdot\|}). \tag{2.1.60}$$

Proof. Assume $T : (H, \tau_w) \to (H, \tau_w)$ is linear and continuous. For each $u \in H$, consider the linear functional

$$f_u : H \to \mathbb{C}, \qquad f_u(x) := \langle Tx, u\rangle. \tag{2.1.61}$$

Since T is continuous on (H, τ_w), it follows that f_u is continuous from (H, τ_w) into $\mathbb{C}$ equipped with the standard topology. The continuous linear functionals on (H, τ_w)

are precisely those coming from H^* (the norm dual), so by the Riesz representation (cf. Theorem 1.5.11) there exists a unique $Su \in H$ such that

$$\langle Tx, u\rangle = \langle x, Su\rangle \quad \text{for all} \quad x \in H. \tag{2.1.62}$$

This defines a map $S : H \to H$. Linearity of S follows from linearity of $u \mapsto f_u$ and the uniqueness in the Riesz representation. Indeed, for any $u, v \in H$ and $\alpha, \beta \in \mathbb{C}$,

$$\begin{aligned}\langle x, S(\alpha u + \beta v)\rangle &= \langle Tx, \alpha u + \beta v\rangle \\ &= \alpha\langle Tx, u\rangle + \beta\langle Tx, v\rangle = \alpha\langle x, Su\rangle + \beta\langle x, Sv\rangle \\ &= \langle x, \alpha Su + \beta Sv\rangle \end{aligned} \tag{2.1.63}$$

which, in light of the arbitrariness of $x \in H$, forces $S(\alpha u + \beta v) = \alpha Su + \beta Sv$, as wanted.

We claim that S is bounded operator on $(H, \tau_{\|\cdot\|})$. To this end, consider its graph

$$\mathcal{G} := \{(u, Su) : u \in H\} \subset H \times H. \tag{2.1.64}$$

Suppose $u_n \to u$ and $Su_n \to w$ in $(H, \tau_{\|\cdot\|})$, as $n \to \infty$. Then, for all $x \in H$,

$$\langle x, w\rangle = \lim_{n\to\infty} \langle x, Su_n\rangle = \lim_{n\to\infty} \langle Tx, u_n\rangle = \langle Tx, u\rangle = \langle x, Su\rangle. \tag{2.1.65}$$

By the Riesz Representation Theorem, $w = Su$. Hence $\mathcal{G}$ is closed in $(H \times H, \tau_{\|\cdot\|} \times \tau_{\|\cdot\|})$. Since H is Banach, by the Closed Graph Theorem (cf., e.g., [39, Theorem 2.15, p. 51]) the operator S is continuous on $(H, \tau_{\|\cdot\|})$, hence also bounded (cf. Theorem 2.1.14).

From (2.1.62) we have $\langle Tx, u\rangle = \langle x, Su\rangle$ for all $x, u \in H$, which goes to show that $T = S^*$ (the adjoint of S). As the adjoint of a bounded operator on a Hilbert space is bounded, T is bounded. In light of Theorem 2.1.14, the linear operator T is therefore continuous in the norm topology $\tau_{\|\cdot\|}$.

Moving on, assume $T : (H, \tau_{\|\cdot\|}) \to (H, \tau_{\|\cdot\|})$ is a linear and continuous operator, hence also bounded (cf. Theorem 2.1.14). As a consequence, the adjoint T^* of T is also bounded on $(H, \tau_{\|\cdot\|})$ (cf. Exercise 1.5.12). The present goal is to show that $T : (H, \tau_w) \to (H, \tau_w)$ is continuous. For this, pick a finite set $F := \{x_1, \dots, x_m\} \subseteq H$ and some $\varepsilon > 0$, and note that if we set $\widetilde{F} := \{T^*x_1, \dots, T^*x_m\} \subseteq H$ then

$$T(N(\widetilde{F}, \varepsilon)) \subseteq N(F, \varepsilon)). \tag{2.1.66}$$

According to the definition of τ_w from Proposition 1.6.22 and item *(1)* in Definition 1.1.7, we conclude from (2.1.66) that $T : (H, \tau_w) \to (H, \tau_w)$ is continuous. This finishes the proof of the theorem. □

We shall next discuss the connection between boundedness and eigenvalues. Recall that on a p-normed space, the set of eigenvalues of each continuous (i.e., bounded) operator is bounded (Proposition 2.1.10). A trivial, but crucial, step involved normalizing the lengths of eigenvectors. A similar argument is needed in the general case and this makes the subject of our next result.

Lemma 2.1.18 (Normalization). *Suppose A is a circled, absorbing subset of a linear topological space X, and suppose $x \in X$ is not in A. Then there exists $r \in (0, 1)$ such that rx belongs to A, but not to $\frac{1}{2}A$.*

Proof. Let $x \in X \setminus A$. Then since A is absorbing, $\{\delta > 0 \,:\, tx \in A,\ \forall\, t \in (0, \delta)\}$ is a nonempty subset of $(0, 1)$, and hence the quantity

$$\eta(x) := \sup\{\delta > 0 \,:\, tx \in A,\ \forall\, t \in (0, \delta)\} \tag{2.1.67}$$

is a well-defined number belonging to $(0, 1]$. Pick $r \in (\frac{1}{2}\eta(x), \eta(x))$ and observe that this forces

$$rx \in A. \tag{2.1.68}$$

Indeed, having $r < \eta(x)$ entails, in light of (2.1.67), that there exists $\delta > r$ such that $tx \in A$ for all $t \in (0, \delta)$. Taking $t := r$ (which is a permissible choice) then proves (2.1.68). We claim that

$$rx \notin \frac{1}{2}A. \tag{2.1.69}$$

To see this, suppose to the contrary that $2rx \in A$. Thus the circledness of A yields

$$tx = \frac{t}{2r}2rx \in A, \qquad \forall\, t \in (0, 2r). \tag{2.1.70}$$

Upon appealing to (2.1.67), we deduce from (2.1.70) that $2r \leq \eta(x)$, contradicting the initial choice of r. This establishes (2.1.69) and completes the proof of the lemma. □

We are now in a position to discuss the boundedness of the point spectrum of a bounded linear operator on a Hausdorff linear topological space.

Theorem 2.1.19. *Suppose T is a bounded linear operator on a Hausdorff linear topological space X. Then the eigenvalues of T form a bounded set.*

Proof. Since T is bounded, there is a zero-neighborhood U such that $T(U)$ is a bounded subset of X. By Proposition 1.6.19 and the fact that every subset of a bounded set is itself bounded, we may assume, without loss of generality, that U is circled. All in all, there exists a circled zero-neighborhood U such that $T(U)$ is bounded, and there exists $r > 0$ such that $T(U) \subseteq rU$.

We claim that the eigenvalues of T lie in the disk $\{\lambda \in \mathbb{C} \,:\, |\lambda| < 2r\}$. To see this, suppose λ is a non-zero eigenvalue of T, and suppose x is an eigenvector for λ; i.e., $x \neq 0$ and $Tx = \lambda x$. Since $T(U)$ is bounded it cannot contain the entire ray $\{t\lambda x \,:\, t > 0\}$ passing through the point λx. Indeed, if $\{t\lambda x \,:\, t > 0\} \subseteq T(U)$, then

$$\{t\lambda x \,:\, t \in \mathbb{R}\} \subseteq T(U), \tag{2.1.71}$$

for $0 \in T(U)$ and $T(U) = -T(U)$ (by the linearity of T and item *(2)* in Proposition 1.6.16). Also, the linearity of T, the fact that U is circled, and (2.1.71) entail

$$\text{span}\{\lambda x\} := \{\alpha\lambda x : \alpha \in \mathbb{C}\} \subseteq T(U) + T(U). \tag{2.1.72}$$

Since $T(U)$ is bounded, (2.1.72), in conjunction with part *(a)* of Theorem 1.6.40 and the fact that every subset of a bounded set inherits the boundedness feature, yields that $\text{span}\{\lambda x\}$ is a bounded subspace of X. However,

$$\lambda x \neq 0 \implies \{0\} \subsetneq \text{span}\{\lambda x\}. \tag{2.1.73}$$

Thusly $\text{span}\{\lambda x\}$ is a nontrivial bounded subspace of the Hausdorff linear topological space X, which is a preposterous conclusion by Proposition 1.6.43. As a result,

$$\{t\lambda x : t > 0\} \not\subset T(U), \tag{2.1.74}$$

and so U does not contain the entire ray through x. Now applying Lemma 2.1.18 to a suitable multiple of x we see that some other positive multiple of x lies in U but not in $\frac{1}{2}U$. Replace the original eigenvector x by this multiple.

Therefore, our normalization argument yields an eigenvector x that lies in U but not in $\frac{1}{2}U$. Now the argument proceeds much like the p-normed version. We have from our choice of x and r:

$$\lambda x = Tx \in T(U) \subseteq rU, \tag{2.1.75}$$

hence, since $\lambda \neq 0$ and U is circled,

$$x \in \frac{r}{\lambda}U = \frac{r}{|\lambda|}U. \tag{2.1.76}$$

Since $x \notin \frac{1}{2}U$ we must have $r/|\lambda| > 1/2$ (again using the fact that U is circled). Thus $|\lambda| < 2r$, as advertised. □

2.1.4 *Compactly Perturbing the Identity*

If K is a compact operator on a linear topological space X we call the operator $I - K$ that occurred in the proof of Theorem 2.1.8 a *compact perturbation of the identity*. So that theorem can be restated as:

in a Hausdorff linear topological space, every compact perturbation of the identity has a finite-dimensional null space. (2.1.77)

We phrase the rest of the "local part" of the Riesz theory in terms of compact perturbations of the identity. The next major steps in the theory are to show that compact perturbations of the identity are relatively open maps (cf. Definition 1.8.14), and have closed range. For F-spaces, the Open Mapping Theorem (recalled in Theorem 1.6.7) provides this relative openness once we know the range is closed. In the general case it seems more efficient to derive the closed range property from relative openness.

Theorem 2.1.20 (Relative Openness). *Every compact perturbation of the identity on a Hausdorff linear topological space is a relatively open mapping (in the sense of Definition 1.8.14).*

Proof. We are assuming $T = I - K$ where K is a compact operator on X. By Definition 2.1.1 and Proposition 1.6.19, there exists a circled zero-neighborhood V_0 in X for which $K(V_0)$ is relatively compact.

Now suppose, reasoning by contradiction, that T is *not* relatively open. This implies that there exists an open set O in X such that $T(O)$ is not open in $\operatorname{ran} T$. This further entails that there exist $x \in X$, a neighborhood N of x with the property that $T(N)$ is not a neighborhood of $T(x)$ in $\operatorname{ran} T$. Since T is linear, we may assume, without loss of generality, that $x = 0$. Appealing again to Proposition 1.6.19, it follows that there exists a circled zero-neighborhood V_1 in X whose image $T(V_1)$ is not a zero-neighborhood in $\operatorname{ran} T$. Thus $V = V_0 \cap V_1$ is a circled zero-neighborhood in X with the following properties:

(a) $K(V)$ is relatively compact, and
(b) $T(V)$ is not a zero-neighborhood in $\operatorname{ran} T$.

By part (b) above, $T(V) \not\supseteq W \cap \operatorname{ran} T$, for every zero-neighborhood W in X. From this and Proposition 1.6.19, it follows that for every circled zero-neighborhood ω in X there is a point $y_\omega \in \omega \cap \operatorname{ran} T$ with $y_\omega \notin T(V)$. Since $T(V)$ is circled and absorbing in $\operatorname{ran} T$, the Normalization Lemma (see Lemma 2.1.18) provides, for each ω, a number $a_\omega \in (0, 1)$ such that $a_\omega y_\omega$ lies in $T(V)$, but *not* in $\frac{1}{2}T(V)$. Moreover, since each zero-neighborhood ω is circled, we have $a_\omega y_\omega \in \omega$.

Upon relabeling each $a_\omega y_\omega$ simply as y_ω, we obtain a net (y_ω), indexed by the circled zero-neighborhoods in X, equipped with the binary relation "$\supseteq$", having these properties:

$$y_\omega \in \omega \text{ and } y_\omega \in T(V) \quad \text{but} \quad y_\omega \notin \frac{1}{2}T(V). \tag{2.1.78}$$

Since each $y_\omega \in T(V)$, there exists a vector $x_\omega \in V$ such that

$$Tx_\omega = y_\omega. \tag{2.1.79}$$

Using next that $K(V)$ is is a relatively compact set, the net $(K(x_\omega))$ has a convergent subnet (see Exercise 1.1.50) which, in the interest of conserving notation, we continue to denote by $(K(x_\omega))$ (by cofinality, the resulting subcollection of indices ω still forms a basis for the zero-neighborhoods of X). Suppose u denotes the limit of this net, i.e., $u = \lim K(x_\omega)$. Since

$$Kx_\omega = x_\omega - Tx_\omega = x_\omega - y_\omega \tag{2.1.80}$$

and (by Lemma 1.1.46) $y_\omega \to 0$, we have $x_\omega \to u$, so using the Hausdorffness of X, we obtain

$$0 = \lim y_\omega = \lim Tx_\omega = Tu. \tag{2.1.81}$$

Thus $u \in \ker T$, so the net (x_ω), which we have just seen converges to u, must eventually lie in $\frac{1}{2}V + u \subseteq \frac{1}{2}V + \ker T$. Thus the T-image of this net, which by (2.1.79) above is the net (y_ω), eventually lies in $T\left(\frac{1}{2}V\right) = \frac{1}{2}T(V)$. But this contradicts the original choice of the vectors y_ω, which demanded that *none of them* belongs to $\frac{1}{2}T(V)$. This contradiction proves that the mapping T is relatively open. □

Theorem 2.1.21 (Closed Range). *On a Hausdorff linear topological space X, every compact perturbation of the identity has closed range.*

Proof. Suppose $T = I - K$ where K is a compact operator on X. Thus there is a zero-neighborhood V for which $K(V)$ is relatively compact. We wish to show that $\operatorname{ran} T$ is closed in X. To this end, suppose (x_α) is a net in X for which the image net (Tx_α) converges to a point $y \in X$. We aim to show that $y \in \operatorname{ran} T$, i.e., $y = Tx$ for some $x \in X$. If some subnet of (x_α) is convergent –say to x– then there is no problem: $y = Tx$ by the continuity of T (cf. Theorem 1.1.47), and the fact that X is Hausdorff.

We will show that, even if (x_α) has no convergent subnet, it can be replaced by a new net which has a convergent subnet, and whose image still converges to y. This will complete the proof.

To produce the replacement net, we begin with the fact that T is a relatively open map (by Theorem 2.1.20), hence $T(V)$ is a zero-neighborhood in $\operatorname{ran} T$ (as may be seen from Definition 1.8.14). Since the net (Tx_α) is Cauchy in $\operatorname{ran} T$ (see Definition 1.8.4 and Proposition 1.8.6), this provides us with an index γ such that:

$$\alpha, \beta \geq \gamma \Rightarrow Tx_\alpha \in Tx_\beta + T(V) = T(x_\beta + V). \tag{2.1.82}$$

In particular, the net (Tx_α) is eventually in $T(x_\gamma + V)$.

Also, for any zero-neighborhood W, this same net (Tx_α), being convergent to y, is eventually in $y + W$. In particular, some member of the original net (x_α), which we denote z_W, has these properties:

$$Tz_W \in T(x_\gamma + V) \tag{2.1.83}$$

and

$$Tz_W \in y + W. \tag{2.1.84}$$

Let $\mathcal{W}$ denote the collection of all zero-neighborhoods in X, ordered by reverse inclusion. According to (2.1.84), the net $(Tz_W : W \in \mathcal{W})$, converges to y. Now (2.1.83) shows that

$$z_W \in T^{-1}(T(x_\gamma + V)) = x_\gamma + V + \ker T, \tag{2.1.85}$$

where T^{-1} denotes the preimage. In other words, we have the decomposition

$$z_W = x_\gamma + v_W + k_W, \tag{2.1.86}$$

where $v_W \in V$ and $k_W \in \ker T$.

We claim that the net $(x_\gamma + v_W : W \in \mathcal{W})$, whose image under T converges to y, contains a convergent subnet. Once this is done, we will know that y is the image of this subnet's limit, and our proof will be completed.

To prove the claim, note that relative compactness of $K(V)$ in X insures that the net (Kv_W) has a convergent subnet –call it (Kv_ω). Since also

$$Tv_\omega = Tz_\omega - Tx_\gamma \to y - Tx_\gamma \tag{2.1.87}$$

we see from the equation $I = T + K$ that the subnet (v_ω) itself converges in X and hence so does $(x_\gamma + v_\omega)$. □

Here is a summary of what we have proved so far:

Theorem 2.1.22 (Compact perturbation). *Suppose X is a Hausdorff linear topological space, and assume $T \in \mathcal{L}(X)$ is a compact perturbation of the identity. Then:*

(a) $\ker T$ *is finite dimensional;*
(b) $\operatorname{ran} T$ *is closed;*
(c) T is a relatively open map (in the sense of Definition 1.8.14).

We close this subsection with an exercise that is a harbinger of things to come. Let us agree to call a continuous linear operator on a linear topological space X *invertible* if it is a linear homeomorphism of X onto itself. The next result says that compact perturbations of an invertible operator are "almost one-to-one."

Exercise 2.1.23 (Weyl's Theorem). Suppose X is a Hausdorff linear topological space and assume $T \in \mathcal{L}(X)$ is invertible. Then for every compact operator K on X the null space of $T + K$ is finite dimensional.

Hint: Use (2.1.77).

2.2 Algebraic Interlude: Range and Null-Space Sequences

The complete Riesz theory of compact operators involves a detailed analysis of the *generalized eigenspaces* of the operator $T = \lambda I - K$, where K is a compact operator on a linear topological space and λ a non-zero eigenvalue of K. The generalized eigenspaces of T that correspond to λ are the subspaces $\ker(\lambda I - K)^n$, where n is a non-negative integer. These subspaces are T-invariant, and the Riesz theory ultimately provides a useful decomposition of the underlying space into a topological direct sum of a generalized eigenspace and another T-invariant subspace (a "generalized range"). As we observed in the previous section, when we work with one non-zero eigenvalue at a time, there is no harm in assuming that eigenvalue to be 1, hence the objects of our interest will be null spaces of non-negative integer powers of compact perturbations of the identity.

In this section we will temporarily put aside considerations of topology in order to emphasize the purely algebraic aspects of the situation. Thus for the remainder of the section:

1. X will be a complex (or real, if you wish) vector space, with no topology assumed.
2. T will be a linear operator on X, with no continuity assumed.

For each non-negative integer k we write:

$$N_k := \ker(T^k) = (T^k)^{-1}(\{0\}) \text{ and } R_k := \operatorname{ran}(T^k) = T^k(X). \tag{2.2.88}$$

Also, set $T^0 := I$, $N_0 := \ker(T^0) = \{0\}$ and $R_0 := \operatorname{ran}(I) = X$, where I denotes the identity map on X.

Exercise 2.2.1. Prove that whenever $0 \leq i \leq j$ one has $T^i(N_j) = N_{j-i} \cap R_i$.

We call $(R_k : k \geq 0)$ the *range sequence* of T, and $(N_k : k \geq 0)$ the `null-space sequence` associated with T. Clearly, as k increases the range sequence decreases, and the null-space sequence increases, i.e.,

$$N_0 \subseteq N_1 \subseteq N_2 \subseteq \cdots \text{ and } R_0 \supseteq R_1 \supseteq R_2 \supseteq \cdots \tag{2.2.89}$$

The work of this section concerns the *stability* of these sequences. For motivation, note first that if we have equality in the first two terms of the range sequence ($R_0 = R_1$) then we have equality ever after. This is trivial: the equation $R_0 = R_1$ means that T maps X onto itself, hence the same is true of T^k for every non-negative integer k, and therefore the range sequence is constant: $R_k \equiv X$. Similar considerations apply to the null-space sequence. These observations constitute the case $k = 0$ of the following result.

Theorem 2.2.2 (First Stabilization). *Let X be a complex (or real) vector space and suppose that $T : X \longrightarrow X$ is a linear operator. For each $k \in \mathbb{N} \cup \{0\}$, let N_k and R_k be as in* (2.2.88). *Then the following properties hold true:*

(a) $R_k = R_{k+1} \Rightarrow R_{k+1} = R_{k+2}$, and consequently $R_m = R_k$, for each $m \geq k$.
(b) $N_k = N_{k+1} \Rightarrow N_{k+1} = N_{k+2}$, and consequently $N_m = N_k$, for each $m \geq k$.

Proof. To treat *(a)*, suppose $R_k = R_{k+1}$. Since we know that $(R_k : k \geq 0)$ is decreasing, it suffices to show that $R_{k+1} \subseteq R_{k+2}$ (because the other inclusion is automatic). Suppose $y \in R_{k+1}$, i.e., there exists $x \in X$ with $y = T^{k+1}x = T(T^k x)$. Given that $T^k x \in R_k = R_{k+1}$ we know that $T^k x = T^{k+1}x'$ for some $x' \in X$. Thus,

$$y = T(T^{k+1}x') = T^{k+2}x' \in R_{k+2}, \tag{2.2.90}$$

as wanted. This shows that $R_{k+1} = R_k$ and an iteration argument immediately yields that $R_m = R_k$, for every $m \geq k$.

To deal with item *(b)*, suppose $N_k = N_{k+1}$. This time, based on the fact that $(N_k : k \geq 0)$ is increasing, there remains to show that $N_{k+2} \subseteq N_{k+1}$. To this end, fix $x \in N_{k+2}$. Then $0 = T^{k+2}x = T^{k+1}(Tx)$, so that $Tx \in N_{k+1} = N_k$, whereupon $0 = T^k(Tx) = T^{k+1}x$. Thus $x \in N_{k+1}$, and therefore $N_{k+1} = N_{k+2}$, as desired. Finally, the last part in *(b)* follows again by iteration. □

Convention 2.2.3 *Theorem 2.2.2 asserts that for the range sequence* $(R_k : k \geq 0)$ *there are only these possibilities:*

(a) The sequence is constant: $R_k \equiv X$ *for all* $k \in \mathbb{N} \cup \{0\}$.
(b) The sequence decreases strictly up to some index $k \in \mathbb{N}$, *after which it is constant* $(\equiv R_k)$.
(c) The sequence decreases strictly, forever.

We say the range sequence stabilizes at $k \in \mathbb{N}$ *in case (b), and at zero in case (a). In case (c) we say that the range sequence never stabilizes. Similar terminology, whose formulation we leave to the reader, applies to the null-space sequence.*

Example 2.2.4. Let $X := \ell^2$ and consider the `forward` and `backward` shift operators defined on X, respectively, by:

$$S(\xi_1, \xi_2, \dots) := (0, \xi_1, \xi_2, \dots),$$
$$B(\xi_1, \xi_2, \dots) := (\xi_2, \xi_3, \xi_4, \dots), \tag{2.2.91}$$

for vectors $(\xi_1, \xi_2, \dots) \in \ell^2$. Clearly, both operators act linearly on ℓ^2, and:

1. the null-space sequence of S is $\equiv \{0, 0, \dots\}$, while the range sequence of S never stabilizes;
2. the range sequence of B is $\equiv \ell^2$, while the null-space sequence never stabilizes.

The next result shows that when stabilization occurs, it leads to the kind of behavior generally associated with finite-dimensional vector spaces, when injectivity is equivalent to surjectivity.

Theorem 2.2.5 (Algebraic Fredholm Alternative). *Suppose* T *is a linear transformation of a vector space* X.

(a) If T *is one-to-one and its range sequence stabilizes, then* T *maps* X *onto itself; i.e., the range sequence stabilizes at zero.*
(b) If T *maps* X *onto itself, and its null-space sequence stabilizes, then* T *is one-to-one; i.e., the null-space sequence stabilizes at zero.*

Proof. To prove *(a)*, suppose T is one-to-one and $R_k = R_{k+1}$ for some $k \in \mathbb{N} \cup \{0\}$. Fix $y \in X$. We wish to show that $y \in T(X)$. Since $T^k y \in R_k = R_{k+1}$, it follows that $T^k y = T^{k+1} y'$ for some $y' \in X$. Using that T^k is one-to-one (because T is), we obtain $y = Ty' \in T(X)$, as desired.

Moving on to *(b)*, suppose $T(X) = X$ and $N_k = N_{k+1}$ for some $k \in \mathbb{N} \cup \{0\}$. We wish to show that $\ker T = \{0\}$. To this end, suppose $x \in \ker T$. Since T maps X *onto* itself the same is true of T^k, so we have $x = T^k z$ for some $z \in X$. Thus $0 = Tx = T^{k+1} z$, i.e., $z \in N_{k+1} = N_k$, and therefore $0 = T^k z = x$, which completes the proof. □

We have already seen in Example 2.2.4 that the range and null-space sequences need not always stabilize. Of course they *will* stabilize if the underlying space X is finite dimensional, in which case Theorem 2.2.5 reduces to the fact that a linear operator on a finite-dimensional vector space is one-to-one if and only if it is onto.

In the next section we prove that both range and null-space sequences stabilize for *compact perturbations of the identity*. This yields the usual Fredholm Alternative for compact operators: *A compact perturbation of the identity is one-to-one if and only if it is onto.*

First, however, we refine Theorem 2.2.5 to provide important information on the structure of operators whose range and null-space sequences both stabilize.

Theorem 2.2.6 (Simultaneous Stabilization). *Suppose T is a linear operator on a vector space X. If both the range and null-space sequences of T stabilize, then they both stabilize at the same index $k \in \mathbb{N} \cup \{0\}$, and induce the algebraic direct sum decomposition*

$$X = N_k \boxplus R_k. \tag{2.2.92}$$

In this decomposition:

(a) T is an isomorphism of R_k onto itself;
(b) $T^i(N_k) = N_{k-i} \cap R_i$, for every $i \in \{0, 1, \ldots, k\}$;
(c) $\operatorname{ran} T^i = (N_{k-i} \cap R_i) \boxplus R_k$, *for every $i \in \{0, 1, \ldots, k\}$.*

Proof. The argument proceeds in several steps, the crucial one being:

`Step 1`: *If T is any linear transformation, then for each $k \in \mathbb{N}$ one has*

$$N_k = N_{k+1} \Leftrightarrow N_k \cap R_k = \{0\}. \tag{2.2.93}$$

To prove this, note that if $N_k = N_{k+1}$ and $x \in N_k \cap R_k$ then $T^k x = 0$ and $x = T^k z$ for some $z \in X$. Thus $T^{2k} z = 0$ which places z in $N_{2k} = N_k$ (the last equality coming item *(b)* in Theorem (2.2.2)). So $0 = T^k z = x$, as desired. Conversely, if $N_k \cap R_k = \{0\}$ and $x \in N_{k+1}$ then

$$T^k x \in N_1 \cap R_k \subseteq N_k \cap R_k = \{0\}, \tag{2.2.94}$$

so $x \in N_k$. Hence $N_{k+1} \subseteq N_k$. Since always $N_k \subseteq N_{k+1}$, this completes the proof of the claim made in Step 1.

`Step 2`: *If the range and null-space sequences of a linear transformation stabilize, then they stabilize at the same index $k \in \mathbb{N} \cup \{0\}$.*

To justify the claim above, let us first treat the case when $N_k = N_{k+1}$, for some $k \in \mathbb{N}$. Then by Step 1, the restriction of T from R_k into R_k, denoted by T_k, is a one-to-one mapping of R_k into itself, with range sequence $R_k, R_{k+1}, \ldots$ which, by our hypothesis, stabilizes. So by Theorem 2.2.5, T_k maps R_k *onto* itself, i.e., $R_k = R_{k+1}$. Thus the range sequence stabilizes no later than the null-space sequence. Now if $N_0 = N_1$, then T is injective on X and appealing again to Theorem 2.2.5, $T(X) = X$, and as such the range sequence stabilizes at 0.

Suppose now that $R_k = R_{k+1}$ for some $k \in \mathbb{N}$, and let T_k continue to denote the restriction of T to R_k. In this scenario, T_k maps R_k onto itself and the null-space sequence of T_k is

$$\{0\}, R_k \cap N_1, R_k \cap N_2, \ldots, \tag{2.2.95}$$

which stabilizes (because the original null-space sequence $(N_j)_{j \in \mathbb{N} \cup \{0\}}$ is assumed to stabilize). Using again Theorem 2.2.5 shows that T_k is one-to-one, hence so is $(T_k)^k$, i.e., $R_k \cap N_k = \{0\}$. Thus, $N_k = N_{k+1}$ by Step 1, hence the null-space sequence stabilizes no later than the range sequence. Next, if $R_0 = R_1$, then T is onto on X. Hence, by Theorem 2.2.5, we obtain that T is injective and therefore the null-space sequence also stabilizes at 0.

In summary, if the null-space sequence of T stabilizes at some $k \in \mathbb{N} \cup \{0\}$ then the range sequence of T also stabilizes at k. Moreover, if the range sequence of T stabilizes at some $k \in \mathbb{N} \cup \{0\}$ then the null-space sequence of T also stabilizes at k. Consequently, if both the range and null-space sequences of T stabilize, then they both stabilize at the same index $k \in \mathbb{N} \cup \{0\}$.

`Step 3`: *Suppose both null-space and range sequences stabilize at* $k \in \mathbb{N} \cup \{0\}$. *Then* $X = N_k \boxplus R_k$.

First we assume $k \in \mathbb{N}$. From Step 1 we have that $R_k \cap N_k = \{0\}$, so we need only prove that X is the algebraic sum of R_k and N_k. For this, take $x \in X$ and note that $T^k x = T^{2k} y$ for some $y \in X$ (because $R_k = R_{2k}$). Thus, $T^k(x - T^k y) = 0$, i.e.,

$$z := x - T^k y \in N_k, \tag{2.2.96}$$

whereupon

$$x = z + T^k y \in N_k + R_k, \tag{2.2.97}$$

as desired. If $k = 0$, then trivially $X = \{0\} + X = N_0 + R_0$. This completes the proof of Step 3.

`Step 4`: *Proof of (a)-(c).*

From Steps 1-3 we know that both the range and null-space sequences stabilize at the same $k \in \mathbb{N} \cup \{0\}$, and the algebraic direct sum decomposition (2.2.92) holds. Then the fact that T maps R_k onto itself follows immediately from the fact that $R_k = R_{k+1}$. To finish *(a)*, it remains to show that $T : R_k \longrightarrow R_k$ is one-to-one. However, $\ker (T|_{R_k}) = R_k \cap N_1 \subseteq R_k \cap N_k = \{0\}$, by (2.2.92). Moving on to the proof of *(b)*, that T^i maps N_k onto $N_{k-i} \cap R_i$ for all $i \in \{0, 1, \ldots, k\}$ is seen directly from definitions (cf. Exercise 2.2.1).

To justify *(c)*, assume $k \geq 1$ (as the statement when $k = 0$ is trivially true). Fix $i \in \{0, 1, \ldots, k\}$ and use (2.2.92), what we proved in *(b)*, and the fact that $T^i(R_k) = R_k$ (since the range sequence stabilizes at k) to write

$$T^i(X) = T^i(N_k) + T^i(R_k) = (N_{k-i} \cap R_i) + R_k. \tag{2.2.98}$$

In addition, if $x \in (N_{k-i} \cap R_i) \cap R_k$ then there exists $y \in X$ such that $x = T^k y$, and also $x = T^i z$ for some $z \in N_k$, by *(b)*. Thus, $T^k x = T^{k+i} z = T^i(T^k z) = 0$ which implies $x \in N_k$. Since we are presently assuming $k \geq 1$, from (2.2.93) we see that $x = 0$. This establishes the algebraic direct sum decomposition claimed in *(c)*. The proof of Theorem 2.2.6 is therefore complete. □

Exercise 2.2.7. Retain the context of Theorem 2.2.6 and continue to denote by k the index of stabilization for the operator T. Prove that $R_k \cap N_\ell = \{0\}$ for each integer $\ell \geq 0$, and that one has $R_i = \operatorname{ran} T^i = R_j \boxplus (N_{j-i} \cap R_i)$ whenever $j \geq k$ and $0 \leq i \leq j$.

Hint: The first claim in the statement is seen directly from definitions and the stability of the sequence of null-spaces. To prove the second claim, recall that

(a) $j \geq k$ and $0 \leq i \leq j$;
(b) $N_\ell \subseteq N_{\ell+1}$ and $R_{\ell+1} \subseteq R_\ell$ for each $\ell \in \mathbb{N} \cup \{0\}$;
(c) $X = N_k \boxplus R_k$;
(d) $N_\ell = N_k$ and $R_\ell = R_k$ whenever $\ell \geq k$;
(e) $R_k \cap N_\ell = \{0\}$ for each integer $\ell \geq 0$.

First, as $R_j \subseteq R_i$ by (a)-(b), and trivially $N_{j-i} \cap R_i \subseteq R_i$, we have $R_j + (N_{j-i} \cap R_i) \subseteq R_i$. Also, thanks to the fact that $R_j = R_k$ by (a) and (d), that, trivially, $N_{j-i} \cap R_i \subseteq N_{j-i}$, and (e) above, we conclude that $R_j \cap (N_{j-i} \cap R_i) = \{0\}$. Thus, so far

$$R_j \boxplus (N_{j-i} \cap R_i) \subseteq R_i. \tag{2.2.99}$$

To prove the opposite inclusion in (2.2.99), start with an arbitrary $x \in R_i$ and use (c) to (uniquely) decompose $x = n + r$ with $n \in N_k$ and $r \in R_k$. Then $r \in R_j$ by (a) and (d), so it remains to show that $n \in N_{j-i} \cap R_i$. For one thing, $n = x - r \in R_i + R_j = R_i$ by (a)-(b), so $n \in N_k \cap R_i$. We are therefore left with proving that

$$N_k \cap R_i \subseteq N_{j-i}. \tag{2.2.100}$$

To this end, pick $y \in N_k \cap R_i$ and note that this entails

$$T^k y = 0 \text{ and } y = T^i z \text{ for some } z \in X. \tag{2.2.101}$$

In particular, $T^{k+i} z = 0$ so $z \in N_{k+i} = N_k$ by (d). As such,

$$T^{j-i} y = T^{j-i}(T^i z) = T^j z = T^{j-k}(T^k z) = 0, \tag{2.2.102}$$

so $y \in N_{j-i}$. This establishes (2.2.100), so all claims have now been justified.

Here is some terminology that allows a catchy summary of part of Theorem 2.2.6.

Definition 2.2.8. *Call a linear operator T on a vector space X* `nilpotent` *provided $T^n = 0$ for some positive integer n. The least such integer is called the* `index of nilpotence` *of the linear operator T.*

Definition 2.2.9. *Let T be a linear operator on a vector space X, and let Y and Z be T-invariant subspaces of X such that $X = Y \boxplus Z$. Then the pair of subspaces (Y, Z)*

is said to reduce *the operator T. If A and B are the restrictions of T to Y and Z respectively, write*

$$T = A \boxplus B \tag{2.2.103}$$

and say T is the algebraic direct sum of A and B.

Corollary 2.2.10. *Suppose T is a linear operator on a vector space X, such that both the range sequence and null-space sequence of T stabilize at $k \in \mathbb{N} \cup \{0\}$. Then T decomposes into the algebraic direct sum of an invertible operator and a nilpotent operator of index k.*

For finite-dimensional vector spaces X, the range sequence and null-space sequence have to stabilize. In such a scenario, Corollary 2.2.10 is the foundation of the Jordan Decomposition Theorem. More specifically, given an eigenvalue λ of the linear operator T, one performs the above decomposition on the operator $T - \lambda I$, then further decomposes the nilpotent part into cyclic suboperators, and having done that, moves on to the isomorphic part, shows the underlying subspace to be T-invariant, and then repeats the process.

We conclude this section with a couple of results, which will play an important role in the future.

Proposition 2.2.11. *Let X be a vector space and assume $T : X \to X$ is a linear operator. Set*

$$R_j := \operatorname{ran} T^j \ \textit{ for each } j \in \mathbb{N}, \textit{ and } \ R_\infty := \bigcap_{j \in \mathbb{N}} R_j, \tag{2.2.104}$$

and assume that there exists $m \in \mathbb{N}$ such that

$$\ker T \cap R_m = \ker T \cap R_{m+k} \quad \forall\, k \in \mathbb{N}. \tag{2.2.105}$$

Then

$$T(R_\infty) = R_\infty. \tag{2.2.106}$$

Proof. In one direction, start by picking $y \in T(R_\infty)$, hence $y = Tx$ for some $x \in R_\infty$. The latter membership implies that $x \in \operatorname{ran}(T^j)$ for each $j \in \mathbb{N}$. As a consequence, $Tx \in \operatorname{ran}(T^{j+1}) = R_{j+1}$ for each $j \in \mathbb{N}$ and since clearly $Tx \in \operatorname{ran}(T) = R_1$, it follows that $y = Tx \in R_\infty$. This establishes the left-to-right inclusion in (2.2.106).

To show the right-to-left inclusion in (2.2.106), pick now $y \in R_\infty$ and let $m \in \mathbb{N}$ be as in the statement. Then for each $k \in \mathbb{N}$ we have $y \in R_{m+k}$, and consequently there exists $x_k \in X$ such that

$$y = T^{m+k} x_k. \tag{2.2.107}$$

Going further, introduce

$$y_k := T^m x_1 - T^{m+k-1} x_k, \qquad \forall\, k \in \mathbb{N}. \tag{2.2.108}$$

Upon recalling the definition of the subspaces R_j, $j \in \mathbb{N}$, from (2.2.104) it follows that $T^m x_1 \in R_m$ and $T^{m+k-1} x_k = T^m(T^{k-1} x_k) \in R_m$, for each $k \in \mathbb{N}$. Using this and

(2.2.108) we may therefore conclude that

$$y_k \in R_m, \quad \forall\, k \in \mathbb{N}. \tag{2.2.109}$$

From (2.2.108) and the definition of the x_j's we also deduce that

$$Ty_k = T^{m+1}x_1 - T^{m+k}x_k = y - y = 0, \tag{2.2.110}$$

i.e., $y_k \in \ker T$. Combining this with (2.2.109), the hypothesis (2.2.105), and the nesting property of the subspaces R_j's, we obtain that

$$y_k \in \ker T \cap R_m = \ker T \cap R_{m+k} \subseteq R_{m+k} \subseteq R_{m+k-1}, \quad \forall\, k \in \mathbb{N}. \tag{2.2.111}$$

This and (2.2.108) yield that $T^m x_1 \in R_{m+k-1}$ for each $k \in \mathbb{N}$. Since in addition we clearly have that $T^m x_1 = T^j(T^{m-j}x_1) \in R_j$ for each $j \in \{1, \dots, m\}$ it follows that $T^m x_1 \in R_\infty$ and that

$$y = T^{m+1}x_1 = T\,(T^m x_1) \in T(R_\infty), \tag{2.2.112}$$

as desired. This completes the proof of the right-to-left inclusion in (2.2.106) and finishes the proof of the proposition. □

An immediate consequence of Proposition 2.2.11 is presented next.

Theorem 2.2.12. *Let X be a vector space and let T be a linear operator on X. If with the operator T one associates $R_\infty := \bigcap_{j\in\mathbb{N}} \operatorname{ran} T^j$, then the following implication holds:*

$$\dim(\ker T) < \infty \implies T(R_\infty) = R_\infty. \tag{2.2.113}$$

Proof. The key observation is that since the subspaces R_j, labeled by $j \in \mathbb{N}$, are nested, whenever $\dim(\ker T) < \infty$ the sequence

$$\Big\{\dim\big(\ker T \cap R_j\big)\Big\}_{j\in\mathbb{N}} \subseteq \mathbb{N}, \tag{2.2.114}$$

is decreasing. Thus, this must be constant from a certain rank on, i.e., there exists $m \in \mathbb{N}$ such that (2.2.105) holds. Then Proposition 2.2.11 applies and yields the desired conclusion. □

2.3 Advanced Riesz Theory

The main result of this section is that on linear topological spaces, compact perturbations of the identity behave like linear operators on finite-dimensional vector spaces in that their range and null-space sequences stabilize. After proving this remarkable fact we investigate its consequences for the spectral properties of compact operators. As a finale, we show that for $0 < p < 1$ the p-Banach space $L^p([0,1])$, which we previously observed (in part *(b)* of Remark 1.8.30) has trivial dual space, has an even stronger property: The only compact operator it supports is the zero operator.

Recall that $\mathcal{L}(X)$ denotes the algebra of continuous linear operators on X and $\mathcal{K}(X)$ the ideal of compact operators. We begin with a simple observation about compact perturbations of the identity.

Proposition 2.3.1. *If T is a compact perturbation of the identity on a linear topological space X, then so is T^k for every non-negative integer k.*

Proof. This is, of course, trivial if $k = 0$. For arbitrary powers $k \in \mathbb{N}$ the Binomial Theorem shows that $T^k = I + KA$, where $A \in \mathcal{L}(X)$. Now KA is compact since by part *(b)* of Theorem 2.1.3 we know that the compact operators form a right ideal in $\mathcal{L}(X)$, and the result follows. □

Corollary 2.3.2. *Let X be a Hausdorff linear topological space and T be a compact perturbation of the identity on X. Then, for every non-negative integer k,*

(a) $R_k = \operatorname{ran} T^k$ *is closed in X;*
(b) $N_k = \ker T^k$ *is finite dimensional.*

Proof. Assertions *(a)* and *(b)* follow readily from Proposition 2.3.1 and Theorem 2.1.22. □

The range-stabilization result requires a variant of the Lemma 1.6.50. This variant gives new information only in case the ambient X is *not* locally bounded (see Theorem 1.8.40), so a good strategy for the first reading might be to assume X is locally bounded, and proceed directly to the stabilization results.

Lemma 2.3.3 (Second Riesz). *Suppose T is a linear operator on a linear topological space X that is a bounded perturbation of the identity; that is, $T = I - K$ for some bounded operator K on X. Let V be a zero-neighborhood in X such that $K(V)$ is bounded. If Y is a proper closed subspace of X that contains the range of T, then there exists a vector x in V with $(x + Y) \cap \frac{1}{2}V = \varnothing$.*

Proof. Proceeding as in the proof of the First Riesz (Lemma 1.6.50), note that we may rephrase the conclusion by: *there exists $x \in V$ such that $x \notin Y + \frac{1}{2}V$*. Arguing by contradiction as in the proof of the First Riesz, if no such vector x exists then $V \subseteq Y + \frac{1}{2}V$. Hence upon iteration:

$$V \subseteq Y + 2^{-n}V, \qquad \forall\, n \in \mathbb{N} \cup \{0\}. \tag{2.3.115}$$

Since $Y \supseteq \operatorname{ran} T$, the subspace Y is T-invariant, and therefore also K-invariant. So upon applying K to both sides of (2.3.115) we obtain for all $n \in \mathbb{N} \cup \{0\}$

$$K(V) \subseteq K(Y) + 2^{-n}K(V) \subseteq Y + 2^{-n}K(V). \tag{2.3.116}$$

Since the set $K(V)$ is *bounded*, it is absorbed by every zero-neighborhood in X. Consequently, given any zero-neighborhood W in X we have $2^{-n}K(V) \subseteq W$ for some $n \in \mathbb{N}$. Thus (2.3.116) above implies that

$$K(V) \subseteq \bigcap_{W \in \mathcal{N}(0;\tau_X)} (Y + W) \tag{2.3.117}$$

where τ_X is the linear topology on X. This, along with Corollary 1.6.32, shows that $K(V)$ is contained in the closure of Y. But Y is closed, so $K(V) \subseteq Y$.

We claim that V itself is contained in Y. This will complete the proof, since V is absorbing (being a zero-neighborhood; cf. part *(a)* in Theorem 1.6.18), hence any subspace that contains it must be all of X by part *(a)* of Theorem 1.6.18, contrary to the fact that Y is assumed to be a *proper* subspace of X. Indeed, the fact that $V \subseteq Y$ follows from $K(V) \subseteq Y$, the fact that $I = T + K$, and the hypothesis that $T(X) \subseteq Y$, by writing:

$$V = (T + K)(V) \subseteq T(V) + K(V) \subseteq Y + Y = Y. \tag{2.3.118}$$

The proof of Lemma 2.3.3 is therefore complete. □

Proposition 2.3.4. *Let X be a Hausdorff linear topological space and $T : X \to X$ be a compact perturbation of the identity on X. Then the range sequence of T stabilizes.*

Proof. Suppose by contradiction that the range sequence $(R_k : k \geq 0)$ of T does *not* stabilize, and therefore strictly decreases forever. We shall show that the operator $K := I - T$ is not compact. This will certainly be the case if some R_k is not closed (cf. Corollary 2.3.2). For that reason, suppose every R_k is closed.

If the image under K of every zero-neighborhood V in X is not bounded then, by Definition 2.1.1 and part *(c)* of Theorem 1.6.40, the operator K cannot be compact on X so, again, we are done. There remains to show that

$$\begin{gathered} K(V) \text{ is not a relatively compact set in } X \\ \text{for every zero-neighborhood } V \text{ in } X \text{ such that } K(V) \text{ is bounded.} \end{gathered} \tag{2.3.119}$$

With this goal in mind, fix a zero-neighborhood V in X with the property that $K(V)$ is bounded in X. Upon recalling Proposition 1.6.19 and Proposition 1.6.39, we may assume, without loss of generality, that V is circled. Fix a non-negative integer k. Since $K(V)$ is bounded in X, it is immediate (from definitions and Proposition 1.6.39) to see that $K(V \cap R_k)$ is bounded in R_k. This permits us to apply Lemma 2.3.3 with the T-invariant subspace R_k in place of X, the proper closed subspace R_{k+1} in place of Y, and the zero-neighborhood $V \cap R_k$ of R_k. There results a vector $x_k \in X$ with

$$x_k \in V \cap R_k \text{ but } (x_k + R_{k+1}) \cap \frac{1}{2}(V \cap R_k) = \varnothing. \tag{2.3.120}$$

From (2.3.120) and the fact that $x_k \in R_k$ and $R_{k+1} \subseteq R_k$, it follows that

$$x_k \in V \cap R_k \text{ but } (x_k + R_{k+1}) \cap \frac{1}{2}V = \varnothing. \tag{2.3.121}$$

Thus the sequence $(Kx_j)_{j \in \mathbb{N}}$ lies in $K(V)$, and we claim that it has no limit point in X. To this end, consider $j, k \in \mathbb{N}$, with $j > k$, and write

$$\begin{aligned} Kx_k - Kx_j &= x_k - x_j + Tx_j - Tx_k \\ &\in x_k + R_j + R_{j+1} + R_{k+1} \\ &\subseteq x_k + R_{k+1}, \end{aligned} \tag{2.3.122}$$

thanks to the fact that $R_{j+1} \subseteq R_j \subseteq R_{k+1}$.

Thus, using (2.3.121), we infer that if $j > k$ then $Kx_k - Kx_j \notin \frac{1}{2}V$ and hence

$$Kx_k \notin Kx_j + \frac{1}{2}V. \tag{2.3.123}$$

In turn, from this we obtain

$$j \neq k \implies Kx_j \notin Kx_k - \frac{1}{2}V = Kx_k + \frac{1}{2}V, \tag{2.3.124}$$

the last equality holding because V is circled. Thus the sequence $\{Kx_j\}_{j\in\mathbb{N}}$ of $K(V)$, satisfies the hypotheses of Proposition 1.8.41, whose conclusion guarantees that $K(V)$ is not relatively compact. This shows (2.3.119) and completes the proof of Proposition 2.3.4. □

Proposition 2.3.5. *Let X be a Hausdorff linear topological space and $T : X \to X$ be a compact perturbation of the identity on X. Then the null-space sequence of T stabilizes.*

Proof. The proof is similar to the one in the proposition above. Suppose that T is a linear transformation on X for which the null-space sequence $(N_k : k \geq 0)$ increases strictly, forever. We wish to show that the operator $K := I - T$ is not compact. This will certainly be the case if some N_k is infinite dimensional (cf. Corollary 2.3.2), so we may assume that N_k is finite dimensional, for any $k \in \mathbb{N} \cup \{0\}$.

Fix V a circled zero-neighborhood in X. The goal is to prove that $K(V)$ is not relatively compact. To this end, we may assume $K(V)$ is bounded – otherwise we are finished. Consider $k \in \mathbb{N}$ and T, viewed as an operator on its invariant subspace N_{k+1}, and let $Y := N_k$. Thus Y is finite dimensional and thus by Corollary 1.8.33 we have that Y is closed. Using that $(N_j : j \geq 0)$ is strictly increasing, Y is a proper subspace of N_{k+1}. In this context, applying Lemma 2.3.3 gives the existence of $x_k \in X$ such that

$$x_k \in V \cap N_{k+1} \quad \text{but} \quad (x_k + N_k) \cap \frac{1}{2}V = \varnothing. \tag{2.3.125}$$

Now we proceed just as in the last proof. Fix a pair of non-negative integers $j < k$ and observe that

$$Kx_k - Kx_j = x_k - x_j + Tx_j - Tx_k \tag{2.3.126}$$

$$\in x_k + N_{j+1} + N_j + N_k \tag{2.3.127}$$

$$\subseteq x_k + N_k + N_k + N_k \tag{2.3.128}$$

$$= x_k + N_k. \tag{2.3.129}$$

Thus, if $j < k$, our choice of x_k, guarantees that

$$Kx_k \notin Kx_j + \frac{1}{2}V. \tag{2.3.130}$$

Once again the circledness of V makes the restriction $j < k$ irrelevant, so the sequence $\{Kx_j\}_{j\in\mathbb{N}}$ satisfies the assumptions of Proposition 1.8.41, whose conclusion ensures that $K(V)$ is not relatively compact in X. □

Theorem 2.3.6 (Riesz Theory). *If T is a compact perturbation of the identity on a Hausdorff linear topological space X, then:*

(a) The operator T is a relatively open map (in the sense of Definition 1.8.14) with finite-dimensional kernel and closed range. Also, $\ker T$ *is topologically complemented in X, and* $\operatorname{ran} T$ *has finite codimension, so it too is topologically complemented in X.*
(b) Each generalized range space $R_j := \operatorname{ran} T^j$, with $j \in \mathbb{N} \cup \{0\}$, is closed and of finite codimension in X, and hence topologically complemented in X. Also, each generalized kernel $N_j := \ker T^j$, with $j \in \mathbb{N} \cup \{0\}$, is finite dimensional, closed, and topologically complemented in X.
(c) The range and null-space sequences of the operator T both stabilize at the same index $k \in \mathbb{N} \cup \{0\}$, for which we have the topological direct sum decompositions

$$X = R_k \oplus N_k \ \ \textit{as well as} \ \ T(X) = T(R_k) \oplus T(N_k) = R_k \oplus T(N_k). \tag{2.3.131}$$

(d) For any $j \geq k$, where k is the index of stabilization in (c), the following are true: T is a linear homeomorphism of R_j onto itself, $T^i(N_j) = N_{j-i} \cap R_i$, and $R_i = \operatorname{ran} T^i = R_j \oplus (N_{j-i} \cap R_i)$, for every $0 \leq i \leq j$.
(e) The Fredholm Alternative holds: T is one-to-one if and only if T is onto (that is, $\operatorname{ran} T = X$*).*

Proof. The relative openness of T, the finite-dimensionality of its kernel, and the closedness of its range were worked out in Theorem 2.1.22; this is relevant for part *(a)*. Next, Corollary 2.3.2 implies that R_j is closed in X and

$$\dim(N_j) < \infty, \qquad \forall\, j \in \mathbb{N} \cup \{0\}. \tag{2.3.132}$$

This is relevant for part *(b)*.

Turning our attention to part *(c)*, the stabilization statement follows from Proposition 2.3.4, Proposition 2.3.5, and Theorem 2.2.6. Theorem 2.2.6 also establishes that $X = N_k \boxplus R_k$. Using (2.3.132) and Theorem 1.8.36 (the Finite-Codimension

Theorem) yield the topological direct sum decomposition $X = R_k \oplus N_k$. To obtain the other topological direct sum decompositions promised in *(c)*, start by applying the operator T to $X = N_k \boxplus R_k$ to obtain $T(X) = T(N_k) + T(R_k)$, and note that $T(R_k) = R_{k+1} = R_k$. Since we also know that $T(N_k) \subseteq N_{k-1} \subseteq N_k$ and $N_k \cap R_k = \{0\}$, it follows that

$$T(X) = T(R_k) \boxplus T(N_k) = R_k \boxplus T(N_k). \tag{2.3.133}$$

Finally, since N_k is finite dimensional, so is $T(N_k)$. The idea is now to invoke Theorem 1.8.36 (the Finite-Codimension Theorem) with the Hausdorff linear topological space $T(X)$ playing the role of ambient, R_k being the closed linear subspace of $T(X)$ of finite codimension (with the latter property implied by Exercise 1.3.37), and $T(N_k)$ playing the role of the algebraic complement of R_k in $T(X)$. Ultimately, this shows that $T(X) = T(R_k) \oplus T(N_k) = R_k \oplus T(N_k)$. This completes the proof of part *(c)*.

Going further, that $\ker T$ is topologically complemented in X follows from the fact that it is a topologically complemented subspace of N_k (by finite dimensionality and Theorem 1.8.36), part *(c)* and Proposition 1.8.23. The finite codimensionality of $\operatorname{ran} T$ follows from *(c)*, the fact that $\operatorname{ran} T \supseteq R_k$, and Exercise 1.3.32. This finishes the treatment of part *(a)*.

If $j \geq k$, where $k \in \mathbb{N} \cup \{0\}$ is the stabilization index for the range and the null-space sequences, then by *(c)* and (2.3.132), it follows that R_j has finite codimension in X. When $j \in \mathbb{N} \cup \{0\}, j < k$, then $R_j \supseteq R_k$, and by Exercise 1.3.32 it follows that R_j has finite codimension in X. Moving on, since each R_j, for $j \in \mathbb{N} \cup \{0\}$, is closed and has finite-dimensional algebraic complement in X, by Theorem 1.8.40, we obtain that R_j is topologically complemented in X. Also, since (2.3.132) holds, Corollary 1.8.33 gives that N_j is closed for every $j \in \mathbb{N} \cup \{0\}$. By *(c)*, it immediately follows that N_j is topologically complemented in X, for any $j \geq k$. If $j < k$, then N_j is topologically complemented in N_k (by finite-dimensionality and Theorem 1.8.36), whereas N_k is topologically complemented in X by *(c)*. Therefore, invoking Proposition 1.8.23, we obtain that N_j is topologically complemented in X, as desired. This completes the proof of part *(b)*.

As regards part *(d)*, fix $j \geq k$ and abbreviate $T_j := T|_{R_j}$. That is (cf. (2.1.1)-(2.1.2)), define $T_j : R_j \to R_j$ by setting $T_j x := Tx$ for each $x \in R_j$. Given that $R_j = R_k$, part *(a)* in Theorem 2.2.6 gives that T is an isomorphism of R_j onto itself. Since its restriction to R_j is precisely T_j, it follows that $T_j : R_j \to R_j$ is an isomorphism. Moreover, the fact that $T : X \to X$ is continuous implies that T_j is also continuous. If we now set $K_j := I_j - T_j : R_j \to R_j$, where I_j is the identity on R_j, then $K_j x = (I - T)x$ for each $x \in R_j$ is a compact operator on R_j (since R_j is closed in X and $I - T$ is compact on X; see part *(d)* in Theorem 2.1.3). Granted this, Theorem 2.1.20 (stating that compact perturbations of identity on Hausdorff linear topological space are relatively open) applies and gives that T_j is a relatively open mapping on R_j. This further implies that T_j^{-1} is continuous, so $T_j : R_j \to R_j$ is actually a linear homeomorphism. Thus, T is a linear homeomorphism of R_j onto itself.

To proceed, assume $0 \leq i \leq j$. In view of Exercise 2.2.1 and Exercise 2.2.7 we are left with showing that the algebraic direct sum $R_i = R_j \boxplus (N_{j-i} \cap R_i)$, for each $0 \leq i \leq j$, is in fact topological. This, in turn, follows from *(b)*, the fact that $R_j \subseteq R_i$, for $0 \leq i \leq j$, and Theorem 1.8.36.

Finally, part *(e)* follows directly from the stabilization of range and null-space sequences, and Theorem 2.2.5. □

2.3.1 Operator Direct Sums

The decomposition in part *(c)* of Theorem 2.3.6 splits X topologically into the direct sum of closed subspaces R_k and N_k, both of which are invariant subspaces[1] of T. In this case we say the operator T is the *topological direct sum* of its restrictions to the two invariant subspaces. We have already introduced the algebraic version of this concept in Definition 2.2.9. For the record, here is a formal definition:

Definition 2.3.7. *Suppose $T \in \mathcal{L}(X)$, with X a linear topological space that decomposes as a topological direct sum $X = Y \oplus Z$, where both factors Y and Z that are T-invariant linear subspaces. In this scenario, the subspaces Y and Z are said to* `reduce` *T, and T is called the* `topological direct sum` *of its restrictions to Y and Z. If A and B, respectively, stand for these restrictions, this shall be denoted as $T = A \oplus B$.*

Note that in Theorem 2.3.6, the direct sum decomposition of T relative to the decomposition $X = R_k \oplus N_k$ of X splits T into the topological direct sum of a linear homeomorphism on R_k (cf. item *(d)* in Theorem 2.3.6) and an operator that is *nilpotent of index k* on N_k (cf. Definition 2.2.8). However, in the topological direct sum decomposition of X into $\ker T$ and a complementary subspace (always possible by part *(a)* of Theorem 2.3.6) it may *not* be possible to find a T-invariant complement to $\ker T$.

We shall discuss next an "external" version of the topological direct sum which provides a useful way of constructing examples.

Definition 2.3.8. *Given two linear topological spaces X_1 and X_2 along with two linear operators $T_1 \in \mathcal{L}(X_1)$ and $T_2 \in \mathcal{L}(X_2)$, write $X_1 \oplus X_2$ for the topological Cartesian product of the sets X_1, X_2 (a small abuse of notation) and define*

$$\begin{aligned} &T_1 \oplus T_2 : X_1 \oplus X_2 \longrightarrow X_1 \oplus X_2 \;\; \textit{acting according to} \\ &(T_1 \oplus T_2)(x_1, x_2) := (T_1 x_1, T_2 x_2) \;\; \textit{for all} \;\; (x_1, x_2) \in X_1 \oplus X_2. \end{aligned} \tag{2.3.134}$$

This is clearly a continuous linear operator on the product space. The "external" notion of topological direct sum introduced in (2.3.134) can be reconciled with the

[1] Recall that an invariant subspace of T is a linear subspace that is mapped by T into (not necessarily onto) itself.

previously defined "internal" one by identifying X_1 with $X_1 \oplus \{0\}$ in the product space, identifying T_1 with the operator $T_1 \oplus 0$, and making similar identifications relative to the subscript "2." Of course these direct sum constructions—both "internal" and "external" –can be carried out with any finite number of factors. We leave the details to the reader.

2.3.2 Invertibility and Spectrum

Let X be a linear topological space, and denote by I the identity operator on X. Recall that an operator $T \in \mathcal{L}(X)$ is said to be `invertible` if it is a linear homeomorphism of X onto itself.

Definition 2.3.9. *Let X be a linear topological space. The* `spectrum` *of $T \in \mathcal{L}(X)$, denoted by $\sigma(T; X)$, is the set of complex numbers λ such that $\lambda I - T$ is not invertible*[2] *on X.*

Finally, define the `resolvent` *of $T \in \mathcal{L}(X)$ as the complement of its spectrum:*

$$\rho(T; X) := \mathbb{C} \setminus \sigma(T; X) = \{\lambda \in \mathbb{C} : \ \lambda I - T \ \ \text{is invertible}\}. \tag{2.3.135}$$

In the context of the above definition, we wish to notice that if $\lambda \in \mathbb{C}$ is an eigenvalue of T, then $\lambda \in \sigma(T; X)$.

Since a compact operator on an infinite-dimensional Hausdorff linear topological space is not invertible (cf. item *(e)* in Theorem 2.1.3), the spectrum of every such operator contains zero. Upon uniting the Fredholm Alternative (part *(e)* of Theorem 2.3.6) with the Global Eigenvalue Theorem of §2.1 (see Theorem 2.1.9), we obtain the following description of the rest of the spectrum.

Theorem 2.3.10. *Suppose K is a compact operator on an infinite-dimensional Hausdorff linear topological space X. Then the spectrum of K consists of zero, and a (possibly empty) set of finite-multiplicity eigenvalues that is either finite or forms a sequence that converges to zero.*

Proof. By item *(e)* in Theorem 2.1.3, $\lambda = 0$ is in the spectrum of K. Suppose next $\lambda \neq 0$ is not an eigenvalue of K, and let $T := \lambda I - K$. Since T is a non-zero scalar multiple of a compact perturbation of the identity, by part *(a)* of Theorem 2.3.6 we may deduce that T is a relatively open mapping, to which the Fredholm Alternative (part *(e)* of Theorem 2.3.6) applies. Specifically, T is one-to-one (here we used that λ is not an eigenvalue of K), ergo T must map X onto itself. Thus, T is an open, one-to-one linear map of X onto itself, i.e., it is invertible. This shows that λ is not in the spectrum of K.

Thus the non-zero part of the spectrum of T can consist only of eigenvalues. The rest of the theorem now follows immediately from the Finite-Multiplicity Theorem

---

[2] That is, there is no $R \in \mathcal{L}(X)$ such that $R(\lambda I - T) = (\lambda I - T)R = I$

and the Global Eigenvalue Theorem (Theorem 2.1.8 and Theorem 2.1.9, respectively). □

Exercise 2.3.11. Weyl's Theorem again. Show that if two bounded linear operators defined on a Hausdorff linear topological space differ by a compact operator, then their spectra differ by at most finite-multiplicity eigenvalues.

Hint: This is a restatement of Exercise 2.1.23.

We shall return to the task of further elucidating the nature of the spectrum linear continuous operators (introduced in Definition 2.3.9) in §3.5.1.

Chapter 3
Fredholm Theory in Topological Vector Spaces

Abstract In this chapter we explore the algebraic and topological properties of a remarkable class of operators, whose definition is inspired by the very foundations of Riesz theory. Our starting point is to clarify what constitutes a natural notion of a Fredholm operator on a Hausdorff linear topological space. This task is undertaken in §3.1, building on insights from our earlier work on Riesz theory. Central to this discussion is the Fundamental Fredholm Decomposition (Theorem 3.1.7), which asserts that a Fredholm operator is essentially a linear homeomorphism augmented by finite-dimensional behavior. This perspective permeates the subject and is particularly evident in Theorem 3.1.10, the analogue of the classical Atkinson Theorem for topological vector spaces. In this sense, it may rightly be regarded as the Fundamental Theorem of Fredholm Theory in the setting of topological vector spaces.

In §3.2 we introduce and study the Calkin algebra of continuous linear operators modulo compact operators on topological vector spaces. The Fredholm index is then introduced in §3.3, with the goal of demonstrating its striking algebraic and topological properties, which deepen our understanding of linear operators and their spectra. We begin with the case of index zero in §3.3.1, followed by the scenario of negative index in §3.3.2. The crucial additivity of the index for Fredholm operators on topological vector spaces is stated in §3.3.4. This result is highly consequential, and immediate applications – such as the invariance of the Fredholm index under compact perturbations – are explored in §3.3.5. The proof of the additivity theorem is completed in §3.3.6.

The continuity of the Fredholm index is examined in §3.4. Here we restrict attention to the most commonly occurring class of linear topological spaces: the p-Banach spaces, with $0 < p \leq 1$. For such spaces, Banach-space techniques provide a "p-norm topology" for the space of continuous linear operators. Within this setting, we show that the Fredholm index, viewed as a mapping from the space of Fredholm operators (equipped with the operator norm topology) into the integers with the discrete topology, is continuous. In fact, we prove a slightly stronger but equivalent statement: Fredholm operators of a fixed index form an open subset of the space of

D. Mitrea et al., *Fredholm Theory in Topological Vector Spaces*, Springer Monographs in Mathematics, https://doi.org/10.1007/978-3-032-24738-4_3

continuous linear operators. Significantly, we also provide an example showing that continuity of the index may fail for Fredholm operators on Fréchet spaces. In §3.5 we apply Fredholm theory to the study of the spectrum of a linear operator. In §2, dealing with Riesz theory in topological vector spaces, we established that the spectrum of a compact operator on an infinite-dimensional Hausdorff linear topological space consists of the origin together with a (possibly empty) set of finite-multiplicity eigenvalues which, if infinite, form a sequence converging to zero. This basic spectral configuration motivates the pursuit of further spectral variants here.

Our efforts culminate in §3.6, where we develop a general Fredholm theory applicable to operators between two (possibly distinct) linear topological spaces. The treatment in this chapter has thus far focused on the "diagonal" case, where a linear operator acts from a space into itself. However, it is of considerable interest to consider the "off-diagonal" case, allowing operators to act between different spaces. Extending Fredholm theory to this broader setting presents new challenges: in particular, operators may no longer be composable with themselves, a key feature in earlier proofs. Our approach is to use the previously developed theory as a stepping stone, showing that it essentially self-extends from the diagonal to the off-diagonal case via a matrix formalism. The core results are established in §3.6.1 for linear topological spaces and in §3.6.2 for quasi-Banach spaces, while §3.6.3 gathers various auxiliary results related to the development of a semi-Fredholm theory on quasi-Banach spaces.

3.1 Fredholm Operators

Recall from the last chapter that if X is a Hausdorff linear topological space and $T \in \mathcal{L}(X)$ is a compact perturbation of the identity, then:

(i) $\operatorname{ran} T$ *is a closed subspace of* X *that has finite codimension in* X.
(ii) $\ker T$ *is finite dimensional and topologically complemented in* X.
(iii) T *is a relatively open map (i.e.,* $T(V)$ *is open in* $\operatorname{ran} T$ *whenever* V *is open in* X*).*

Inspired by the format of this result, we make the following definition:

Definition 3.1.1. *Given a Hausdorff linear topological space* X*, call* $T \in \mathcal{L}(X)$ *a* `Fredholm operator` *on* X *provided it verifies the following three axioms:*

(F1) $\operatorname{ran} T$ *is a closed subspace of* X *that has finite codimension in* X.
(F2) $\ker T$ *is finite dimensional and topologically complemented in* X.
(F3) T *is a relatively open map (i.e.,* $T(V)$ *is open in* $\operatorname{ran} T$ *whenever* V *is open in* X*).*

For each Fredholm operator T *on* X*, define the* `Fredholm index` *(or simply* `index`*) of* T *by setting*

$$i(T) := \dim(\ker T) - \operatorname{codim}(\operatorname{ran} T) \in \mathbb{Z}. \tag{3.1.1}$$

In relation to this definition, some comments are in order.

Remark 3.1.2. Regarding condition *(F2)* recall that a finite-dimensional subspace of a Hausdorff linear topological space must be closed (see Corollary 1.8.33), but it need not always be topologically complemented (cf. part *(b)* of Remark 1.8.30). On the other hand, if X has enough continuous linear functionals to separate points (cf. Definition 1.6.12; this is the case if, e.g., X is a normed space, or more generally, X is a Hausdorff locally convex topological vector space – cf. Theorem 1.12.2) then every finite-dimensional subspace *is* topologically complemented (cf. part *(c)* of Remark 1.8.30, and Corollary 1.12.3).

Regarding condition *(F3)*, we wish to note that for F-spaces the Open Mapping Theorem (recalled in Theorem 1.6.7) provides this relative openness property once we know the range is closed.

Example 3.1.3. Let X be a Hausdorff linear topological space. The reader will have no trouble checking that the following classes of operators in $\mathcal{L}(X)$ are Fredholm:

(a) Compact perturbations of the identity –in particular *every* linear mapping from X into X if $\dim X < +\infty$.
(b) Invertible operators.
(c) Direct sums (Definition 2.3.8) of Fredholm operators.
(d) The forward and backward shifts of Example 2.2.4.
(e) Restrictions of Fredholm operators to *reducing* subspaces (see Definition 2.3.7).

In relation to item *(e)* of Example 3.1.3, it is illuminating to point out that the restriction of a Fredholm operator to an *invariant* subspace need not be Fredholm; see below.

Exercise 3.1.4. Give an example of a Fredholm operator on a Hilbert space whose restriction to some closed invariant subspace fails to be Fredholm.

Hint: Let H be an infinite-dimensional Hilbert space, and consider the "two-sided" Hilbert space $\ell^2_{\mathbb{Z}}(H)$, consisting of all H-valued sequences $x = (\xi_n)_{n\in\mathbb{Z}}$ with $\|x\| := \big(\sum_{n\in\mathbb{Z}} \|\xi_n\|_H^2\big)^{1/2}$. Let T denote the "forward shift" $(Tx)_n := \xi_{n-1}$ for each $x = (\xi_m)_{m\in\mathbb{Z}} \in \ell^2_{\mathbb{Z}}(H)$ and $n \in \mathbb{Z}$. Let M denote the subspace of $\ell^2_{\mathbb{Z}}(H)$ consisting of those $x = (\xi_n)_{n\in\mathbb{Z}} \in \ell^2_{\mathbb{Z}}(H)$ for which $\xi_n = 0$ whenever $n < 0$. Hence, M is a closed subspace of $\ell^2_{\mathbb{Z}}(H)$, which is invariant under the action of T. Show that T is an invertible (hence Fredholm) operator on $\ell^2_{\mathbb{Z}}(H)$, but that its restriction to the closed invariant subspace M is not Fredholm. More specifically, the range of $T : M \to M$ consists of those sequences in M with zero 0-th coordinate, which is a closed subspace of codimension $\dim H$. The quotient $M/\operatorname{ran}(T|_M)$ is naturally isomorphic to H via the map $x \mapsto \xi_0$. Since H is infinite-dimensional, $\dim \operatorname{coker}(T|_M) = \dim H = \infty$. Therefore $T : M \to M$ fails the finite-codimension requirement and is not Fredholm. (For comparison, in the scalar case when $H := \mathbb{C}$, the unilateral shift has range of codimension 1 and, hence, is Fredholm of index -1.)

Some elementary but useful insights are collected below.

Exercise 3.1.5. Show that:

(a) The forward and backward shifts of Examples 2.2.4 can be defined on other sequence spaces (e.g., ℓ^p for any value of p). Show that they are Fredholm operators on any "reasonable" sequence space (where the definition of "reasonable" is up to you). Give some examples.
(b) Which diagonal operators on ℓ^2 as in Proposition 2.1.6 (or, more generally, on any "reasonable" sequence space) are Fredholm?
(c) Which projections on a Hausdorff linear topological space are Fredholm?

Proposition 3.1.6. *Let X be a Hausdorff topological vector space and let $Y \subseteq X$ be a closed subspace of finite codimension in X. Let Z be a topological complement of Y, i.e., a finite-dimensional subspace of X with the property that*

$$X = Y \oplus Z \tag{3.1.2}$$

(with item (d) of Theorem 1.8.1 guaranteeing the existence of such a space Z). Finally, denote by π the projection of X onto Y, i.e.,

$$\begin{aligned} &\pi : X \longrightarrow X, \quad \pi(x) := y \ \text{ whenever} \\ &x = y + z \ \text{ with } \ y \in Y \ \text{ and } \ z \in Z. \end{aligned} \tag{3.1.3}$$

Then the operator $\pi : X \to X$ is a Fredholm operator on X and $i(\pi) = 0$.

Proof. Clearly,

$$\operatorname{ran} \pi = Y \ \text{ and } \ \ker \pi = Z. \tag{3.1.4}$$

Proposition 1.8.25 then ensures that π is a relatively open mapping. Also $\ker \pi = Z$ is finite dimensional and topologically complemented (cf. (3.1.2)), while $\operatorname{ran} \pi = Y$ is closed and has finite codimension in X by hypothesis. Consequently, π is a Fredholm operator on X (cf. Definition 3.1.1). Finally, based on the definition of the index, (3.1.4), and (3.1.2) we may compute

$$\begin{aligned} i(\pi) &= \dim(\ker \pi) - \operatorname{codim}(\operatorname{ran} \pi) \\ &= \dim Z - \operatorname{codim} Y = 0, \end{aligned} \tag{3.1.5}$$

thanks to (3.1.2) and (1.3.179). □

Theorem 3.1.7 (Fundamental Fredholm Decomposition). *If T is a Fredholm operator on a Hausdorff linear topological space X then there exist subspaces Y and M of X, where Y is closed and of finite codimension, and M is finite dimensional (hence closed), such that:*

(a) $X = Y \oplus \ker T$, and
(b) $X = \operatorname{ran} T \oplus M$.

In addition, if Y and ran T *are equipped with the relative topology induced by X, they become Hausdorff linear topological spaces themselves and T maps Y homeomorphically onto* ran T.

Proof. This is mostly a restatement of Definition 3.1.1. Since $\operatorname{ran} T$ is closed and has finite codimension, it is topologically complemented in X (cf. Theorem 1.8.36), so M can be taken to be any topological complement of $\operatorname{ran} T$. Similarly we can take Y to be any topological complement of $\ker T$ (which exists by *(F2)*). These subspaces are automatically closed in X by Corollary 1.8.18. Also, Proposition 1.6.8 tells us that Y and $\operatorname{ran} T$ become Hausdorff linear topological spaces themselves when equipped with the relative topology induced by X.

It is clear from the decomposition *(a)* that T maps Y one-to-one and onto $\operatorname{ran} T$. To complete the proof, it remains to show that $T : Y \longrightarrow \operatorname{ran} T$ is open. To this end, let U be an open subset of Y. Then the decomposition *(a)* together with Proposition 1.8.13 imply that $U + \ker T$ is an open subset of X, and *(F3)* guarantees that $T(U) = T(U + \ker T)$ is open in $\operatorname{ran} T$. □

Comment 3.1.8 *For F-spaces, the continuity of the inverse of $T : Y \to \operatorname{ran} T$ in the proof above follows from the Open Mapping Theorem (recalled in Theorem 1.6.7) because* $\operatorname{ran} T$ *is itself an F-space.*

If X is a finite-dimensional linear topological space, then every linear operator on X is Fredholm, and part *(a)* of the result above translates into a fundamental result of linear algebra:

Corollary 3.1.9. *If X is a finite-dimensional vector space and T is a linear mapping from X into X then*

$$\dim X = \dim(\ker T) + \dim(\operatorname{ran} T). \tag{3.1.6}$$

The Fundamental Fredholm Decomposition (cf. Theorem 3.1.7) tells us that: *A Fredholm operator is a linear homeomorphism to which some finite-dimensional behavior has been added.* This point of view pervades the subject. It is particularly evident in the next result, which should be viewed as the "Fundamental Theorem of Fredholm Theory."

Theorem 3.1.10 (Atkinson's Theorem). *For X a Hausdorff linear topological space and $T \in \mathcal{L}(X)$, the following five statements are equivalent:*

(a) T is a Fredholm operator (cf. Definition 3.1.1).
(b) T is invertible modulo finite rank projections in the sense that there exists some $S \in \mathcal{L}(X)$ such that both $I - ST$ and $I - TS$ are finite-rank projections.
(c) T is invertible modulo finite rank operators in the sense that there exists $S \in \mathcal{L}(X)$ such that both $I - ST$ and $I - TS$ have finite-dimensional range.
(d) T is invertible modulo compact operators in the sense that there exists $S \in \mathcal{L}(X)$ such that both $I - ST$ and $I - TS$ are compact.
(e) There exist $S_1, S_2 \in \mathcal{L}(X)$ such that both $I - S_1T$ and $I - TS_2$ are compact.

Moreover, if $T \in \mathcal{L}(X)$ is Fredholm then any operators $S, S_1, S_2 \in \mathcal{L}(X)$ doing the job in items (b)-(e) above are also Fredholm.

Proof. *(a)*$\Rightarrow$*(b)*: Suppose T is Fredholm. We use part *(b)* of Theorem 3.1.7 to define $S : X \to X$ as follows:

1. On $\operatorname{ran} T$ we define the operator S to be the inverse of the linear homeomorphism $T : Y \to \operatorname{ran} T$.
2. On M we set $S \equiv 0$.

Since S is continuous on both $\operatorname{ran} T$ and M, it is continuous on their topological direct sum (cf. Proposition 1.8.13) which is X. Moreover, ST is the identity on Y, and is $\equiv 0$ on $\ker T$, so it is the natural projection of X onto Y. This makes $I - ST$ be the natural projection of X onto the finite-dimensional subspace $\ker T$ and in particular a finite-rank continuous operator.

Note also that TS is the identity on $\operatorname{ran} T$, and is $\equiv 0$ on M. Thus (referring to part *(b)* of Theorem 3.1.7) TS is the natural projection of X onto $\operatorname{ran} T$, hence $I - TS$ is the natural projection of X onto the finite-dimensional subspace M. These arguments establish conclusion *(b)*.

$(b) \Rightarrow (c)$: Immediate.

$(c) \Rightarrow (d)$: Trivial, by part *(c)* of Theorem 2.1.3.

$(d) \Rightarrow (a)$: Here the full strength of the Riesz Theory shall be employed. First suppose only that $I - ST$ is compact for some $S \in \mathcal{L}(X)$. Since ST is a compact perturbation of the identity, the Riesz Theory (cf. Theorem 2.3.6) gives that $\ker(ST)$ is both finite dimensional and topologically complemented. Consequently, the same is true of $\ker T$, because it is a subspace of $\ker(ST)$, hence complemented therein by finite dimensionality, and therefore, using Proposition 1.8.23, it is topologically complemented in X. Thus,

$$\text{if } I - ST \in \mathcal{K}(X) \text{ then } \ker T \text{ is finite dimensional and topologically complemented in } X. \tag{3.1.7}$$

Next suppose only that $I - TS$ is compact for some $S \in \mathcal{L}(X)$, so that now TS is a compact perturbation of the identity. According to Theorem 2.3.6, $\operatorname{ran}(TS)$ is closed in X, and of finite codimension in X. Now $\operatorname{ran} T \supseteq \operatorname{ran}(TS)$, so $\operatorname{ran} T$ also has finite codimension in X, by Exercise 1.3.32. Invoking now Proposition 1.3.46, it follows that $\operatorname{ran}(TS)$ has finite codimension in $\operatorname{ran} T$. The latter, combined with the fact that $\operatorname{ran}(TS)$ is closed in X, and the Finite-Dimensional Extension Theorem (see Theorem 1.8.35) implies $\operatorname{ran} T$ is also closed in X. In conclusion:

$$\text{if } I - TS \in \mathcal{K}(X)\text{, then } \operatorname{ran} T \text{ is a closed subspace of } X \text{ of finite codimension.} \tag{3.1.8}$$

The argument presented so far shows that if *(d)* holds, then T satisfies conditions *(F1)-(F2)* in Definition 3.1.1. There remains to prove that T satisfies condition *(F3)* in Definition 3.1.1, i.e., to show that $T : X \to X$ is a relatively open map. If X is an F-space then this follows immediately from the Open Mapping Theorem (cf. Theorem 1.6.7). In general, however, one needs additional effort—the argument below comes from the work of B. Gramsch (cf. [19]). The idea is to apply the Advanced Riesz Theory of §2.3 – particularly the stabilization of range and null-space sequences – to the operator ST, and then decode the relative openness of T from the information thus obtained about ST.

For each $k \in \mathbb{N} \cup \{0\}$, let R_k and N_k denote respectively the range and null-space of the operator $(ST)^k$. Since ST is a compact perturbation of the identity, by part *(b)* of Theorem 2.3.6, R_k is closed in X, and N_k is finite dimensional. Also, using parts *(c)* and *(d)* of Theorem 2.3.6, upon choosing k large enough we obtain the topological decomposition

$$X = R_k \oplus N_k \tag{3.1.9}$$

where

$$ST \text{ is a homeomorphism of } R_k \text{ onto itself,} \tag{3.1.10}$$

$$ST(N_k) = N_{k-1} \cap R_1, \tag{3.1.11}$$

and there is the corresponding range-decomposition

$$\operatorname{ran}(ST) = R_k \oplus (N_{k-1} \cap R_1). \tag{3.1.12}$$

Fix such an index k for the rest of the proof. We claim that

$$\operatorname{ran} T = T(R_k) \oplus T(N_k). \tag{3.1.13}$$

The fact that $\operatorname{ran} T$ is the sum of $T(R_k)$ and $T(N_k)$ is clear from (3.1.9), so to check the *algebraic* version of (3.1.13) we need only show $T(R_k) \cap T(N_k) = \{0\}$. If x belongs to $T(R_k) \cap T(N_k)$ then

$$Sx \in ST(R_k) \cap ST(N_k) \subseteq R_k \cap N_{k-1} \subseteq R_k \cap N_k = \{0\}, \tag{3.1.14}$$

where the first inclusion in (3.1.14) is justified by (3.1.10), (3.1.11), (2.2.89), the second inclusion comes from (2.2.89), and where the last equality uses (3.1.9). In turn, (3.1.14) forces $Sx = 0$. Yet S is one-to-one on $T(R_k)$ (because ST is one-to-one on R_k; cf. (3.1.10)), and since $x \in T(R_k)$ we conclude that $x = 0$, as desired.

The argument so far shows that

$$\operatorname{ran} T = T(R_k) \boxplus T(N_k). \tag{3.1.15}$$

Consequently, $T(R_k)$ has the finite-dimensional subspace $T(N_k)$ as an algebraic complement in $\operatorname{ran} T$, so (3.1.13) will follow from the Finite-Codimension Theorem (cf. Theorem 1.8.36) once we show that $T(R_k)$ is closed in $\operatorname{ran} T$. For this, we start by observing that

$$T(R_k) = T(ST)^k(X) = (TS)^k(T(X)) \supseteq (TS)^k(TS(X)) = \operatorname{ran}(TS)^{k+1}, \tag{3.1.16}$$

where the second equality uses the fact that $T(ST)^k = (TS)^k T$. By hypothesis, TS is a compact perturbation of the identity on X, so is $(TS)^{k+1}$. Thus, by using part *(b)* of Theorem 2.3.6, $\operatorname{ran}(TS)^{k+1}$ is a closed subspace of X having finite codimension, and the computation above shows that $T(R_k)$ is a finite-dimensional extension of this closed subspace. Another appeal to the Finite-Dimensional Extension Theorem (Theorem 1.8.35) shows that $T(R_k)$ is closed in X, hence also closed in $\operatorname{ran} T$ (since the latter is closed; cf. (3.1.8)). This completes the proof of (3.1.13).

With (3.1.13) in hand, we may rely on Corollary 1.8.17 (presently used with $X_0 := N_k$, $X_1 := R_k$, $Y_0 := T(N_k) = T(X_0)$, $Y_1 := T(R_k) = T(X_1)$, and $Y := X$) to establish the relative openness of the map $T : X \to X$ by proving the result separately for $T : R_k \to T(R_k)$ and $T : N_k \to T(N_k)$. Openness for the restriction of T to N_k follows from finite dimensionality: the spaces involved are linearly homeomorphic to Euclidean spaces (cf. Theorem 1.8.31), so the desired openness can be derived from the Open Mapping Theorem.

As for the action of T on R_k, recall that ST is a linear homeomorphism of R_k onto itself, by (3.1.10). Denote by S_k the restriction of S to $T(R_k)$, that is, the operator $S_k : T(R_k) \to R_k$ given by $S_k x := Sx$ for each $x \in T(R_k)$. Since $S : X \to X$ is continuous, Exercise 1.1.11 guarantees that $S_k : T(R_k) \to R_k$ is continuous. Also, since ST is an isomorphism of the space R_k onto itself (cf. (3.1.10)) we conclude that S_k is a one-to-one map of $T(R_k)$ onto R_k. Thus if V is an open subset of R_k, we know that $W := S_k T(V) = (ST)V$ is also open in R_k (because ST is a homeomorphism when acting on R_k), hence $T(V) = S_k^{-1}(W)$ is open in $T(R_k)$ by the continuity of S_k. Thus $T : R_k \to T(R_k)$ is an open map, and the proof of *(d)* ⇒ *(a)* is now complete.

At this stage, we have shown that *(a)*-*(d)* are all equivalent. There remains to include item *(e)* in this circle of equivalences. Obviously, *(d)* ⇒ *(e)*, so we focus on *(e)* ⇒ *(d)*. Suppose there exist $S_1, S_2 \in \mathcal{L}(X)$ such that $K_1 := I - S_1 T$ together with $K_2 := I - TS_2$ are compact on X. Composing the terms in the first equality with S_2 to the right and then taking into account the second equality gives

$$\begin{aligned} K_1 S_2 &= S_2 - (S_1 T) S_2 = S_2 - S_1 (T S_2) \\ &= S_2 - S_1 (I - K_2) = S_2 - S_1 + S_1 K_2 \end{aligned} \tag{3.1.17}$$

thus,

$$S_1 - S_2 = K := S_1 K_2 - K_1 S_2 \in \mathcal{K}(X), \tag{3.1.18}$$

with the membership provided by item *(b)* in Theorem 2.1.3. Consequently,

$$I - TS_1 = I - T(S_2 + K) = (I - TS_2) - TK = K_2 - TK \in \mathcal{K}(X), \tag{3.1.19}$$

again by item *(b)* in Theorem 2.1.3. Hence, we have found $S_1 \in \mathcal{L}(X)$ with the property that $I - S_1 T \in \mathcal{K}(X)$ and $I - TS_1 \in \mathcal{K}(X)$, finishing the justification of *(e)* ⇒ *(d)*.

There remains to show that, under the assumption that $T \in \mathcal{L}(X)$ is Fredholm, then any operators $S, S_1, S_2 \in \mathcal{L}(X)$ doing the job in items *(b)*-*(e)* are also Fredholm. Indeed, the current working assumption plus the fact that *(b)* ⇒ *(a)*, *(c)* ⇒ *(a)*, and *(d)* ⇒ *(a)* imply that any operator $S \in \mathcal{L}(X)$ doing the job in items *(b)* and *(c)* is also Fredholm. Let's look at $S_1, S_2 \in \mathcal{L}(X)$ doing the job in item *(d)*. Bring in $S \in \mathcal{L}(X)$ as in item *(d)* which, from what we have proved so far, is known to be a Fredholm operator on X. Running the argument that has produced (3.1.18) with S in place of S_2 then gives

$$Q := S_1 - S \in \mathcal{K}(X). \tag{3.1.20}$$

Consequently, in view of (3.1.20), the job S does in the current item *(d)*, and items *(a)-(b)* in Theorem 2.1.3, we see that

$$\begin{aligned} I - TS_1 &= I - T(S + Q) = (I - TS) - TQ \in \mathcal{K}(X) \text{ and} \\ I - S_1 T &= I - (S + Q)T = (I - ST) - QT \in \mathcal{K}(X). \end{aligned} \tag{3.1.21}$$

Granted this, the implication *(d)* $\Rightarrow$ *(a)* then proves that S_1 is indeed Fredholm. Since a similar argument shows that S_2 is also Fredholm, the proof of Theorem 3.1.10 is therefore complete. □

Exercise 3.1.11. Show that on $L^p([0, 1])$, with $0 < p < 1$, the only Fredholm operators are the invertible operators.

Hint: Recall that the space in question has no nontrivial continuous linear functionals, and that on a space with trivial dual any continuous finite-rank operator must be zero. If T is a Fredholm operator on $L^p([0, 1])$ then, according to Theorem 3.1.10, there exists a continuous operator $S : X \to X$ such that both $I - ST$ and $I - TS$ are finite-rank, hence necessarily zero. This makes T invertible.

Exercise 3.1.12. Show that no compact operator on an infinite-dimensional Hausdorff linear topological space is Fredholm.

Hint: Suppose K is compact and Fredholm on an infinite-dimensional Hausdorff linear topological space X. Then Theorem 3.1.10 guarantees the existence of a continuous operator $S : X \to X$ with the property that $KS - I$ is a continuous finite rank operator. Hence, on the one hand, $KS - I$ is a compact operator on X by item *(c)* in Theorem 2.1.3. On the other hand, KS is a compact operator on X by item *(b)* in Theorem 2.1.3. Ultimately, this makes the identity on X a compact operator since $I = KS - (KS - I)$. Use item *(e)* in Theorem 2.1.3 to obtain a contradiction.

3.2 The Calkin Algebra

Let X be a Hausdorff linear topological space and recall the algebra structure on $\mathcal{L}(X)$ where the multiplication is given by composition. Since by part *(b)* of Theorem 2.1.3 we have that $\mathcal{K}(X)$, the collection of compact operators on X, is a two-sided ideal in $\mathcal{L}(X)$, the quotient space $\mathcal{L}(X)/\mathcal{K}(X)$ is also an algebra over the original scalar field (in this case, $\mathbb{C}$). It is called the `Calkin algebra`. Since no topology is yet assumed for $\mathcal{L}(X)$, at this point we also have to regard the Calkin algebra as a purely algebraic entity.

Let π denote the quotient map, i.e., π is the algebra homomorphism of $\mathcal{L}(X)$ onto the Calkin algebra that maps operators from $\mathcal{L}(X)$ into their cosets in $\mathcal{L}(X)/\mathcal{K}(X)$:

$$\mathcal{L}(X) \ni T \longmapsto \pi(T) := T + \mathcal{K}(X) \in \mathcal{L}(X)/\mathcal{K}(X). \tag{3.2.22}$$

With this piece of notation, the equivalence *(a)* $\Leftrightarrow$ *(d)* in the statement of Atkinson's Theorem (see Theorem 3.1.10) may be restated purely in terms of the Calkin algebra:

Corollary 3.2.1. *Let X be a Hausdorff linear topological space and fix an operator $T \in \mathcal{L}(X)$. Then T is a Fredholm operator on X if and only if $\pi(T)$ is an invertible element of the Calkin algebra $\mathcal{L}(X)/\mathcal{K}(X)$.*

In turn, this corollary yields two important stability properties of the class of Fredholm operators.

Corollary 3.2.2. *Let X be a Hausdorff linear topological space. Then the following properties hold.*

(a) If T is a Fredholm operator on X then so is $T + K$ for every $K \in \mathcal{K}(X)$. In other words, the class of Fredholm operators is stable under compact perturbations.
(b) If S and T are Fredholm operators on X, then so is ST.

Proof. Since $\pi(T + K) = \pi(T)$, part *(a)* follows immediately from Corollary 3.2.1. As for part *(b)*, just observe that because the quotient map π is an algebra homomorphism, $\pi(ST) = \pi(S)\pi(T)$, so the result follows from Corollary 3.2.1 and the fact that in any algebra the product of invertible elements is invertible. □

Exercise 3.2.3. Suppose X is a Hausdorff linear topological space, and consider $S, T \in \mathcal{L}(X)$. Then the following results hold:

(a) If ST and TS are both Fredholm on X, then so are S and T.
(b) The conclusion of part *(a)* does not hold if we merely assume that ST is Fredholm.

Hint: For part *(a)*, recall from item *(e)* in Theorem 3.1.10 that there exist two operators, $Q_1, Q_2 \in \mathcal{L}(X)$, such that both $I - Q_1(ST)$ and $I - (TS)Q_2$ are compact. Thus, for $S_1 := Q_1 S \in \mathcal{L}(X)$ and $S_2 := SQ_2 \in \mathcal{L}(X)$ we see that both $I - S_1 T$ and $I - TS_2$ are compact. With this in hand, we may now appeal to item *(e)* in Theorem 3.1.10 to conclude that T is Fredholm. Prove that S is Fredholm reasoning in a similar fashion.

For part *(b)*, let H be an infinite-dimensional Hilbert space, and define ℓ^2_H to be the collection of all sequences $(x_n)_{n\in\mathbb{N}}$ of vectors in H for which $\sum_{n=1}^{\infty} \|x_n\|^2 < \infty$. Check that ℓ^2_H is a Hilbert space. Define forward and backward shifts "of infinite multiplicity" on ℓ^2_H by repeating the definitions we used for ordinary ℓ^2. Show that neither shift is a Fredholm operator. Show that the forward shift so defined is left invertible, having the backward shift as a left inverse.

3.3 Properties of the Fredholm Index

We begin the study of the remarkable integer-valued function attached to the collection of Fredholm operators as in (3.1.1). In the next few sections we show that the

Fredholm index has striking algebraic and topological properties that we can employ to further our understanding of linear operators and their spectra. Some preliminary remarks are in order.

1. Every invertible operator on a linear topological space has index zero.
2. The forward shift S on ℓ^2 has index -1, and the backward shift B on ℓ^2 has index $+1$. More generally, for n a positive integer, $i(S^n) = -n$ and $i(B^n) = n$.
3. Since the identity operator on any linear topological space has index zero, we then conclude that every integer occurs as the index of an operator on some linear topological space.
4. Somewhat less trivially,

$$\text{every linear mapping on a finite-dimensional vector space is Fredholm with index zero.} \tag{3.3.23}$$

Indeed (3.3.23) is a rephrasing of the fact that in finite-dimensional spaces the dimension of the range plus the dimension of the null-space is the dimension of the whole space (cf. Corollary 3.1.9).

The next result shows that the index interacts well with topological direct sums. Its proof is straightforward and we leave it as an exercise to the reader, but its usefulness makes us give it the status of a theorem.

Theorem 3.3.1 (Topological Direct Sum). *If T_1 and T_2 are Fredholm operators on the Hausdorff linear topological spaces X_1 and X_2, respectively, then $T_1 \oplus T_2$ (cf. Definition 2.3.8) is a Fredholm operator on the space $X_1 \oplus X_2$, and one has*

$$i(T_1 \oplus T_2) = i(T_1) + i(T_2). \tag{3.3.24}$$

In the converse direction, if X_1 and X_2 are Hausdorff linear topological spaces and $T_j \in \mathcal{L}(X_j)$ for $j = 1, 2$ are such that $T_1 \oplus T_2$ is a Fredholm operator on the space $X_1 \oplus X_2$, then T_j is Fredholm on X_j for $j = 1, 2$, and once again (3.3.24) *holds.*

Exercise 3.3.2. Supply a detailed proof for Theorem 3.3.1.

Corollary 3.3.3. *In a Hausdorff linear topological space X, every compact perturbation of the identity on X is a Fredholm operator with index zero.*

Proof. Let $T \in \mathcal{L}(X)$ be a compact perturbation of the identity on X. From Riesz Theory (Theorem 2.3.6) we know that T is a Fredholm operator on X. Theorem 2.3.6 also provides a non-negative integer k for which we have $X = R_k \oplus N_k$, where $R_k = \operatorname{ran} T^k$ and $N_k = \ker T^k$. Moreover, T is the direct sum of its restrictions to R_k and N_k, i.e.,

$$T = (T|_{R_k}) \oplus (T|_{N_k}) \quad \text{on} \quad R_k \oplus N_k = X. \tag{3.3.25}$$

The restriction to R_k is invertible, and therefore has index zero. Since N_k is a finite-dimensional space, the restriction of T to it also has index zero by (3.3.23). Therefore T has index zero by Theorem 3.3.1. □

Exercise 3.3.4. Suppose H is a Hilbert space and T a Fredholm operator on H. Show that the adjoint operator T^* is also Fredholm, and that $i(T^*) = -i(T)$.

Hint: Relate the kernel and range of T^* to the range and kernel of T.

Here is more good news: *the index is a similarity invariant.* This follows immediately from the next result.

Theorem 3.3.5 (Permanence of Index). *If T is a Fredholm operator on a Hausdorff linear topological space X, and Q is an invertible operator on X, then QT and TQ are Fredholm operators on X, satisfying $i(QT) = i(TQ) = i(T)$.*

Proof. From part *(b)* of Example 3.1.3 we know that invertible operators (on linear topological spaces) are Fredholm. Then the Fredholmness of QT and TQ follows from part *(b)* of Corollary 3.2.2. Since $\ker(TQ) = Q^{-1}(\ker T)$ and $\ker(QT) = \ker T$ we see that all the null-spaces under consideration have the same dimension. Since $\operatorname{ran}(TQ) = \operatorname{ran} T$ we obtain $i(T) = i(TQ)$.

As far as $\operatorname{ran}(QT)$ is concerned, apply Q to both sides of the fundamental decomposition $X = \operatorname{ran} T \oplus M$ (cf. Theorem 3.1.7), where M is a finite-dimensional linear subspace of X, to obtain

$$X = Q(\operatorname{ran} T) \oplus QM = \operatorname{ran} QT \oplus QM. \tag{3.3.26}$$

Since M and QM both have the same (finite) dimension, $\operatorname{ran} T$ and $\operatorname{ran}(QT)$ have the same (finite) codimension, hence $i(QT) = i(T)$. □

As a direct consequence of Theorem 3.3.5 we have the following similarity invariance of the Fredholm index:

Corollary 3.3.6. *Index is a "similarity invariant." Specifically, if T is a Fredholm operator on a Hausdorff linear topological space X, and if Q is a linear homeomorphism of X onto itself, then QTQ^{-1} is a Fredholm operator on X and $i(QTQ^{-1}) = i(T)$.*

Corollary 3.3.7 (The Generalized Fredholm Alternative). *Let X be a Hausdorff linear topological space. If $Q \in \mathcal{L}(X)$ is invertible and $K \in \mathcal{K}(X)$, then $Q + K$ is a Fredholm operator on X and $i(Q + K) = 0$. In particular, $Q + K$ is one-to-one if and only if it maps X onto itself.*

Proof. Write $Q + K = Q(I + Q^{-1}K)$. Since by part *(b)* of Theorem 2.1.3 we know that $Q^{-1}K$ is compact, Corollary 3.3.3 implies that $I + Q^{-1}K$ is a Fredholm operator on X and $i(I + Q^{-1}K) = 0$. Theorem 3.3.5 tells us that $Q + K = Q(I + Q^{-1}K)$ is a Fredholm operator on X and $i(Q + K) = 0$. In particular, $\ker(Q + K) = \{0\}$ if and only if the codimension of $\operatorname{ran}(Q + K)$ is zero, i.e., if and only if $(Q + K)(X) = X$. □

3.3.1 Index Zero

A more careful consideration of the proof of Theorem 3.1.10, in particular the implication *(a)* ⇒ *(c)*, allows us to considerably refine Corollary 3.3.7.

Theorem 3.3.8 (Index-Zero). *For a continuous linear operator T on a Hausdorff linear topological space X, the following three conditions are equivalent:*

(a) T is Fredholm and $i(T) = 0$.
(b) There exists a compact operator K on X such that $T + K$ is invertible.
(c) There exists a finite-rank operator $F \in \mathcal{L}(X)$ such that $T + F$ is a linear homeomorphism.

Proof. The implication *(c)* $\Rightarrow$ *(b)* is a consequence of item *(c)* in Theorem 2.1.3, while *(b)* $\Rightarrow$ *(a)* follows from the Generalized Fredholm Alternative, Corollary 3.3.7.

As such, it remains to prove that *(a)* $\Rightarrow$ *(c)*. To this end, recall from Theorem 3.1.7 the fundamental Fredholm decomposition

$$X = \operatorname{ran} T \oplus M = Y \oplus \ker T, \tag{3.3.27}$$

where $\operatorname{ran} T$, M, Y, $\ker T$ are quipped with the relative topology induced by X. According to Proposition 1.6.8, these all become Hausdorff linear topological spaces themselves. With the direct sum decompositions in (3.3.27) we associate the canonical projections

$$P_{\operatorname{ran}T} : X \to \operatorname{ran} T,\ P_M : X \to M, \text{ and } P_{\ker T} : X \to \ker T \tag{3.3.28}$$

as in Remark 1.8.21. Since said sum is actually topological, Proposition 1.8.22 ensures that these projections are continuous. Also, let

$$\iota_M : M \hookrightarrow X,\ \iota_Y : Y \hookrightarrow X, \text{ and } \iota_{\ker T} : \ker T \hookrightarrow X \tag{3.3.29}$$

be canonical inclusions, and recall from Proposition 1.6.8 that these are continuous as well. Pressing on, the last claim in Theorem 3.1.7 ensures that T induces a linear homeomorphism of Y onto $\operatorname{ran} T$, which we denote by

$$T\big|_{Y \to \operatorname{ran} T} : Y \longrightarrow \operatorname{ran} T. \tag{3.3.30}$$

Also, condition $i(T) = 0$ implies that M and $\ker T$ have the same (finite) dimension. If said dimension is zero, then T is already invertible, and we are done. If the dimension is positive, consider a linear isomorphism Φ taking $\ker T$ onto M. Since linear operators on finite-dimensional Hausdorff linear topological spaces are always continuous (Corollary 1.8.34), it follows that Φ is actually a homeomorphism. Then

$$F := \iota_M \circ \Phi \circ P_{\ker T} : X \to X \tag{3.3.31}$$

is a continuous linear map of finite-rank. Clearly, $T + F$ belongs to $\mathcal{L}(X)$, and we claim that this is a homeomorphism. However, this follows by checking (in a straightforward fashion) that

$$S := \iota_Y \circ \Big(T\big|_{Y \to \operatorname{ran} T}\Big)^{-1} \circ P_{\operatorname{ran}T} + \iota_{\ker T} \circ \Phi^{-1} \circ P_M, \tag{3.3.32}$$

which by design is a continuous linear map on X, serves as an inverse for $T + F$. $\square$

In the next two subsections, we turn our attention to analogous characterizations of positive and negative index. To get a feeling for these results, consider the forward and backward shift operators S and B of Examples 2.2.4. Recall that $i(S) = -1$ and $i(B) = +1$. Regarding invertibility, note that $BS = I$ and $I - SB$ is a projection of rank one.

3.3.2 Negative Index

Let X be a vector space equipped with a topology. Let us say that $T \in \mathcal{L}(X)$ is `left-invertible` (or invertible from the left) if there exists $S \in \mathcal{L}(X)$ such that $ST = I$. We shall call S a `left-inverse` for T. Note that S maps X *onto* X, while T is one-to-one. Moreover the operator TS is a continuous *projection* of X onto $\operatorname{ran}(TS) = \operatorname{ran} T$ (just check that $(TS)^2 = TS$), and we also have $\ker(TS) = \ker S$ (because T is one-to-one). So left-invertibility leads to the following topological direct sum decomposition:

$$X = \operatorname{ran} T \oplus \ker S \tag{3.3.33}$$

and TS becomes the natural projection of X onto $\operatorname{ran} T$ (see Theorem 1.8.27). These observations lead to the following result.

Proposition 3.3.9. *Let X be a Hausdorff linear topological space. If T is a left-invertible Fredholm operator on X then $i(T) \leq 0$. Furthermore if S is a left-inverse of T, then S is Fredholm on X and $i(S) = -i(T)$.*

Proof. In view of the fact that "left-invertible" implies "one-to-one," the index of the operator T is $-\operatorname{codim}(\operatorname{ran} T) \leq 0$. By (3.3.33) the codimension of $\operatorname{ran} T$ is the dimension of $\ker S$, so because S maps X onto itself its index is $\dim(\ker S) = -i(T)$. It remains to prove that S is a Fredholm operator. What is at stake here is the issue of openness (recall that for non-F-spaces ontoness does not always imply openness). To this end, suppose V is an open subset of X. Since continuous projections are relatively open (Theorem 1.8.27) we know that $TS(V)$ is open in $\operatorname{ran} T$. Since T is one-to-one, $S(V) = T^{-1}(TS(V))$ is an open set because of the continuity of T. Thus S is a Fredholm operator. □

Theorem 3.3.10 (Negative-Index Prototheorem). *Let X be a Hausdorff linear topological space. If T is a Fredholm operator satisfying $i(T) \leq 0$ then there exists a finite-rank operator $F \in \mathcal{L}(X)$ such that $T + F$ is left-invertible; in particular, $T + F$ is injective.*

Moreover, $T + F$ is a Fredholm operator on X and $i(T + F) = i(T)$.

Proof. Consider the fundamental Fredholm decomposition (see Theorem 3.1.7)

$$X = \ker T \oplus Y = M \oplus \operatorname{ran} T, \tag{3.3.34}$$

where Y is a closed subspace of X, M is a finite-dimensional subspace of X. (Recall that $\operatorname{ran} T$, M, Y, $\ker T$ are equipped with the relative topology induced by X, and that, thanks to Proposition 1.6.8, these all become Hausdorff linear topological spaces themselves.)

With the direct sum decompositions in (3.3.34) come the associated canonical projections

$$P_{\mathrm{ran}T} : X \to \operatorname{ran} T, \quad P_M : X \to M, \quad \text{and} \quad P_{\mathrm{ker}T} : X \to \ker T \tag{3.3.35}$$

as described in Remark 1.8.21. Since these decompositions are topological, Proposition 1.8.22 guarantees that each of these projections is continuous. Next, consider the canonical inclusions

$$\iota_M : M \hookrightarrow X, \quad \iota_Y : Y \hookrightarrow X, \quad \text{and} \quad \iota_{\mathrm{ker}T} : \ker T \hookrightarrow X, \tag{3.3.36}$$

which are continuous by Proposition 1.6.8. The last assertion of Theorem 3.1.7 shows that T induces a linear homeomorphism from Y onto $\operatorname{ran} T$, which we denote by

$$T\big|_{Y\to\operatorname{ran} T} : Y \longrightarrow \operatorname{ran} T. \tag{3.3.37}$$

Since $i(T) \leq 0$, the dimension of $\ker T$ is at most the dimension of M, which is finite. Let Φ be any one-to-one linear map taking $\ker T$ into M. By finite dimensionality, Φ is automatically continuous (Corollary 1.8.34) and, in fact,

$$\Phi\big|_{\ker T\to\operatorname{ran}\Phi} : \ker T \longrightarrow \operatorname{ran}\Phi \tag{3.3.38}$$

is a linear homeomorphism between finite-dimensional linear spaces. We extend Φ to a continuous linear map of finite-rank, defined on all of X, by setting

$$F := \iota_M \circ \Phi \circ P_{\mathrm{ker}T} : X \to X. \tag{3.3.39}$$

Note that $\operatorname{ran} F = \operatorname{ran}\Phi$, so we may adjust notation in (3.3.38) to say that

$$\Phi\big|_{\ker T\to\operatorname{ran} F} : \ker T \longrightarrow \operatorname{ran} F \ \text{ is a linear homeomorphism.} \tag{3.3.40}$$

Also, $T + F$ is a linear and continuous operator on X, mapping X onto a subspace R of X. From (3.3.34) and the fact $F \equiv 0$ on Y (as seen from (3.3.39)), we see that, actually,

$$R = \operatorname{ran} T + \operatorname{ran} F. \tag{3.3.41}$$

Upon recalling that $\operatorname{ran} T$ is closed of finite codimension in X, we may invoke Theorem 1.8.38 to conclude that R is itself closed and of finite codimension in X. Granted these properties, Corollary 1.8.37 guarantees the existence of a finite dimensional linear subspace N of X such that

$$X = R \oplus N. \tag{3.3.42}$$

Since, as seen from (3.3.39), we have

$$\operatorname{ran} F = \operatorname{ran}\Phi \subseteq M, \tag{3.3.43}$$

it follows that $\operatorname{ran} T \cap \operatorname{ran} F = \{0\}$, hence (3.3.41) improves to

$$R = \operatorname{ran} T \boxplus \operatorname{ran} F. \tag{3.3.44}$$

Given that $\operatorname{ran} T$ is closed and $\operatorname{ran} F$ is finite dimensional (since M is so), we may invoke Theorem 1.8.36 to further improve (3.3.44) to

$$R = \operatorname{ran} T \oplus \operatorname{ran} F. \tag{3.3.45}$$

With (3.3.42) and (3.3.45) in hand, we may invoke the result in Exercise 1.8.24 to obtain a projection $P_{\operatorname{ran}F}$ of X onto $\operatorname{ran} F$, i.e.,

$$\begin{aligned} &P_{\operatorname{ran}F} : X \to X \text{ is a linear and continuous map, satisfying} \\ &\quad P^2_{\operatorname{ran}F} = P_{\operatorname{ran}F} \text{ and } \operatorname{ran} P_{\operatorname{ran}F} = \operatorname{ran} F = \operatorname{ran} \Phi. \end{aligned} \tag{3.3.46}$$

Use this to define (bearing in mind that (3.3.37) is a linear homeomorphism, and (3.3.40))

$$S := \iota_Y \circ \left(T\big|_{Y \to \operatorname{ran} T}\right)^{-1} \circ P_{\operatorname{ran}T} + \iota_{\ker T} \circ \left(\Phi\big|_{\ker T \to \operatorname{ran} F}\right)^{-1} \circ P_{\operatorname{ran}F}. \tag{3.3.47}$$

This may be readily seen to be a continuous linear map on X, serving as an inverse from the left for $T + F$. Hence, $T + F$ is indeed left-invertible hence, in particular, injective.

Next, that $T + F$ is Fredholm is a consequence of part *(a)* in Corollary 3.2.2 and part *(c)* in Theorem 2.1.3. Also, from (3.3.44), (3.3.43), and the definition of R we see that

$$\operatorname{ran}(T + F) = \operatorname{ran} T \boxplus \operatorname{ran} \Phi, \tag{3.3.48}$$

while the fact that Φ is a one-to-one linear map defined on the finite dimensional space $\ker T$ implies

$$\dim \operatorname{ran} \Phi = \dim \ker T. \tag{3.3.49}$$

Based on (3.1.1), (3.3.48)-(3.3.49), the injectivity of $T + F$, and Exercise 1.3.40 we then conclude that

$$\begin{aligned} i(T + F) &= \dim \ker (T + F) - \operatorname{codim} \operatorname{ran} (T + F) \\ &= 0 - \big(\operatorname{codim} \operatorname{ran} T - \dim \operatorname{ran} \Phi\big) \\ &= \dim \ker T - \operatorname{codim} \operatorname{ran} T = i(T), \end{aligned} \tag{3.3.50}$$

as wanted. This finishes the proof of Theorem 3.3.10. □

Comment 3.3.11 *We call the last result a "prototheorem" because it is only a down-payment (albeit a substantial one) on the real thing. Based on our experience with Theorem 3.3.8, we are entitled to expect that the following conditions on a Fredholm operator T should be equivalent:*

(a) T has index ≤ 0.

(b) T has the form "left-invertible plus compact."
(c) T has the form "left-invertible plus finite rank."

Our Prototheorem establishes the implication (a)⇒(c), while (c)⇒(b) is a consequence of item (c) in Theorem 2.1.3. Proposition 3.3.9 is a special case of (b)⇒(a) (for which the compact perturbation is zero), and it will imply the full result if we can establish that:

$$\textit{the index remains unchanged under compact perturbation.} \tag{3.3.51}$$

This is indeed true, (Corollary 3.3.16), but its proof requires the Addition Theorem that is established later. In the next subsection we content ourselves with a discussion of positive index that parallels the work just completed for negative index.

3.3.3 Positive Index

Let X be a vector space equipped with a topology. We say that $T \in \mathcal{L}(X)$ is `right-invertible` (or invertible from the right) provided there exists $S \in \mathcal{L}(X)$ such that $TS = I$. We call S a `right-inverse` for T.

This is a mirror-image of the negative-index discussion. In a Hilbert space the results to follow can be derived as corollaries of the corresponding results for negative index by using the adjoint operation (see Exercises 3.3.4).

Proposition 3.3.12. *Let X be a Hausdorff linear topological space. If T is a Fredholm operator on X which is right-invertible, with right-inverse S, then $i(T) \geq 0$ and S is also Fredholm, and $i(S) = -i(T)$.*

Proof. The equation $TS = I$ forces T to map X onto itself, so $i(T) \geq 0$. This also forces S to be one-to-one, so we focus on establishing that $\operatorname{ran} S$ is closed, with codimension equal to $i(T)$, and that S is relatively open. An argument similar to the discussion at the beginning of the subsection §3.3.2 shows that ST is a continuous projection of X onto $\operatorname{ran} S$, and that the complementary subspace is $\ker T$. Thus $\operatorname{ran} S$ is closed and has codimension equal to $\dim(\ker T) = i(T)$. It remains to prove that S is a relatively open map. Let V be an open set in X. Then $W = T^{-1}(V)$ is open in X by the continuity of T, hence $ST(W)$ is open in $\operatorname{ran} S$ by the relative openness of continuous projections (Theorem 1.8.27). But $ST(W) = ST(T^{-1}(V)) = S(V)$, so $S(V)$ is open in $\operatorname{ran} S$, as desired. □

We next establish that Fredholm operators with positive index are right-invertible modulo finite-rank operators.

Theorem 3.3.13 (Positive-Index Prototheorem). *Let X be a Hausdorff linear topological space. If T is a Fredholm operator on X satisfying $i(T) \geq 0$ then there exists a finite-rank operator $F \in \mathcal{L}(X)$ such that $T + F$ is right-invertible; in particular, $T + F$ is surjective.*

Furthermore, $T + F$ is a Fredholm operator on X and $i(T + F) = i(T)$.

Proof. Recall the Fundamental Fredholm Decomposition (see Theorem 3.1.7)

$$X = \ker T \oplus Y = M \oplus \operatorname{ran} T, \tag{3.3.52}$$

where Y is a closed subspace of X, and M is a finite-dimensional subspace of X. (Recall that $\operatorname{ran} T$, M, Y, and $\ker T$ carry the subspace topology inherited from X, and that, by Proposition 1.6.8, each of them is a Hausdorff linear topological space.)

With the direct sum decompositions in (3.3.52) come the corresponding canonical projections

$$P_{\operatorname{ran} T} : X \to \operatorname{ran} T, \quad P_M : X \to M, \quad \text{and} \quad P_{\ker T} : X \to \ker T \tag{3.3.53}$$

as described in Remark 1.8.21. Since these sums are topological, Proposition 1.8.22 ensures that each projection is continuous. Next, consider the canonical inclusions

$$\iota_M : M \hookrightarrow X, \quad \iota_Y : Y \hookrightarrow X, \quad \text{and} \quad \iota_{\ker T} : \ker T \hookrightarrow X, \tag{3.3.54}$$

which are continuous by Proposition 1.6.8. The final statement of Theorem 3.1.7 shows that T induces a linear homeomorphism from Y onto $\operatorname{ran} T$, which we denote by

$$T\big|_{Y \to \operatorname{ran} T} : Y \longrightarrow \operatorname{ran} T. \tag{3.3.55}$$

Since $i(T) \geq 0$, it follows that the (finite) dimension of $\ker T$ is at least as large as that of M, so we can find a linear map Φ taking $\ker T$ onto M. By finite dimensionality, Φ is automatically continuous (Corollary 1.8.34). Let W be a topological complement of $\ker \Phi$ in the finite dimensional space $\ker T$:

$$\ker T = \ker \Phi \oplus W. \tag{3.3.56}$$

In particular, the restriction

$$\Phi\big|_W : W \longrightarrow M \tag{3.3.57}$$

is a linear homeomorphism between finite-dimensional linear spaces. We extend Φ to a continuous linear map of finite-rank, defined on all of X, by setting

$$F := \iota_M \circ \Phi \circ P_{\ker T} : X \to X. \tag{3.3.58}$$

As seen from (3.3.58),

$$\operatorname{ran} F = \operatorname{ran} \Phi = M \quad \text{and} \quad F \equiv 0 \quad \text{on} \quad Y. \tag{3.3.59}$$

Also, $T + F$ is a linear and continuous operator on X, mapping X onto

$$\operatorname{ran}(T + F) = \operatorname{ran} T + \operatorname{ran} F = \operatorname{ran} T + M = X, \tag{3.3.60}$$

thanks to (3.3.52) and (3.3.59). Hence, $T + F$ is surjective. Note that the canonical inclusion ι_W of W into X is a continuous linear map. Use this to define (bearing in mind that (3.3.55) and (3.3.57) are linear homeomorphisms)

$$S := \iota_Y \circ \left(T\big|_{Y\to \operatorname{ran} T}\right)^{-1} \circ P_{\operatorname{ran}T} + \iota_W \circ \left(\Phi\big|_W\right)^{-1} \circ P_M. \tag{3.3.61}$$

This may be readily seen to be a continuous linear map on X, serving as an inverse from the right for $T + F$. Hence, $T + F$ is indeed right-invertible.

Next, that $T + F$ is Fredholm is a consequence of part *(a)* in Corollary 3.2.2 and part *(c)* in Theorem 2.1.3. Also, from (3.3.52), the fact that T induces a linear homeomorphism from Y onto $\operatorname{ran} T$, and that Φ maps $\ker T$ into M, on the one hand we obtain

$$\ker(T + F) = \ker \Phi. \tag{3.3.62}$$

On the other hand, since actually Φ is a linear operator mapping $\ker T$ onto M, the First Isomorphism Theorem for Vector Spaces tells us that

$$\dim\ker\Phi = \dim\ker T - \dim M = \dim\ker T - \operatorname{codim}\operatorname{ran} T = i(T) \tag{3.3.63}$$

where the second equality comes from (3.1.1), and the last equality uses (3.1.1). From (3.1.1), (3.3.62)-(3.3.63), and the surjectivity of $T + F$ we then conclude that

$$i(T + F) = \dim\ker(T + F) - \operatorname{codim}\operatorname{ran}(T + F) = i(T) - 0 = i(T), \tag{3.3.64}$$

as wanted. This finishes the proof of Theorem 3.3.13. □

We are now in a position to provide the proof of the following important result.

Corollary 3.3.14. *Suppose n is an integer, and X is a Hausdorff linear topological space on which there is a Fredholm operator of index n. Then X also supports a Fredholm operator of index $-n$.*

Proof. If $n = 0$ there is nothing to prove. If $n \neq 0$, Propositions 3.3.9 and 3.3.12 assert that the "partial inverse" operators S that show up in Theorem 3.3.10 and Theorem 3.3.13 do the job. □

3.3.4 Additivity of the Index

We have already seen (cf. Theorem 3.3.1) that the index possesses an elementary additivity property: *If S and T are Fredholm operators on linear topological spaces, then $S \oplus T$ is also Fredholm and $i(S \oplus T) = i(S) + i(T)$.* We devote this subsection to a much deeper result:

Theorem 3.3.15 (Addition). *If S and T are Fredholm operators on a Hausdorff linear topological space X, then ST is also Fredholm on X and*

$$i(ST) = i(S) + i(T). \tag{3.3.65}$$

From part *(b)* in Corollary 3.2.2 we already know ST is a Fredholm operator on X. We also know some special cases of Theorem 3.3.15. For example, if S is an invertible operator on X then $i(S) = 0$ and $i(ST) = i(T)$ (cf. Theorem 3.3.5). Next, if S and T are Fredholm operators on X and $ST = I$, then our discussion of left and right invertibility (respectively, Proposition 3.3.9 and Proposition 3.3.12) showed that $i(S) = -i(T)$. Finally there is the forward shift on, for example, ℓ^2 (see Example 2.2.4). By part *(d)* of Example 3.1.3, we know that S is Fredholm on ℓ^2, and that $i(S^2) = -2 = 2i(S)$, giving yet another instance of the Addition Formula. Similar comments apply, of course, to the backward shift (see Example 2.2.4).

The proof of Theorem 3.3.15 is given at the very end of §3.3.6. To fortify ourselves for this endeavor, we first present some applications in the next subsection.

3.3.5 Applications of Theorem 3.3.15

First, let us tie up the loose ends concerning right and left invertibility that were left hanging in Comment 3.3.11. Recall that the results we wished to prove depended on the index remaining constant under compact perturbations (cf. (3.3.51)), something we have already encountered for invertible operators under the name of "The Generalized Fredholm Alternative" (Corollary 3.3.7). So we begin with this; here is the formal statement of (3.3.51):

Corollary 3.3.16. *If T is a Fredholm operator on a Hausdorff linear topological space X, and K is a compact operator on X, then $T + K$ s a Fredholm operator on X and $i(T + K) = i(T)$. Succinctly put, the index of a Fredholm operator is constant under compact perturbations. As a consequence, the result recorded in* (2.1.77) *as well as Corollary 3.3.3 may be strengthened to:*

in a Hausdorff linear topological space, every compact perturbation of an invertible operator is a Fredholm operator with index zero. (3.3.66)

Proof. From part *(a)* in Corollary 3.2.2 we already know that $T + K$ is a Fredholm operator on X. By Corollary 3.3.14 there exists a Fredholm operator S on X with $i(S) = -i(T)$. Then part *(b)* in Corollary 3.2.2 guarantees that ST is a Fredholm operators on X, and we may write

$$\begin{aligned} i(T + K) - i(T) &= i(T + K) + i(S) = i(S(T + K)) \\ &= i(ST + SK) = i(ST + \text{compact}) \end{aligned} \tag{3.3.67}$$

where we have used Theorem 3.3.15 in the second equality and the fact that $\mathcal{K}(X)$ is a two-sided ideal in $\mathcal{L}(X)$ (cf. part *(b)* of Theorem 2.1.3) in the last equality. Since $i(ST) = i(S) + i(T) = 0$ (again by Theorem 3.3.15), Theorem 3.3.8 provides a compact operator F (which we could even take to have finite rank), such that $ST + F$ is invertible. Thus, we may continue (3.3.67) by writing

$$\begin{aligned} i(T+K) - i(T) &= i((ST+F) - F + \text{ compact}) \\ &= i(\text{ invertible } + \text{ compact}) \\ &= 0, \end{aligned} \tag{3.3.68}$$

where at the last step we have used the Generalized Fredholm Alternative (Corollary 3.3.7). Hence, $i(T+K) = i(T)$, as wanted. □

Exercise 3.3.17. If X is an infinite-dimensional Hausdorff linear topological space, show that there are no compact operators on X that are Fredholm.

Hint: If T is a compact operators on X that happens to be Fredholm, then $0 = T - T$ becomes (as a compact perturbation of Fredholm operator) a Fredholm operator itself. Reach a contradiction by looking at its kernel.

According to the discussion following our Negative-Index Prototheorem (Theorem 3.3.10) we now have the promised "right" and "left" versions of the Index-Zero Theorem (Theorem 3.3.8). For reference we give official statements of both results, and repeat the proof of the "left-handed" theorem.

Theorem 3.3.18 (Negative Index). *For T a Fredholm operator on a Hausdorff linear topological space X, the following statements are equivalent:*

(a) $i(T) \leq 0$.
(b) There is a finite-rank operator $F \in \mathcal{L}(X)$ such that $T + F$ is left-invertible.
(c) There is a compact operator K on X such that $T + K$ is left-invertible.

Proof. The implication *(a)* ⇒ *(b)* is Theorem 3.3.10, while the implication *(b)* ⇒ *(c)* is seen from part *(c)* in Theorem 2.1.3. Note that if $K = 0$, the implication *(c)* ⇒ *(a)* is precisely Proposition 3.3.9. In the general case, this implication follows from Proposition 3.3.9 and Corollary 3.3.16. □

A similar argument gives the following companion result:

Theorem 3.3.19 (Positive Index). *For T a Fredholm operator on a Hausdorff linear topological space X, the following statements are equivalent:*

(a) $i(T) \geq 0$.
(b) There is a finite-rank operator $F \in \mathcal{L}(X)$ such that $T + F$ is right-invertible.
(c) There is a compact operator K on X such that $T + K$ is right-invertible.

There is a subtle difference between the syntax of the theorems on positive and negative index and that of the Index-Zero Theorem (Theorem 3.3.8). In the latter theorem the operator becomes Fredholm once it is recognized to be a compact perturbation of an invertible one, whereas in the right and left-hand versions above, Fredholmness is an overriding assumption. The following exercise shows that this difference is unavoidable.

Exercise 3.3.20. Find a continuous linear operator on a Hilbert space that is left-invertible but not Fredholm.

Hint: Try a forward shift of "infinite multiplicity," (cf. item *(b)* in Exercise 3.2.3).

In a lighter vein, here is a generalization of the fact that you cannot shift coordinates by half a unit.

Exercise 3.3.21. Let S denote the forward shift on ℓ^2 (see Examples 2.2.4). Show that S has no square root, i.e., that there is no continuous linear operator T on ℓ^2 such that $T^2 = S$. Generalize this result to a larger class of Fredholm operators.

Hint: You may wish to review item *(a)* in Exercise 3.2.3.

We conclude this subsection by presenting a restriction result for Fredholm operators in the proposition below.

Proposition 3.3.22. *Let X, Y be two Hausdorff topological vector spaces with the property that Y is a dense subset of X. Consider a Fredholm operator $T : X \to X$ for which Y is an invariant subspace and such that its restriction to Y, namely*

$$T|_Y : Y \to Y \text{ defined as } (T|_Y)(x) := Tx \text{ for each } x \in Y, \tag{3.3.69}$$

is also a Fredholm operator. Then

$$\begin{gathered} i(T) = i(T|_Y) \text{ if and only if} \\ \ker T = \ker(T|_Y) \text{ and } \operatorname{codim}(\operatorname{ran} T) = \operatorname{codim}(\operatorname{ran}(T|_Y)). \end{gathered} \tag{3.3.70}$$

Proof. From (3.3.69) it is clear that

$$T(Y) = \operatorname{ran}(T|_Y). \tag{3.3.71}$$

Since $\operatorname{ran}(T|_Y)$ has finite codimension in Y, there exists a finite-dimensional subspace M of Y such that $T(Y) \boxplus M = Y$. We claim that

$$T(X) + M = X. \tag{3.3.72}$$

To prove the claim, observe that $Y = T(Y) + M \subseteq TX + M$. Hence, since Y is a dense subset of X, it follows that $T(X) + M$ is also a dense subset of X. Moreover, because $T(X) = \operatorname{ran} T$ is a closed subspace of X and M is a finite-dimensional subspace of X, Theorem 1.8.35 implies that $T(X) + M$ is closed in X. Combining these results, the claim in (3.3.72) follows. In concert with Corollary 1.3.44, (3.3.72) implies

$$\dim\Big(\frac{Y}{TY}\Big) = \dim M \geq \dim\Big(\frac{X}{TX}\Big). \tag{3.3.73}$$

From this, (1.3.158), and (3.3.71) we then conclude that

$$\operatorname{codim}(\operatorname{ran}(T|_Y)) \geq \operatorname{codim}(\operatorname{ran} T). \tag{3.3.74}$$

Assume now $i(T) = i(T|_Y)$. Then (3.3.74) forces $\dim(\ker(T|_Y)) \geq \dim(\ker T)$. Given that $\ker(T|_Y)$ is a subspace of $\ker T$, itself a finite-dimensional linear space, the opposite inequality also holds, and we ultimately deduce that $\ker T = \ker (T|_Y)$. With this in hand, we also conclude that $\operatorname{codim}(\operatorname{ran} T) = \operatorname{codim}(\operatorname{ran}(T|_Y))$, since $i(T) = i(T|_Y)$. This proves the direct implication in (3.3.70), and the converse is clear from definitions. □

3.3.6 The Proof of Theorem 3.3.15

The arguments involved in the proof of Theorem 3.3.15 are purely algebraic. To emphasize this we assume for the rest of this subsection that X, Y are simply vector spaces over $\mathbb{C}$ (or $\mathbb{R}$ if you prefer) with no topology assumed, and let

> $\Phi^{\text{alg}}(X, Y)$ denote the set of all linear transformations $T : X \to Y$ with finite dimensional null-space and finite codimensional range. (3.3.75)

We shall refer to elements in $\Phi^{\text{alg}}(X, Y)$ as `algebraic Fredholm operators` from X to Y. Also, for each $T \in \Phi^{\text{alg}}(X, Y)$ we shall define

$$i(T) := \dim(\ker T) - \operatorname{codim}(\operatorname{ran} T). \tag{3.3.76}$$

Exercise 3.3.23. Let X be a vector space and consider a subspace $Y \subseteq X$ of finite codimension in X. Denote by ι the canonical inclusion of Y into X, i.e.,

$$\iota : Y \longrightarrow X, \quad \iota(x) := x \text{ for each } x \in Y. \tag{3.3.77}$$

Then $\iota \in \Phi^{\text{alg}}(Y, X)$ and $i(\iota) = -\operatorname{codim} Y$.

We are going to prove that the product (aka composition) of algebraic Fredholm operators is again algebraic Fredholm, and that the addition formula (3.3.65) holds. The argument comprises three separate computations which yield formulas for, respectively, the dimension of $\ker ST$, the codimension of $\operatorname{ran} ST$, and the codimension of the sum of two subspaces. Here they are!

Theorem 3.3.24 (Dimension of $\ker ST$). *Let X, Y, Z be three vector spaces and let $T : X \to Y$ and $S : Y \to Z$ be linear transformations, both of which with a finite-dimensional null-space. Then ST also has finite-dimensional null-space and*

$$\dim(\ker(ST)) = \dim(\ker T) + \dim(\ker S \cap \operatorname{ran} T). \tag{3.3.78}$$

Proof. Since $\ker(ST) \supseteq \ker T$, Theorem 1.3.29 ensures the existence of a subspace M of X such that

$$\ker(ST) = \ker T \boxplus M. \tag{3.3.79}$$

Consequently, we have (cf. Exercise 1.3.28)

$$\dim(\ker(ST)) = \dim(\ker T) + \dim M, \tag{3.3.80}$$

so the issue is to show that

$$\dim M = \dim(\ker S \cap \operatorname{ran} T). \tag{3.3.81}$$

For this, observe that $\ker(ST) = T^{-1}(\ker S)$, so by (3.3.79) we have

$$T(M) = T(\ker(ST)) = T(T^{-1}(\ker S)) = \ker S \cap \operatorname{ran} T. \tag{3.3.82}$$

Since T is one-to-one on M, both M and $T(M)$ have the same dimension, so the last equation implies (3.3.81). □

Theorem 3.3.25 (Codimension of $\operatorname{ran} ST$**).** *Let X, Y, Z be vector spaces and suppose $T : X \to Y$ and $S : Y \to Z$ are linear transformations, both of which with a finite codimensional range. Then ST has finite codimensional range and*

$$\operatorname{codim}(\operatorname{ran}(ST)) = \operatorname{codim}(\operatorname{ran} S) + \operatorname{codim}(\operatorname{ran} T + \ker S). \tag{3.3.83}$$

Proof. To get started, observe that the result is obvious if the last term on the right-hand side of (3.3.83) is zero. For then $\operatorname{ran} T + \ker S$ is all of Y, which in turn allows us to compute

$$\operatorname{ran} S = S(Y) = S(\operatorname{ran} T + \ker S) = S(\operatorname{ran} T) = \operatorname{ran}(ST). \tag{3.3.84}$$

In general, we know that $\operatorname{ran}(ST) \subseteq \operatorname{ran} S$, so (1.3.170) permits us to write (keeping (1.3.158) in mind)

$$\operatorname{codim}(\operatorname{ran}(ST)) = \operatorname{codim}(\operatorname{ran} S) + (\text{codim of } \operatorname{ran}(ST) \text{ in } \operatorname{ran} S). \tag{3.3.85}$$

The issue, then, is to prove that

$$\operatorname{codim}(\operatorname{ran} T + \ker S) = \text{codim of } \operatorname{ran}(ST) \text{ in } \operatorname{ran} S. \tag{3.3.86}$$

For this, let N be an algebraic complement to $\operatorname{ran} T + \ker S$ in Y (which always exists, thanks to Theorem 1.3.29), so

$$Y = (\operatorname{ran} T + \ker S) \boxplus N. \tag{3.3.87}$$

Apply S to both sides of the last equation to obtain

$$\operatorname{ran} S = \operatorname{ran}(ST) + S(N). \tag{3.3.88}$$

Since $\ker S$ lies in a subspace complementary to N (as seen from (3.3.87)), the operator S is one-to-one on N, hence

$$\dim S(N) = \dim N = \operatorname{codim}(\operatorname{ran} T + \ker S), \tag{3.3.89}$$

where the last equality uses (3.3.87) and Exercise 1.3.41. Then (3.3.86) follows upon showing that $\dim S(N)$ is the codimension of $\operatorname{ran}(ST)$ in $\operatorname{ran} S$. This in turn will follow from (3.3.88) if we can show that the sum on the right-hand side of that equation is an algebraic direct sum, i.e., that $\operatorname{ran}(ST) \cap S(N) = \{0\}$.

To this end, assume y belongs to the intersection, so

$$y = STx \quad \text{for some} \quad x \in X, \quad \text{and} \quad y = Sz \quad \text{for some} \quad z \in N. \tag{3.3.90}$$

This forces

$$Tx - z \in \ker S, \tag{3.3.91}$$

thus

$$z \in \ker S + Tx \subseteq \ker S + \operatorname{ran} T. \tag{3.3.92}$$

Since also $z \in N$ we have $z = 0$ by (3.3.87). Hence (3.3.91) reduces to $Tx \in \ker S$, and therefore $y = STx = 0$, as wanted. □

The reader is reminded that the class of algebraic Fredholm operators has been introduced in (3.3.75).

Theorem 3.3.26 (Additivity of index for algebraic Fredholm operators). *Let X, Y, Z be vector spaces and consider $T \in \Phi^{\text{alg}}(X, Y)$ along with $S \in \Phi^{\text{alg}}(Y, Z)$. Then $ST \in \Phi^{\text{alg}}(X, Z)$ and*

$$i(ST) = i(S) + i(T). \tag{3.3.93}$$

Proof. That ST belongs to $\Phi^{\text{alg}}(X, Z)$ follows from Theorem 3.3.24 and Theorem 3.3.25. Upon substituting the results of (3.3.78) and (3.3.83) in the definition of $i(ST)$ (cf. (3.3.76)) we obtain:

$$\begin{aligned} i(ST) &= \dim(\ker(ST)) - \operatorname{codim}(\operatorname{ran}(ST)) \\ &= \dim\ker T + \dim(\ker S \cap \operatorname{ran} T) \\ &\quad - \operatorname{codim}(\operatorname{ran} S) - \operatorname{codim}(\operatorname{ran} T + \ker S). \end{aligned} \tag{3.3.94}$$

To compute the codimension of $\operatorname{ran} T + \ker S$ in Y we use formula (1.3.183) with $Y := \operatorname{ran} T$ and $N := \ker S$:

$$\operatorname{codim}(\operatorname{ran} T + \ker S) = \operatorname{codim}(\operatorname{ran} T) - \dim(\ker S) + \dim(\operatorname{ran} T \cap \ker S). \tag{3.3.95}$$

Upon substituting this into the previous equation we arrive at

$$\begin{aligned} i(ST) &= \dim(\ker T) + \dim(\ker S \cap \operatorname{ran} T) - \operatorname{codim}(\operatorname{ran} S) \\ &\quad - \operatorname{codim}(\operatorname{ran} T) + \dim(\ker S) - \dim(\operatorname{ran} T \cap \ker S) \\ &= i(T) + i(S), \end{aligned} \tag{3.3.96}$$

as desired. □

At last, we are now in a position to give a formal proof of Theorem 3.3.15.

Proof of Theorem 3.3.15. That ST is Fredholm on X follows from part *(b)* in Corollary 3.2.2, while formula (3.3.65) is a consequence of Theorem 3.3.26 (bearing in mind Definition 3.1.1 and (3.3.75)). □

3.4 Continuity of the Index

In this section we restrict our attention to the most commonly occurring class of linear topological spaces—the *p-Banach spaces* ($0 < p \leq 1$). For such spaces X, Banach-space techniques carry over *verbatim* to provide a "p-norm-topology" for the space $\mathcal{L}(X)$ of continuous linear operators on X. We show that within this setting the Fredholm index—viewed as a mapping of the Fredholm operators space, equipped with the induced topology of $\mathcal{L}(X)$, into the set of integers with the discrete topology—is continuous. We actually prove something that is seemingly a little stronger although ultimately equivalent: *for each integer n the Fredholm operators of index n form an open subset of* $\mathcal{L}(X)$. An example showing that the continuity of the index may fail for Fredholm operators on Fréchet spaces is outlined in Exercise 3.4.16.

3.4.1 The Space $\mathcal{L}(X)$ as a p-Banach Algebra

Let us start with the observation that if X is a p-normed space for some $0 < p \leq 1$, with p-norm $\|\cdot\|$, then—just as for ordinary normed spaces—a linear transformation T on X is continuous if and only if it is bounded on the unit ball of X (cf. Theorem 2.1.14). Using the piece of notation introduced in Definition 1.4.26, for each $T \in \mathcal{L}(X)$ we therefore have

$$\|T\| = \sup\{\|Tx\| : x \in X \text{ and } \|x\| \leq 1\} < \infty. \tag{3.4.97}$$

Exercise 3.4.1. Show that if X is a p-normed space for some $0 < p \leq 1$, then

$$\|ST\| \leq \|S\| \cdot \|T\| \quad \text{for all} \quad S, T \in \mathcal{L}(X). \tag{3.4.98}$$

Moreover, if X is actually a p-Banach space, verify that $\mathcal{L}(X)$ equipped with the functional $\|\cdot\|$ recalled in (3.4.97) becomes a p-Banach space itself.

Definition 3.4.2. *Suppose A is a p-normed vector space that is also an algebra with identity, and that*

$$\|xy\| \leq \|x\| \cdot \|y\| \quad \textit{for each} \quad x, y \in A. \tag{3.4.99}$$

Call A a p-`normed algebra` *and the p-norm on A a p-*`algebra norm`*. If A is complete in a p-algebra norm, we call it a p-*`Banach algebra`.

As a point of reference, we record here that, as seen from Definition 3.4.2 and Exercise 3.4.1,

if X is a p-Banach space for some $0 < p \leq 1$, then $\mathcal{L}(X)$, equipped with the operator p-norm (3.4.97), becomes, in fact, a p-Banach algebra. (3.4.100)

3.4.2 Invertibility in p-Banach Algebras

Suppose A is a p-Banach algebra, for some $p \in (0, 1]$, with identity. In particular, there exists $e \in A$ (called `identity`) for which

$$ex = xe = x, \quad \text{for every } x \in A. \tag{3.4.101}$$

Clearly, from (3.4.101), there can be at most one identity element, and when it exists, we assume further that its norm is 1. In fact, a slightly more economical (although equivalent) demand on the identity element $e \in A$) satisfying (3.4.101) is by asking $\|e\| \leq 1$ while, in addition, requesting that $A \neq \{0\}$. The fact that $\|e\| \geq 1$ is then seen from

$$\|e\| = \|ee\| \leq \|e\| \cdot \|e\| = \|e\|^2, \tag{3.4.102}$$

where the inequality follows from (3.4.99), and that $\|e\| \neq 0$, which is implied by the assumption $A \neq \{0\}$.

Going further, as part (b) of the exercise below shows, if $\|e\| \neq 1$, then the algebra can be equivalently renormed so that the identity element is a unit vector in the new norm.

Exercise 3.4.3. (a) Show that in any non-trivial p-Banach algebra (where $p \in (0, 1]$) with identity e, the norm of the identity is at least 1.

(b) Suppose $(A, \|\cdot\|)$ is a p-Banach algebra (where $p \in (0, 1]$) with identity e, but do *not* assume the norm of the identity to be 1. Then

$$\|x\|_* := \sup\{\|xy\| : y \in A, \ \|y\| \leq 1\} \tag{3.4.103}$$

defines a p-norm on A (in fact, $\|\cdot\|_*$ is the p-norm A inherits from $\mathcal{L}(A)$ when each element of A is viewed as a left-multiplication operator on A). Show that $(A, \|\cdot\|_*)$ is also a p-Banach algebra, that $\|e\|_* = 1$, and that $\|\cdot\|$ and $\|\cdot\|_*$ are equivalent.

Hint: Show that the new p-norm is $\leq$ the old one at every point of A. Then show that A, viewed as a set of left-multiplication operators on itself, is closed in $\mathcal{L}(A)$. Finally, use the Open Mapping Theorem (in the version recalled in Theorem 1.6.7) to show that the old and new p-norms are equivalent.

Continue to assume that A is a p-Banach algebra, for some $p \in (0, 1]$, with identity. Recall that an element $x \in A$ is `invertible` if there exists $y \in A$ such that $xy = yx = e$. When this happens, y is the unique element with this property, and we write $x^{-1} := y$. Also, find it useful to single out the collection of invertible elements of A:

$$\mathcal{G}(A) := \{x \in A : x \ \text{ is invertible}\}. \tag{3.4.104}$$

Then just as for Banach algebras, we have the following result:

Theorem 3.4.4 (Invertibility). *If A is a p-Banach algebra with identity then $\mathcal{G}(A)$ is an open subset of A. More precisely, if $x_0 \in \mathcal{G}(A)$ and $x \in A$ with*

$$\|x - x_0\| < \|x_0^{-1}\|^{-1}, \tag{3.4.105}$$

then $x \in \mathcal{G}(A)$.

Proof. The proof is similar to the one for the Banach algebras, and we include it here for the convenience of the reader.

`Step 1.` Suppose $x_0 = e$, the identity element of A, and $x \in A$ is such that (3.4.105) is satisfied. Our goal is to show that $x \in \mathcal{G}(A)$. Thanks to Exercise 3.4.3 we may assume that $\|e\| = 1$. As such, (3.4.105) gives

$$\|e - x\| < 1. \tag{3.4.106}$$

Since

$$\|e - x\| < 1 \quad \text{and} \quad \|(e - x)^n\| \leq \|e - x\|^n, \tag{3.4.107}$$

for each $n \in \mathbb{N} \cup \{0\}$, the completeness of $(A, \|\cdot\|)$ guarantees that the series

$$\sum_{n=0}^{\infty} (e - x)^n \tag{3.4.108}$$

converges to an element $z \in A$. We claim that z is the inverse of x. To see this, let S_N be the N-th partial sum of the series (3.4.108). Then it is straightforward that

$$xS_N = S_N x = e - (e - x)^{N+1}. \tag{3.4.109}$$

By (3.4.106) and (3.4.107), the sequence of elements $(e - x)^{N+1}$ converges to zero as $N \to \infty$, hence $xS_N = S_N x \to e$ as $N \to \infty$. Upon recalling that $S_N \to z$ as $N \to \infty$, we obtain, by appealing to (3.4.99), that $zx = xz = e$. Thus $x \in \mathcal{G}(A)$ and $z = x^{-1}$. This completes the treatment of Step 1.

`Step 2.` Let $x_0 \in \mathcal{G}(A)$ be arbitrary. Then for $x \in A$ we have

$$\|x_0^{-1} x - e\| = \|x_0^{-1}(x - x_0)\| \leq \|x_0^{-1}\| \|x - x_0\|. \tag{3.4.110}$$

Hence, if x satisfies (3.4.105) above, then the right-hand side of the last inequality is < 1. By Step 1, $x_0^{-1} x$ is invertible in A. Thus $x = x_0(x_0^{-1} x)$, being the product of two invertible elements, is also invertible. $\square$

The special case treated in Step 1 of the above proof deserves special mention:

Corollary 3.4.5. *Let* $p \in (0, 1]$. *In a p-Banach algebra with identity, the open ball of radius one centered at the identity consists entirely of invertible elements.*

3.4.3 Topologizing the Calkin Algebra

Let X be a p-Banach space, for some $p \in (0, 1]$. We wish to turn the Calkin algebra $\mathcal{L}(X)/\mathcal{K}(X)$, introduced in §3.2, into a p-Banach algebra. The first step is to show that the subspace $\mathcal{K}(X)$ is closed in $\mathcal{L}(X)$. Then the construction is exactly the same as the one done for the quotient of a Banach algebra modulo a closed ideal, and, for completeness, we present it here.

Proposition 3.4.6. *Let X be a p-Banach space, for some $p \in (0, 1]$. The set $\mathcal{K}(X)$ is a closed subspace of $\mathcal{L}(X)$.*

Proof. This is a direct consequence of Proposition 2.1.5. □

The next definition introduces a p-algebra norm on the Calkin algebra.

Definition 3.4.7. *Let X be a p-Banach space for some $p \in (0, 1]$. If $T \in \mathcal{L}(X)$ define the* `quotient p-norm` *of the coset $T + \mathcal{K}(X)$ in $\mathcal{L}(X)/\mathcal{K}(X)$ by:*

$$\|T + \mathcal{K}(X)\| := \inf\{\|T + K\| : K \in \mathcal{K}(X)\}. \tag{3.4.111}$$

A few comments are in order here. First, a combination of (3.4.111) and Proposition 3.4.6 shows that $\|T + \mathcal{K}(X)\|$ is the distance in $\mathcal{L}(X)$ from T to the closed subspace $\mathcal{K}(X)$. In particular, if $T + \mathcal{K}(X)$ is not the zero element of $\mathcal{L}(X)/\mathcal{K}(X)$, then $T \notin \mathcal{K}(X)$, and consequently we have $\|T + \mathcal{K}(X)\| > 0$. This shows that the quotient p-norm from (3.4.111) satisfies the non-degeneracy condition. As for the homogeneity condition, if $\alpha \in \mathbb{C}$ and $T \in \mathcal{L}(X)$, then

$$\begin{aligned} \|\alpha(T + \mathcal{K}(X))\| &= \|\alpha T + \mathcal{K}(X)\| \\ &= \inf\Big\{\|\alpha T + K\| : K \in \mathcal{K}(X)\Big\} \\ &= |\alpha|^p \cdot \inf\Big\{\|T + K\| : K \in \mathcal{K}(X)\Big\} \\ &= |\alpha|^p \cdot \|T + \mathcal{K}(X)\|. \end{aligned} \tag{3.4.112}$$

Next, we verify that the quotient p-norm defined in (3.4.111) satisfies the sub-additivity condition. Let $T, L \in \mathcal{L}(X)$ and fix some arbitrary $\varepsilon > 0$. Choose $K_T, K_L \in \mathcal{K}(X)$ satisfying

$$\|T + K_T\| < \|T + \mathcal{K}(X)\| + \frac{\varepsilon}{2} \quad \text{and} \quad \|L + K_L\| < \|L + \mathcal{K}(X)\| + \frac{\varepsilon}{2}. \tag{3.4.113}$$

Thus, since $K_T + K_L \in \mathcal{K}(X)$, we obtain

$$\begin{aligned} \|(T + L) + \mathcal{K}(X)\| &\le \|(T + L) + (K_T + K_L)\| \\ &= \|(T + K_T) + (L + K_L)\| \\ &\le \|T + K_T\| + \|L + K_L\| \\ &< \|T + \mathcal{K}(X)\| + \|L + \mathcal{K}(X)\| + \varepsilon. \end{aligned} \tag{3.4.114}$$

By letting $\varepsilon \to 0^+$, we obtain the desired conclusion. The foregoing discussion shows that:

the quotient p-norm defined in (3.4.111) is a genuine p-norm and, thus, $\mathcal{L}(X)/\mathcal{K}(X)$ is a p-normed vector space when equipped with this p-norm. (3.4.115)

Moving on, recall the quotient map $\pi : \mathcal{L}(X) \to \mathcal{L}(X)/\mathcal{K}(X)$, defined by

$$\pi(T) := T + \mathcal{K}(X) \quad \text{for each} \quad T \in \mathcal{L}(X). \tag{3.4.116}$$

When the Calkin algebra is equipped with the quotient p-norm, π is clearly a contraction; i.e.,

$$\|\pi(T)\| \leq \|T\| \quad \text{for every } T \in \mathcal{L}(X). \tag{3.4.117}$$

In particular, π is continuous.

Proposition 3.4.8. *Let X be a p-Banach space for some $p \in (0, 1]$. The quotient p-norm introduced in* (3.4.111) *turns the Calkin algebra into a p-Banach algebra.*

Proof. That the Calkin algebra $\mathcal{L}(X)/\mathcal{K}(X)$ becomes a p-Banach space when it is equipped with the quotient p-norm defined in (3.4.111) is seen from (3.4.100), Lemma 1.11.6, and Proposition 3.4.6. To check (3.4.99) in the current context, start by fixing $T_1, T_2 \in \mathcal{L}(X)$. The goal is to show that

$$\|(T_1 + \mathcal{K}(X)) \circ (T_2 + \mathcal{K}(X))\| \leq \|T_1 + \mathcal{K}(X)\| \, \|T_2 + \mathcal{K}(X)\|. \tag{3.4.118}$$

To justify this, given $\varepsilon > 0$, choose $K_1, K_2 \in \mathcal{K}(X)$ satisfying

$$\|T_1 + K_1\| < \|T_1 + \mathcal{K}(X)\| + \varepsilon \text{ and } \|T_2 + K_2\| < \|T_2 + \mathcal{K}(X)\| + \varepsilon. \tag{3.4.119}$$

Then,

$$\begin{aligned} \|(T_1 + \mathcal{K}(X)) \circ (T_2 + \mathcal{K}(X))\| &= \|T_1 \circ T_2 + \mathcal{K}(X)\| \\ &\leq \|T_1 \circ T_2 + (T_1 \circ K_2 + K_1 \circ T_2 + K_1 \circ K_2)\| \\ &= \|(T_1 + K_1) \circ (T_2 + K_2)\| \\ &\leq \|T_1 + K_1\| \, \|T_2 + K_2\| \\ &< (\|T_1 + \mathcal{K}(X)\| + \varepsilon) \, (\|T_2 + \mathcal{K}(X)\| + \varepsilon), \end{aligned} \tag{3.4.120}$$

where the first inequality holds due to the fact that

$$T_1 \circ K_2 + K_1 \circ T_2 + K_1 \circ K_2 \in \mathcal{K}(X), \tag{3.4.121}$$

(cf. parts *(a)* and *(b)* of Theorem 2.1.3), whereas the second inequality holds by virtue of the fact that $\mathcal{L}(X)$ is itself a p-Banach algebra (cf. (3.4.100)). Finally, (3.4.118) is obtained by letting $\varepsilon \to 0^+$. □

Comment 3.4.9 *For the quotient p-norm of the Calkin algebra identity $I + \mathcal{K}(X)$, note that it is ≤ 1 by* (3.4.117). *However, part (a) of Exercise 3.4.3 shows that* *the norm of the identity in a non-trivial p-Banach algebra is always* ≥ 1. *As a consequence,* $\|\pi(I)\| = 1$.

Recall that Corollary 3.2.1 exhibits the collection of Fredholm operators on a Hausdorff linear topological space X as $\pi^{-1}(\mathcal{G}(A))$, where A denotes the Calkin algebra and $\mathcal{G}(A)$ denotes the collection of invertible elements of A. If X is a p-Banach space then the continuity of π and the openness of $\mathcal{G}(A)$ from Theorem 3.4.4 combine to give the following result.

Corollary 3.4.10. *The Fredholm operators on a p-Banach space X with* $p \in (0, 1]$ *form an open subset of* $\mathcal{L}(X)$.

3.4.4 Fredholm Operators of Index n

The goal of this subsection is to discuss the continuity of the (Fredholm) index in the class of Fredholm operators on a p-Banach space with $0 < p \leq 1$. For the road ahead, it helps to introduce some notation.

Definition 3.4.11. *For a linear topological space X let:*

(a) $\Phi(X)$ *denote the collection of Fredholm operators on X;*
(b) $\Phi_n(X)$ *denote the collection of Fredholm operators on X of index* $n \in \mathbb{Z}$*;*
(c) $\mathcal{G}(X)$ *denote the collection of invertible operators on X.*

Clearly

$$\Phi(X) = \bigcup_{n\in\mathbb{Z}} \Phi_n(X). \tag{3.4.122}$$

Also, when X is a p-Banach space for some $p \in (0, 1]$, it follows that $\mathcal{L}(X)$ is a p-Banach algebra and

$$\mathcal{G}(X) = \mathcal{G}(\mathcal{L}(X)), \tag{3.4.123}$$

where for a p-Banach algebra A, the collection $\mathcal{G}(A)$ has been introduced in the discussion preceding Theorem 3.4.4. Thus by Theorem 3.4.4 and Corollary 3.4.10 both $\mathcal{G}(X)$ and $\Phi(X)$ are open subsets of $\mathcal{L}(X)$. The next order of business is to show that the same is true of $\Phi_n(X)$, for each $n \in \mathbb{Z}$.

Theorem 3.4.12. *If X is a p-Banach space for some* $p \in (0, 1]$, *then*

$$\text{for each } n \in \mathbb{Z} \text{ the set } \Phi_n(X) \text{ is an open subset of } \mathcal{L}(X). \tag{3.4.124}$$

In view if (3.4.122), *this gives another proof of Corollary 3.4.10, i.e., of the fact that the set of Fredholm operators* $\Phi(X)$ *is open in* $\mathcal{L}(X)$.

Proof. We begin with the case $\dim X < \infty$. In this scenario,

$$\mathcal{L}(X) = \mathcal{K}(X) = \Phi(X) \text{ and } \Phi(X) = \Phi_0(X). \tag{3.4.125}$$

Consequently $O \in \Phi(X)$, where O denotes the zero operator, and as such for each $T \in \Phi(X)$ we may write

$$0 = i(O) = i(OT) = i(O) + i(T), \tag{3.4.126}$$

from which it follows that $i(T) = 0$. Consequently

$$\Phi(X) = \Phi_0(X) \text{ and } \Phi_n = \varnothing, \quad \forall\, n \in \mathbb{Z} \setminus \{0\}. \tag{3.4.127}$$

Combined with the fact that $\Phi(X)$ is open in $\mathcal{L}(X)$ (cf. Corollary 3.4.10), (3.4.127) completes the proof of the theorem in the case when $\dim X < \infty$.

Assume next that $\dim X = \infty$ and let first $n = 0$. By Theorem 3.3.8 an operator $T \in \mathcal{L}(X)$ belongs to $\Phi_0(X)$ if and only if T has the form "invertible + compact." Thus

$$\Phi_0(X) = \mathcal{G}(X) + \mathcal{K}(X) = \bigcup_{K \in \mathcal{K}(X)} (\mathcal{G}(X) + K). \tag{3.4.128}$$

Since every element of the union in (3.4.128) is an open subset of $\mathcal{L}(X)$, the same is true of $\Phi_0(X)$.

Suppose now that $n \in \mathbb{Z}$ is such that $n \neq 0$ and notice that if $\Phi_n(X) = \varnothing$ there is nothing to prove. We remind the reader that this is a possible scenario. Indeed, as seen in Exercise 3.1.11 the only Fredholm operators on $L^p([0,1])$ for $0 < p < 1$ are the invertible ones. We shall consider next the case when $\Phi_n(X) \neq \varnothing$ and we start by fixing an operator $T_0 \in \Phi_n(X)$. Appealing to the openness of $\Phi(X)$ in $\mathcal{L}(X)$, given by Corollary 3.4.10, we may conclude that

$$\exists\, \delta_0 > 0 \text{ such that } B(T_0, \delta_0) \subseteq \Phi(X), \tag{3.4.129}$$

where, for $T \in \mathcal{L}(X)$ and $\delta > 0$ the set $B(T, \delta)$ denotes the open ball in $\mathcal{L}(X)$ of radius δ centered at T. In addition, by Corollary 3.3.14 there exists $S_0 \in \Phi_{-n}(X)$, and based on Theorem 3.3.15 we have that $i(S_0 T_0) = 0$. Since we have already shown that $\Phi_0(X)$ is open in $\mathcal{L}(X)$, we may infer that

$$\exists\, \delta_1 > 0 \text{ such that } B(S_0 T_0, \delta_1) \subseteq \Phi_0(X). \tag{3.4.130}$$

Notice that $S_0 \in \Phi_{-n}(X)$ and $\dim X = \infty$ guarantee that $\|S_0\| \neq 0$. Thus, it is meaningful to define $\delta > 0$ by setting

$$\delta := \min\{\delta_0, \delta_1/\|S_0\|\}. \tag{3.4.131}$$

Consider next $T \in B(T_0, \delta)$, thus $\|T - T_0\| < \delta$ and $T \in \Phi(X)$, by (3.4.129) and (3.4.131). Consequently,

$$\|S_0 T - S_0 T_0\| \leq \|S_0\| \cdot \|T - T_0\| < \|S_0\| \cdot \delta \leq \delta_1. \tag{3.4.132}$$

This and (3.4.130) imply $S_0 T \in B(S_0 T_0, \delta_1) \subseteq \Phi_0(X)$. With this in hand, Theorem 3.3.15 yields

$$i(T) = -i(S_0) = n, \tag{3.4.133}$$

i.e., $T \in \Phi_n(X)$. This shows that $B(T_0, \delta) \subseteq \Phi_n(X)$ finishing the proof of the fact that $\Phi_n(X)$ is open in $\mathcal{L}(X)$.

Finally, $\Phi(X) = \bigcup_{n\in\mathbb{Z}} \Phi_n(X)$, and consequently this is open in $\mathcal{L}(X)$ since each $\Phi_n(X)$ is. The proof of Theorem 3.4.12 is now complete. □

Several notable consequences are singled out below.

Corollary 3.4.13. *If X is a p-Banach space for some $p \in (0, 1]$, then the following hold:*

(a) The index function

$$i : \Phi(X) \longrightarrow \mathbb{Z} \tag{3.4.134}$$

is continuous on $\Phi(X)$ (endowed with the relative topology induced from $\mathcal{L}(X)$), where $\mathbb{Z}$ is equipped with the discrete topology.

(b) For each $n \in \mathbb{Z}$, the set $\Phi_n(X)$ of Fredholm operators of index n is both open and closed in $\Phi(X)$, i.e., is a clopen subset of $\Phi(X)$. In particular, since $\Phi(X)$ is open in $\mathcal{L}(X)$ (cf. Theorem 3.4.12), it follows that $\Phi_n(X)$ is open in $\mathcal{L}(X)$ for each $n \in \mathbb{Z}$ (thus recovering Theorem 3.4.12).

Proof. Fix $n \in \mathbb{Z}$. Then the pre-image of the open set $\{n\} \subset \mathbb{Z}$ of the function i in (3.4.134) is $\Phi_n(X)$, which we have just seen in Theorem 3.4.12 to be open in $\mathcal{L}(X)$, hence also open in $\Phi(X)$ equipped with the relative topology. Since every open set in $\mathbb{Z}$ equipped with the discrete topology is a union of singletons, this establishes the continuity of the Fredholm index claimed in item *(a)*. As far as item *(b)* is concerned, each singleton $\{n\}$ is closed in $\mathbb{Z}$ and i is continuous, therefore $\Phi_n(X)$ is also closed in $\Phi(X)$. □

Another way of proving Theorem 3.4.12 is outlined in the next exercise; the idea is to reduce the whole problem to just proving that $\Phi_0(X)$ is open in $\mathcal{L}(X)$, which is more manageable.

Exercise 3.4.14. Suppose X is an infinite-dimensional p-Banach space, for some $p \in (0, 1]$.

(1) Show that $\Phi_0(X)$ is open in $\mathcal{L}(X)$.

(2) Suppose next that $n \in \mathbb{Z}$ is such that $n \geq 0$ and $\Phi_n(X) \neq \varnothing$. Show that there exists a left-invertible operator $S \in \Phi_{-n}(X)$. Use this to define the assignment

$$\Phi_n(X) \ni T \longmapsto TS \in \Phi_0(X). \tag{3.4.135}$$

Use the addition formula (cf. Theorem 3.3.15) to prove that the above map is indeed well defined. Also, show that the map (3.4.135) is continuous in the norm topology on $\mathcal{L}(X)$.

(3) If L is any left-inverse of the operator S then $L \in \Phi_n(X)$, and for every $A \in \Phi_0(X)$ we have $A = (AL)S$, with $AL \in \Phi_n(X)$. Thus, $\mathcal{L}(X) \ni T \mapsto TS \in \mathcal{L}(X)$ is a continuous map and takes $\Phi_n(X)$ *onto* $\Phi_0(X)$. At this stage, conclude (based on item *(1)* and Proposition 1.1.8) that $\Phi_n(X)$ is an open set.

(4) Formulate a similar argument to *(1)-(3)* above that works for each integer $n < 0$.

Hint: For item (1) use Theorem 3.3.8, Theorem 3.4.4 with $A := \mathcal{L}(X)$, and (3.4.100). For the first claim in item (2), combine Theorem 3.3.13 with Proposition 3.3.12. In item (3), use Proposition 3.3.9 to show that $L \in \Phi_n(X)$.

Here is another exercise shedding light on the nature of the collection of Fredholm operators.

Exercise 3.4.15. (a) Let X be a nontrivial p-Banach space for some $p \in (0, 1]$. Suppose $T \in \mathcal{L}(X)$ is left-invertible, i.e., there exists some $L \in \mathcal{L}(X)$ such that $LT = I$ where I is the identity operator on X. Show that if $R \in \mathcal{L}(X)$ satisfies $\|T-R\| < \|L\|^{-1}$, then necessarily $\operatorname{ran} R \cap \ker L = \{0\}$.

(b) Suppose S is the forward shift of infinite multiplicity on a Hilbert space. That is, if H is a given infinite dimensional Hilbert space and one considers

$$X := \left\{(x_1, x_2, \dots) : \ x_n \in H \ \text{ for each } n \in \mathbb{N} \text{ and } \sum_{n=1}^{\infty} \|x_n\|_H^2 < \infty\right\} \tag{3.4.136}$$

(which becomes a Hilbert space itself when equipped with the natural inner product), then S is the operator

$$\begin{aligned} S : X \to X \ \text{ defined by } \ Sx &:= (0, x_1, x_2, \dots) \\ \text{for each } \ x &= (x_1, x_2, \dots) \in X. \end{aligned} \tag{3.4.137}$$

Prove that the open unit ball in $\mathcal{L}(X)$ centered at S contains no Fredholm operators.

(c) Show that $\Phi(X)$ may not be dense in $\mathcal{L}(X)$ even when X is a Hilbert space. If X is an infinite-dimensional p-Banach space for some $p \in (0, 1]$, is $\Phi(X)$ closed in $\mathcal{L}(X)$?

Hint: To deal with part (a), let $x \in \operatorname{ran} R \cap \ker L$. Then $Lx = 0$ and there exists $y \in X$ such that $x = Ry$. Use $LRy = Lx = 0$ and $LT = I$ to write $y = L(T - R)y$. Taking norms, we get

$$\|y\| \leq \|L\| \cdot \|T - R\| \cdot \|y\|. \tag{3.4.138}$$

But $\|T - R\| < \|L\|^{-1}$, so $\|L\| \cdot \|T - R\| < 1$. Use this back in (3.4.138) to conclude that $y = 0$.

For the current part (b), bring in the backward shift operator

$$\begin{aligned} B : X \to X \ \text{ defined by } \ Bx &:= (x_2, x_3, \dots) \\ \text{for each } \ x &= (x_1, x_2, \dots) \in X. \end{aligned} \tag{3.4.139}$$

Show that $\|B\| = 1$ and $BS = I$ on X. If the open unit ball in $\mathcal{L}(X)$ centered at S contains a Fredholm operator T, then according to part *(a)* we have $\operatorname{ran} T \cap \ker B = \{0\}$. However, $\ker B$ is infinite dimensional (since it is isomorphic to the infinite dimensional Hilbert space H), and this prevents T for having a range of finite codimension (according to Exercise 1.3.40).

For the first claim in part (c), use part (b). For the last claim in part (c), consider the sequence given by $T_n := n^{-1}I$ for each $n \in \mathbb{N}$, where I is the identity operator on X.

We close this subsection with a discussion which shows that the index is *not* always continuous for Fredholm operators on Fréchet spaces.

Exercise 3.4.16. Let X denote the space of all scalar (= $\mathbb{R}$, or $\mathbb{C}$) sequences, endowed with the topology of coordinatewise convergence (i.e., $X = \mathbb{R}^{\mathbb{N}}$, or $X = \mathbb{C}^{\mathbb{N}}$, with the product topology).

(a) Show that X is a topological vector space which is locally convex (i.e., its topology has a base of convex neighborhoods).
(b) Show that the topology of X is given by the translation-invariant metric

$$d(x, y) := \sum_{n=1}^{\infty} \frac{1}{2^n} \frac{|\xi_n - \eta_n|}{1 + |\xi_n - \eta_n|} \tag{3.4.140}$$

for each pair $x = (\xi_n)_{n \in \mathbb{N}}$ and $y = (\eta_n)_{n \in \mathbb{N}}$ of vectors in X, and that X is complete in this metric (i.e., $\{x_j\}_j \subseteq X$ converges to $x \in X$ coordinatewise if and only if $d(x_j, x) \to 0$ as $j \to \infty$; thus, X is a Fréchet space[1]).
(c) Let $T : X \to X$ denote the forward shift operator, defined by

$$T((\xi_1, \xi_2, \ldots)) := (0, \xi_1, \xi_2, \ldots) \quad \text{for each} \quad (\xi_1, \xi_2, \ldots) \in X. \tag{3.4.141}$$

Show that T is a Fredholm operator on X of index -1, but that $T - \lambda I$ is an isomorphism of X (hence Fredholm of index 0 on X) for each non-zero scalar λ.

Hint: Here is an outline of proof of part (c). The fact that $T - \lambda I$ is injective is clear. To show surjectivity, given any $y \in X$ find $x \in X$ which solves the equation $y = (T - \lambda I)x$ either by brute force, or by showing that the geometric series expansion for the "formal expression" $\frac{1}{T/\lambda - I}$ actually converges in X. The Open Mapping Theorem (available since X is an F-space; cf. Definition 1.6.6 and Theorem 1.6.7) then guarantees that $(T - \lambda I)^{-1}$ is continuous on X. Conclude that the Fredholm index on $\Phi(X)$ is *discontinuous* at T.

3.5 Fredholm Theory and Spectra

The goal of this section is to use the Fredholm theory developed earlier to further study the *spectrum* of a linear operator; cf. Definition 2.3.9. For motivation, recall that, according to Theorem 2.3.10, if K is a compact operator on an infinite-dimensional Hausdorff linear topological space X, then $\sigma(K; X)$ consists of the

[1] recall that a Fréchet space is a locally convex topological vector space, whose topology is metrizable by a complete translation-invariant metric; equivalently, a Fréchet space is a locally convex F-space – see, e.g., [39, §1.8, items *(e)* and *(f)*, p. 9]

origin (since the operator is not invertible) and a (possibly empty) set of finite-multiplicity eigenvalues which, if not finite, can be written as a sequence converging to zero. Furthermore, $K \notin \Phi(X)$ (see Exercise 3.1.12), while if $\lambda \in \sigma(K; X)$, $\lambda \neq 0$, then $K - \lambda I$ is a Fredholm operator of index zero (item *(a)* in Example 3.1.3 and Corollary 3.3.3).

We will see that this spectral configuration is a special case of a beautiful result that applies to all continuous linear operators on p-Banach spaces. We begin by reviewing, in the context of p-Banach algebras, some elementary properties of spectra.

3.5.1 Spectral Basics

In what follows, A is a p-Banach algebra (for some $p \in (0,1]$) with identity. As usual, we assume the identity element e has p-norm one (see the discussion in Exercise 3.4.3), a case in which we shall say that *the identity is normalized.* In our applications, X will be a p-Banach space and A will either be $\mathcal{L}(X)$, the algebra of continuous linear operators on a p-Banach space X, or $\mathcal{L}(X)/\mathcal{K}(X)$, the Calkin algebra of X. In relation to these, the reder is advised to recall Definition 2.3.9, (3.4.100), and Proposition 3.4.8.

Definition 3.5.1. *Let A be a p-Banach algebra, for some $p \in (0,1]$, with identity e which is normalized. The* `spectrum` *of an element $x \in A$, denoted by $\sigma(x; A)$, is defined as*

$$\sigma(x; A) := \{\lambda \in \mathbb{C} : \lambda e - x \ \text{ is not invertible on } A\}. \tag{3.5.142}$$

The complement of the spectrum of x (the set of $\lambda \in \mathbb{C}$ such that $x - \lambda e$ is invertible) is called the `resolvent` *of x and is denoted by $\rho(x; A)$, i.e.,*

$$\rho(x; A) := \{\lambda \in \mathbb{C} : \lambda e - x \ \text{ is invertible on } A\}. \tag{3.5.143}$$

The manner in which (3.5.142) is compatible with notation introduced earlier in Definition 2.3.9 is via the identification of $\sigma(T; X)$ with $\sigma(T; \mathcal{L}(X))$.

The next result is well known for Banach algebras, and the proof for that case works as well for p-Banach algebras. We present the proof in the interest of completeness.

Theorem 3.5.2. *Let A be a p-Banach algebra with an identity which is normalized. Then, for every $x \in A$ the resolvent $\rho(x; A)$ is an open subset of $\mathbb{C}$, and the spectrum $\sigma(x; A)$ is compact in $\mathbb{C}$.*

Proof. Denote by e the identity element in A, which is normalized. Fix $x \in A$ and recall that the set $\mathcal{G}(A)$ of invertible elements of A is open in A (Theorem 3.4.4). Note that the function $\mathbb{C} \ni \lambda \mapsto x - \lambda e \in A$ is continuous (in fact, distance preserving). Since $\rho(x; A)$ is the pre-image under this continuous function of the open set $\mathcal{G}(A)$, it follows that $\rho(x;A)$ is open in $\mathbb{C}$. Its complement, $\sigma(x;A)$, is therefore closed. To show that $\sigma(x;A)$ is compact, it remains to verify that it is bounded and, to this end, we claim that

$$\sigma(x;A) \subseteq \{\lambda \in \mathbb{C} : |\lambda| \leq \|x\|^{1/p}\}. \tag{3.5.144}$$

To see this, let $\lambda \in \mathbb{C} \setminus \{0\}$ and notice that estimate (3.4.105) of Theorem 3.4.4 (The Invertibility Theorem) applied with $x_0 := -\lambda e$ and x replaced by $x-\lambda e$ amounts to having

$$|\lambda| > \|x\|^{1/p}. \tag{3.5.145}$$

Using Theorem 3.4.4 this shows that $x - \lambda e$ is invertible whenever (3.5.145) holds. The inclusion claimed in (3.5.144) now follows from this observation and the definition in (3.5.142). □

Exercise 3.5.3. Prove the claim made in (3.5.144) by showing that whenever (3.5.145) holds, the Neumann series

$$-\frac{1}{\lambda} \cdot \sum_{k=0}^{\infty} \left(\frac{x}{\lambda}\right)^k \tag{3.5.146}$$

converges to an element of A which is the inverse of $x - \lambda e$.

3.5.2 Spectra: Essential Versus Inessential

Assume X is a p-Banach space for some $p \in (0,1]$, and $T \in \mathcal{L}(X)$. The goal is to further parse the spectrum of T into its "essential" and "inessential" parts. These categories are formally introduced below:

Definition 3.5.4. *If X is a p-Banach space for some $p \in (0,1]$ and $T \in \mathcal{L}(X)$, define the* `essential spectrum` *of T, denoted by $\sigma_e(T;X)$, by setting*

$$\sigma_e(T;X) := \big\{\lambda \in \mathbb{C} : \lambda I - T \ \text{ is not Fredholm on } X\big\}. \tag{3.5.147}$$

The complement of $\sigma_e(T;X)$ is called the `essential resolvent` *of the operator T and is denoted by $\rho_e(T;X)$. Thus,*

$$\rho_e(T;X) = \mathbb{C} \setminus \sigma_e(T;X) = \big\{\lambda \in \mathbb{C} : \lambda I - T \ \text{ is a Fredholm operator on } X\big\}. \tag{3.5.148}$$

Finally, call each point in $\sigma(T;X) \setminus \sigma_e(T;X)$ an `inessential point` *of $\sigma(T;X)$.*

Upon recalling from Corollary 3.2.1 that an operator $T \in \mathcal{L}(X)$ is Fredholm if and only if its coset $\pi(T) := T + \mathcal{K}(X)$ is invertible in the Calkin algebra $\mathcal{L}(X)/\mathcal{K}(X)$, it follows that the essential spectrum of T is just the spectrum of $\pi(T)$ in the Calkin algebra, i.e.,

$$\begin{aligned}&\sigma_e(T;X) \text{ from Definition 3.5.4 is precisely}\\ &\sigma(\pi(T);\mathcal{L}(X)/\mathcal{K}(X)) \text{ considered in the sense of Definition 3.5.1.}\end{aligned} \tag{3.5.149}$$

Thus, by Theorem 3.5.2 (cf. also (3.4.100)),

> if X is a p-Banach space for some $p \in (0, 1]$ and $T \in \mathcal{L}(X)$, then the essential spectrum $\sigma_e(T;X)$ is a compact set in $\mathbb{C}$, and the essential resolvent $\rho_e(T;X)$ is an open set in $\mathbb{C}$. (3.5.150)

In addition, since invertible operators are Fredholm, we clearly have

$$\sigma_e(T;X) \subseteq \sigma(T;X), \qquad \forall\, T \in \mathcal{L}(X). \tag{3.5.151}$$

Exercise 3.5.5. (a) Suppose X is an infinite-dimensional p-Banach space for some $p \in (0, 1]$ and suppose that $T \in \mathcal{K}(X)$. Show that $\sigma_e(T;X) = \{0\}$. (**Suggestion**: use (3.3.66) and Exercise 3.3.17.)

(b) Suppose $\mathbb{T}$ is the `unit circle` in $\mathbb{C}$, i.e.,

$$\mathbb{T} := \{z \in \mathbb{C} : |z| = 1\}, \tag{3.5.152}$$

and denote by T the operator of "multiplication by the variable" acting on $C(\mathbb{T})$, the space of continuous functions on $\mathbb{T}$ equipped with the uniform norm. That is, given any $f \in C(\mathbb{T})$, we set $(Tf)(z) := zf(z)$ for every $z \in \mathbb{T}$. Show that

$$\sigma(T;C(\mathbb{T})) = \sigma_e(T;C(\mathbb{T})) = \mathbb{T}, \qquad \forall\, z \in \mathbb{T}. \tag{3.5.153}$$

(c) Suppose T is, again, the operator of multiplication by the variable, but this time operating on $X := A$, the `disk algebra` (i.e., the collection of functions continuous on the closed unit disk $\overline{\mathbb{D}}$ and holomorphic in the open unit disk $\mathbb{D}$, equipped with the uniform norm on the closed disk). Show that, this time, $\sigma(T;X)$ is the closed unit disk, but $\sigma_e(T;X)$ is the unit circle.

(d) Establish similar results as in part (b) for the operator of multiplication by the variable, this time acting on $L^2(\mathbb{T})$.

(e) Establish similar results as in part (c) for the operator of multiplication by the variable, this time acting on the `Hardy space` H^2 of the disk, defined as

$$H^2(\mathbb{D}) := \Big\{ f \text{ holomorphic in } \mathbb{D} : \|f\|_{H^2} := \sup_{0<r<1} \int_0^{2\pi} r|f(re^{i\theta})|^2\, d\theta < \infty \Big\}. \tag{3.5.154}$$

Hint: For part (c), to show that for every $\lambda \in \mathbb{D}$ the operator $\lambda I - T$ is Fredholm on X, argue along the following lines. First, use the fact if $f \in X$ is a function satisfying $(\lambda - z)f(z) = 0$ for every $z \in \mathbb{D}$ then by unique continuation for holomorphic functions one necessarily has $f = 0$ in $\mathbb{D}$, and hence also in $\overline{\mathbb{D}}$ by continuity. Conclude that $\lambda I - T$ is injective on X. Second, show that

$$\operatorname{ran}(\lambda I - T) = \{g \in X : g(\lambda) = 0\}, \tag{3.5.155}$$

then observe that, with constant functions identified with $\mathbb{C}$,

$$\{g \in X : g(\lambda) = 0\} \boxplus \mathbb{C} = X. \tag{3.5.156}$$

To justify (3.5.156), note that if for each $f \in X$ we define

$$F(z) := \begin{cases} \frac{f(z)-f(\lambda)}{z-\lambda} & \text{if } z \in \overline{\mathbb{D}} \setminus \{\lambda\}, \\ f'(\lambda) & \text{if } z = \lambda, \end{cases} \tag{3.5.157}$$

then $F \in X$ and f may be expressed as $f = -(\lambda I - T)F + f(\lambda)$.

Exercise 3.5.6 (Weyl's Theorem). Show that if X is a Hausdorff linear topological space and $T, S \in \mathcal{L}(X)$ differ by a compact operator on X, then their spectra differ by at most Fredholm points of index zero.

Hint: Given $\lambda \in \sigma(T;X) \setminus \sigma(S;X)$, write $\lambda I - T$ as $\lambda I - S + (S - T)$ and use (3.3.66).

3.5.3 Fredholm Decomposition of the Essential Resolvent

We continue to assume that X is a p-Banach space for some $p \in (0, 1]$. As usual, I stands for the identity operator on X. Fix an operator $T \in \mathcal{L}(X)$. Then the Fredholm index associates to each point $\lambda \in \rho_e(T;X)$ the integer

$$i_T(\lambda) := i(T - \lambda I), \tag{3.5.158}$$

called the T-index of λ. The map

$$i_T : \rho_e(T;X) \longrightarrow \mathbb{Z} \ \text{ is continuous,} \tag{3.5.159}$$

since this is the composition of two continuous functions, specifically: the map $\rho_e(T;X) \ni \lambda \mapsto T - \lambda I \in \Phi(X)$, followed by the Fredholm index $i : \Phi(X) \to \mathbb{Z}$ (cf. Corollary 3.4.13). Observe that the essential resolvent of T decomposes into the *disjoint union* of sets

$$\rho_e(T;X) = \bigsqcup_{n\in\mathbb{Z}} \varphi_n(T;X), \tag{3.5.160}$$

where

$$\varphi_n(T;X) := i_T^{-1}(\{n\}) = \{\lambda \in \rho_e(T;X) : i(T - \lambda I) = n\}, \qquad \forall\, n \in \mathbb{Z}. \tag{3.5.161}$$

In the sequel, we shall refer to complex numbers in $\varphi_n(T;X)$ as Fredholm points of index n (for the operator T).

In view of (3.5.159) and the fact that singletons are simultaneously open and closed in $\mathbb{Z}$, we conclude from Proposition 1.1.8 and Exercise 1.1.10 that

$$\text{each set } \varphi_n(T;X) \text{ is both open and relatively closed} \\ \text{(aka "clopen") in the essential resolvent } \rho_e(T;X). \tag{3.5.162}$$

Proposition 3.5.7. *Let X be a p-Banach space for some $p \in (0, 1]$ and let $T \in \mathcal{L}(X)$. Then for each $n \in \mathbb{Z}$ the set $\varphi_n(T;X)$ is a union of connected components of $\rho_e(T;X)$.*

Proof. Since the connected components of any set are its *minimal* clopen subsets, this follows at a glance from (3.5.162). □

The decomposition of the essential resolvent into the sets $\varphi_n(T;X)$, and its relationship with $\sigma(T;X)$, will occupy the rest of this section. We will concentrate on the fundamental differences between the properties of the sets $\varphi_0(T;X)$ and $\varphi_n(T;X)$ for $n \neq 0$. To start, here is one very simple such difference:

$$\text{for each integer } n \neq 0 \text{ the set } \varphi_n(T;X) \text{ is} \\ \text{contained in } \sigma(T;X)\text{, and is therefore bounded.} \tag{3.5.163}$$

In particular, from Proposition 3.5.7 and (3.5.163) we see that

$$\varphi_n(T;X) \text{ with } n \neq 0 \text{ is a union of } \textit{bounded} \text{ components of } \rho_e(T;X). \tag{3.5.164}$$

In stark contrast with the case of non-zero index discussed above,

$$\text{the set } \varphi_0(T;X) \text{ is always } \textit{unbounded.} \tag{3.5.165}$$

Indeed, it follows from (2.3.135) and the discussion pertaining to (3.5.145) in the proof of Theorem 3.5.2 that

$$\{z \in \mathbb{C} : |z| > \|T\|^{1/p}\} \subseteq \rho(T;X) \subseteq \varphi_0(T;X). \tag{3.5.166}$$

In particular, $\varphi_0(T;X)$ is never contained entirely in the spectrum $\sigma(T;X)$.

To further shed light on the structure of the sets $\varphi_n(T;X)$, we make the following definition.

Definition 3.5.8. *Let X be a p-Banach space for some $p \in (0, 1]$ and let $T \in \mathcal{L}(X)$. Call a* `hole` *any bounded connected component of $\rho_e(T;X)$.*

An immediate consequence of this definition and (3.5.164) is as follows.

Corollary 3.5.9. *Let X be a p-Banach space for some $p \in (0, 1]$ and let $T \in \mathcal{L}(X)$. Then for each integer $n \in \mathbb{Z} \setminus \{0\}$ the set $\varphi_n(T;X)$ is a union of holes.*

As some of the examples below show, $\varphi_0(T;X)$ may even be completely disjoint from the spectrum $\sigma(T;X)$.

Exercise 3.5.10. (a) Suppose X is an infinite-dimensional p-Banach space, for some $p \in (0, 1]$, and fix some $T \in \mathcal{K}(X)$. Show that $\varphi_0(T;X) \cap \sigma(T;X)$ is either empty, a finite set, or a sequence tending to zero (see Theorem 2.1.9). Also, show that $\varphi_n(T;X)$ is empty for $n \neq 0$ (see Exercise 3.5.5 and Exercise 3.3.17).

(b) Suppose T is the operator of multiplication-by-the-variable on $X := C(\mathbb{T})$, or on the space $X := L^2(\mathbb{T})$, as discussed in items (b) and (d) in Exercise 3.5.5. Show that $\sigma(T;X) = \sigma_e(T;X) = \mathbb{T}$, while $\varphi_0(T;X) = \mathbb{C}\setminus\mathbb{T}$, and $\varphi_n(T;X)$ is empty for every integer $n \neq 0$. (**Suggestion**: For the very last claim, show that for each $\lambda \in \mathbb{T}$ the space $\{f \in C(\mathbb{T}) : f(\lambda) = 0\}$ has infinite codimension in $C(\mathbb{T})$, by bringing in the linearly independent family $\{e^{imz} : m \in \mathbb{Z}\} \subseteq C(\mathbb{T})$ and making use of Exercise 1.3.40.)

(c) Let X be either the disk algebra A (i.e., the collection of functions continuous on the closed unit disk and holomorphic in the open unit disk, equipped with the uniform norm on the closed disk), or the Hardy space $H^2(\mathbb{D})$ defined in (3.5.154). Let T be multiplication-by-the-variable on X. Show that the essential spectrum of T is still the unit circle, but now the spectrum is the closed unit disk, $\varphi_{-1}(T;X)$ is the open unit disk, and $\varphi_0(T;X)$ is the complement of the closed unit disk. Also, show that all the other sets $\varphi_n(T;X)$ are empty. (**Suggestion**: Reason as in the hint given for part *(c)* of Exercise 3.5.5.)

Proposition 3.5.11. *Recall the forward shift operator S on ℓ^2, and the backward shift operator B on ℓ^2 (cf. (2.2.91)). Then*

$$\begin{gathered}\sigma(S;\ell^2) = \overline{\mathbb{D}}, \quad \sigma_e(S;\ell^2) = \mathbb{T},\\ \varphi_0(S;\ell^2) = \mathbb{C}\setminus\overline{\mathbb{D}}, \quad \varphi_{-1}(S;\ell^2) = \mathbb{D},\\ \textit{and } \varphi_n(S;\ell^2) = \varnothing \textit{ for all } n \in \mathbb{Z}\setminus\{-1,0\},\end{gathered} \tag{3.5.167}$$

and

$$\begin{gathered}\sigma(B;\ell^2) = \overline{\mathbb{D}}, \quad \sigma_e(B;\ell^2) = \mathbb{T},\\ \varphi_0(B;\ell^2) = \mathbb{C}\setminus\overline{\mathbb{D}}, \quad \varphi_{1}(B;\ell^2) = \mathbb{D},\\ \textit{and } \varphi_n(B;\ell^2) = \varnothing \textit{ for all } n \in \mathbb{Z}\setminus\{0,1\}.\end{gathered} \tag{3.5.168}$$

Proof. Let $\lambda \in \mathbb{C}$. Then $y = (y_1, y_2, \dots) \in \ell^2$ belongs to $\operatorname{ran}(\lambda I_{\ell^2} - S)$ if and only if there exists $x = (x_1, x_2, \dots) \in \ell^2$ such that $(\lambda I_{\ell^2} - S)x = y$. Explicitly, the latter equality is equivalent to

$$y_1 = \lambda x_1, \tag{3.5.169}$$

$$y_2 = \lambda x_2 - x_1, \tag{3.5.170}$$

$$y_3 = \lambda x_3 - x_2, \tag{3.5.171}$$

$$\cdots\cdots\cdots\cdots\cdots \tag{3.5.172}$$

which is further equivalent to having

$$x_n = \sum_{k=1}^{n} \frac{y_k}{\lambda^{n+1-k}} \quad \text{for each} \quad n \in \mathbb{N}. \tag{3.5.173}$$

In the case when $|\lambda| > 1$, we claim that if for any given $y = (y_1, y_2, \dots) \in \ell^2$ we define the terms of the sequence $x = (x_1, x_2, \dots)$ as in (3.5.173) then $x \in \ell^2$. Indeed,

(3.5.173) implies that for each $n \in \mathbb{N}$ we have

$$
\begin{aligned}
|x_n| &\le \sum_{k=1}^{n} \frac{|y_k|}{|\lambda|^{n+1-k}} = \sum_{k=1}^{n} \Big(\frac{|y_k|}{|\lambda|^{(n+1-k)/2}}\Big)\Big(\frac{1}{|\lambda|^{(n+1-k)/2}}\Big) \\
&\le \left[\sum_{k=1}^{n} \Big(\frac{|y_k|}{|\lambda|^{(n+1-k)/2}}\Big)^2\right]^{1/2} \left[\sum_{k=1}^{n} \Big(\frac{1}{|\lambda|^{(n+1-k)/2}}\Big)^2\right]^{1/2} \\
&= \left[\sum_{k=1}^{n} \frac{|y_k|^2}{|\lambda|^{n+1-k}}\right]^{1/2} \left[\sum_{k=1}^{n} \frac{1}{|\lambda|^{n+1-k}}\right]^{1/2} \\
&= \left[\sum_{k=1}^{n} \frac{|y_k|^2}{|\lambda|^{n+1-k}}\right]^{1/2} \left[\sum_{j=1}^{n} \frac{1}{|\lambda|^{j}}\right]^{1/2} \\
&\le \left[\sum_{k=1}^{n} \frac{|y_k|^2}{|\lambda|^{n+1-k}}\right]^{1/2} \left[\sum_{j=1}^{\infty} \frac{1}{|\lambda|^{j}}\right]^{1/2} \\
&= \frac{1}{|\lambda|-1} \left[\sum_{k=1}^{n} \frac{|y_k|^2}{|\lambda|^{n+1-k}}\right]^{1/2},
\end{aligned} \tag{3.5.174}
$$

by the Cauchy-Schwarz inequality, re-indexing $j := n + 1 - k$, and using the formula for the geometric series with ratio $|\lambda|^{-1} < 1$. Consequently, after squaring and summing up in n,

$$
\begin{aligned}
\sum_{n=1}^{\infty} |x_n|^2 &= \frac{1}{(|\lambda|-1)^2} \sum_{n=1}^{\infty} \left(\sum_{k=1}^{n} \frac{|y_k|^2}{|\lambda|^{n+1-k}}\right) \\
&= \frac{1}{(|\lambda|-1)^2} \sum_{k=1}^{\infty} \left(\sum_{n=k}^{\infty} \frac{|y_k|^2}{|\lambda|^{n+1-k}}\right) \\
&= \frac{1}{(|\lambda|-1)^2} \sum_{k=1}^{\infty} |y_k|^2 \left(\sum_{n=k}^{\infty} \frac{1}{|\lambda|^{n+1-k}}\right) \\
&= \frac{1}{(|\lambda|-1)^2} \sum_{k=1}^{\infty} |y_k|^2 \left(\sum_{j=1}^{\infty} \frac{1}{|\lambda|^{j}}\right) \\
&= \frac{1}{(|\lambda|-1)^3} \sum_{k=1}^{\infty} |y_k|^2 < +\infty,
\end{aligned} \tag{3.5.175}
$$

since $y = (y_1, y_2, \dots) \in \ell^2$. Thus, $x \in \ell^2$ as claimed. In conclusion, the above argument proves that

$$
|\lambda| > 1 \implies \operatorname{ran}(\lambda I_{\ell^2} - S) = \ell^2. \tag{3.5.176}
$$

From the fact that $(\lambda I_{\ell^2} - S)x = y$ is equivalent with having (3.5.173), we also conclude (taking $y := 0$) that

$$\ker(\lambda I_{\ell^2} - S) = \{0\} \quad \text{for each} \quad \lambda \in \mathbb{C}. \tag{3.5.177}$$

For further reference, note that combining (3.5.176) with (3.5.177) and Theorem 1.6.7 gives

$$|\lambda| > 1 \implies \lambda I_{\ell^2} - S \ \text{ invertible on } \ \ell^2. \tag{3.5.178}$$

To study the spectral properties of the backward shift operator $B : \ell^2 \to \ell^2$, start from the realization that we have $\|B\| = 1$. Since for every $\lambda \in \mathbb{C}$ with $|\lambda| > \|B\| = 1$ we can invert the operator $\lambda I_{\ell^2} - B = \lambda\big(I_{\ell^2} - \frac{B}{\lambda}\big)$ via the absolutely convergent Neumann series (cf. Exercise 3.5.3)

$$\lambda^{-1} \sum_{n=0}^{\infty} \Big(\frac{B}{\lambda}\Big)^n \in \mathcal{L}(\ell^2), \tag{3.5.179}$$

we conclude that

$$|\lambda| > 1 \implies B : \ell^2 \to \ell^2 \ \text{ is an invertible operator.} \tag{3.5.180}$$

Next consider the case when $\lambda \in \mathbb{C}$ satisfies $|\lambda| < 1$. Pick a complex sequence $(y_0, y_1, y_2, \dots)$ and consider the sequence $x = (x_1, x_2, \dots)$ whose terms are given by

$$x_n := -\sum_{k=0}^{n-1} y_k \lambda^{n-1-k} \quad \text{for each} \quad n \in \mathbb{N}. \tag{3.5.181}$$

Then, much as we have reasoned in (3.5.174)-(3.5.175) but now with λ replaced by λ^{-1}, we see that if $\sum_{n=0}^{\infty} |y_n|^2 < \infty$ then $x \in \ell^2$. With this in hand, straight from definitions one can then check that $(\lambda I_{\ell^2} - B)x = (y_1, y_2, \dots)$. This argument shows that

$$|\lambda| < 1 \implies \operatorname{ran}(\lambda I_{\ell^2} - B) = \ell^2. \tag{3.5.182}$$

Regarding injectivity, fix $\lambda \in \mathbb{C}$ and assume $x = (x_1, x_2, \dots) \in \ell^2 \setminus \{0\}$ is a sequence with the property that $(\lambda I_{\ell^2} - B)x = 0$. Then

$$x_n = \lambda^{n-1} x_1 \quad \text{for each} \quad n \in \mathbb{N} \tag{3.5.183}$$

hence the nonzero sequence x belongs to the one-dimensional linear span of the the sequence $(1, \lambda, \lambda^2, \dots)$. This forces the latter sequence to be in ℓ^2, hence necessarily $|\lambda| < 1$. Conversely, if $|\lambda| < 1$ then every complex sequence belonging to the one-dimensional linear span of the the sequence $(1, \lambda, \lambda^2, \dots)$ is in the kernel of the operator $\lambda I_{\ell^2} - B : \ell^2 \to \ell^2$. This analysis proves that

$$\dim\ker(\lambda I_{\ell^2} - B) = \begin{cases} 0 & \text{if } |\lambda| \geq 1, \\ 1 & \text{if } |\lambda| < 1, \end{cases} \tag{3.5.184}$$

and, in fact,

$$\ker(\lambda I_{\ell^2} - B) = \operatorname{span}_{\mathbb{C}}\{(1, \lambda, \lambda^2, \dots)\} \text{ if } |\lambda| < 1. \tag{3.5.185}$$

What we have proved up to this stage is enough to conclude that

$$|\lambda| < 1 \implies \lambda I_{\ell^2} - B : \ell^2 \to \ell^2 \text{ is a Fredholm operator with index 1.} \tag{3.5.186}$$

Moving on, a direct computation based on definitions, shows that if $\langle \cdot, \cdot \rangle$ is the inner product in the Hilbert space ℓ^2, then

$$\langle S x, y \rangle = \langle x, B y \rangle \text{ for any two sequences } x, y \in \ell^2. \tag{3.5.187}$$

In other words, $S^* = B$ and $B^* = S$ (cf. Exercise 1.5.12). As such,

$$(\overline{\lambda} I_{\ell^2} - B)^* = \lambda I_{\ell^2} - S \text{ for all } \lambda \in \mathbb{C}. \tag{3.5.188}$$

Granted (3.5.186) and (3.5.188), item *(3)* in Theorem 0.0.1 then guarantees that

$$|\lambda| < 1 \implies \lambda I_{\ell^2} - S : \ell^2 \to \ell^2 \text{ is a Fredholm operator with index } -1. \tag{3.5.189}$$

Incidentally, from (3.5.189) and (3.5.177) we conclude that

$$|\lambda| < 1 \implies \operatorname{codim} \operatorname{ran}(\lambda I_{\ell^2} - S) = 1, \tag{3.5.190}$$

which may then be used to show that, in fact,

$$\operatorname{ran}(\lambda I_{\ell^2} - S) = \big[\operatorname{span}_{\mathbb{C}}\{(1, \lambda, \lambda^2, \dots)\}\big]^{\perp} \text{ if } |\lambda| < 1. \tag{3.5.191}$$

At this stage, we are ready to draw conclusions regarding the nature of the spectrum of S. Specifically,

$$\mathbb{D} \subseteq \sigma(S; \ell^2) \subseteq \overline{\mathbb{D}} \tag{3.5.192}$$

with the first inclusion a consequence of, say, (3.5.191), and the second inclusion implied by (3.5.176)-(3.5.177). In light of the fact that $\sigma(S; \ell^2)$ is a closed set (cf. Theorem 3.5.2), we ultimately conclude from (3.5.192) that

$$\sigma(S; \ell^2) = \overline{\mathbb{D}}. \tag{3.5.193}$$

A similar argument, now relying on (3.5.180) and (3.5.184), gives

$$\sigma(B; \ell^2) = \overline{\mathbb{D}}. \tag{3.5.194}$$

Next, based on (3.5.189) and (3.5.193), as well as on (3.5.186) and (3.5.194), we obtain

$$\sigma_e(S; \ell^2) \subseteq \mathbb{T} \text{ and } \sigma_e(B; \ell^2) \subseteq \mathbb{T}. \tag{3.5.195}$$

Going further, recall from (3.5.184) that for each $\lambda \in \mathbb{T}$ the operator $\overline{\lambda} I_{\ell^2} - B$ is injective on ℓ^2. Via duality (cf. (3.5.188)) we arrive at the conclusion that

$$\text{for each } \lambda \in \mathbb{T} \text{ the operator } \lambda I_{\ell^2} - S \text{ has dense range on } \ell^2. \tag{3.5.196}$$

If we now take $\lambda \in \mathbb{T} \setminus \sigma_e(S\,;\ell^2)$ it follows that $\lambda I_{\ell^2} - S$ is also Fredholm, hence has closed range, on ℓ^2. In concert with (3.5.196) this establishes:

$$\text{for each } \lambda \in \mathbb{T} \setminus \sigma_e(S\,;\ell^2) \text{ the operator } \lambda I_{\ell^2} - S \text{ is surjective on } \ell^2. \tag{3.5.197}$$

Together, (3.5.197), (3.5.177), and Theorem 1.6.7 imply that the operator $\lambda I_{\ell^2} - S$ is actually invertible on ℓ^2 for each $\lambda \in \mathbb{T} \setminus \sigma_e(S\,;\ell^2)$. As such, $\mathbb{T} \setminus \sigma_e(S\,;\ell^2)$ is contained in $\mathbb{C} \setminus \sigma(S\,;\ell^2) = \mathbb{C} \setminus \overline{\mathbb{D}}$, with the last equality coming from (3.5.193). This forces $\mathbb{T} \setminus \sigma_e(S\,;\ell^2)$ to be the empty set, hence we have $\mathbb{T} \subseteq \sigma_e(S\,;\ell^2)$. From this and (3.5.195) we ultimately conclude that

$$\sigma_e(S\,;\ell^2) = \mathbb{T}. \tag{3.5.198}$$

Based on this, (3.5.188), and item *(3)* in Theorem 0.0.1, we also obtain

$$\sigma_e(B;\ell^2) = \mathbb{T}. \tag{3.5.199}$$

Pressing forward, from (3.5.160), (3.5.148), and (3.5.198) we see that, on the one hand,

$$\bigsqcup_{n\in\mathbb{Z}} \varphi_n(S\,;\ell^2) = \rho_e(S\,;\ell^2) = \mathbb{C} \setminus \sigma_e(S\,;\ell^2) = \mathbb{C} \setminus \mathbb{T}. \tag{3.5.200}$$

On the other hand, from (3.5.178) and (3.5.189) we obtain

$$\mathbb{C} \setminus \overline{\mathbb{D}} \subseteq \varphi_0(S\,;\ell^2) \text{ and } \mathbb{D} \subseteq \varphi_{-1}(S\,;\ell^2). \tag{3.5.201}$$

A combination of (3.5.200) and (3.5.201) gives that

$$\begin{gathered} \varphi_0(S\,;\ell^2) = \mathbb{C} \setminus \overline{\mathbb{D}}, \quad \varphi_{-1}(S\,;\ell^2) = \mathbb{D}, \\ \text{and } \varphi_n(S\,;\ell^2) = \varnothing \text{ for all } n \in \mathbb{Z} \setminus \{-1, 0\}. \end{gathered} \tag{3.5.202}$$

Reasoning in a similar fashion (now employing (3.5.199), (3.5.180), and (3.5.186)) one obtains

$$\begin{gathered} \varphi_0(B;\ell^2) = \mathbb{C} \setminus \overline{\mathbb{D}}, \quad \varphi_1(B;\ell^2) = \mathbb{D}, \\ \text{and } \varphi_n(B;\ell^2) = \varnothing \text{ for all } n \in \mathbb{Z} \setminus \{0, 1\}. \end{gathered} \tag{3.5.203}$$

This concludes the proof of Proposition 3.5.11. □

Exercise 3.5.12. Let $T := B \oplus S$ on $X := \ell^2 \oplus \ell^2$, where B and S are respectively the backward and forward shifts on ℓ^2. Show that the essential spectrum of T on X is the unit circle, the spectrum of T on X is the closed unit disk, $\varphi_0(T;X)$ is the complement of the unit circle in the plane, and $\varphi_n(T;X)$ is empty for all $n \neq 0$.

Hint: Write $\lambda I_X - T$ as $(\lambda I_{\ell^2} - B) \oplus (\lambda I_{\ell^2} - S)$ and use the Direct Sum Theorem (cf. Theorem 3.3.1) together with Proposition 3.5.11.

We have already seen that $\varphi_0(T;X)$ need *not* be connected (cf. item (b) in Exercise 3.5.10). The next exercise shows that $\varphi_0(T;X) \cap \sigma(T;X)$, as well as $\varphi_n(T;X)$ for $n \neq 0$, need *not* be connected.

Exercise 3.5.13. Let S be the forward shift on ℓ^2, and let B be the backward shift on ℓ^2. Define

$$T := S \oplus 2B \oplus 3S \oplus 4B \quad \text{on} \quad X := \ell^2 \oplus \ell^2 \oplus \ell^2 \oplus \ell^2. \tag{3.5.204}$$

Determine $\sigma(T;X)$, $\sigma_e(T;X)$, and all the sets $\varphi_n(T;X)$, exhibiting each one that is nonvoid as a union of holes.

3.5.4 The Weyl Spectrum

We now begin to study how the spectrum of an operator changes when the operator undergoes compact perturbations. By a version of Weyl's Theorem we know that the changes must be confined to Fredholm points of index zero (see Exercise 3.5.6). Also, since the class of Fredholm operators is stable under compact perturbations (cf. Corollary 3.2.2), the essential spectrum as a whole is unchanged:

Proposition 3.5.14. *If X is a p-Banach space for some $p \in (0,1]$ and if $T \in \mathcal{L}(X)$, then for every compact operator K on X one has:*

$$\sigma_e(T+K;X) = \sigma_e(T;X). \tag{3.5.205}$$

Going further, continue to assume that X is a p-Banach space for some $p \in (0,1]$ and that $T \in \mathcal{L}(X)$. According to Corollary 3.3.16, for each integer n the set $\varphi_n(T;X)$ of Fredholm points of index n (introduced in (3.5.161)) enjoys a similar stability property:

$$\varphi_n(T+K;X) = \varphi_n(T;X) \text{ for every } K \in \mathcal{K}(X) \text{ and } n \in \mathbb{Z}. \tag{3.5.206}$$

From (3.5.205) and (3.5.206) we see that the following set[2], which is called the `Weyl spectrum` of T,

$$\sigma_\omega(T;X) := \sigma_e(T;X) \bigsqcup \Bigg(\bigsqcup_{n\neq 0} \varphi_n(T;X) \Bigg) \tag{3.5.207}$$

is also unchanged under compact perturbations of T. Since for each integer $n \neq 0$ the set $\varphi_n(T;X)$ is contained in $\sigma(T;X)$ (cf. (3.5.163)), and so is $\sigma_e(T;X)$ (cf. (3.5.151)), we conclude that

$$\sigma_\omega(T;X) \subseteq \sigma(T;X). \tag{3.5.208}$$

---

[2] recall that the symbol $\sqcup$ indicates union of disjoint sets

Also, as is apparent from definitions (cf. (3.5.207), (3.5.147), (3.5.161)) as well as (3.5.160) and (3.5.148), we have

$$\sigma_\omega(T;X) = \mathbb{C} \setminus \varphi_0(T;X). \tag{3.5.209}$$

In particular, the `Weyl resolvent` of T satisfies

$$\rho_\omega(T;X) := \mathbb{C} \setminus \sigma_\omega(T;X) = \varphi_0(T;X). \tag{3.5.210}$$

Going further, we claim that

$$\begin{array}{c}\bigcap_{K\in\mathcal{K}(X)} \sigma(T+K;X) \ \text{ is the } \textit{largest} \text{ subset of } \sigma(T;X) \\ \text{that is invariant under compact perturbation of } T.\end{array} \tag{3.5.211}$$

Indeed, if for each $T \in \mathcal{L}(X)$ we temporarily denote by $s(T;X)$ the largest subset of $\sigma(T;X)$ satisfying $s(T+K;X) = s(T;X)$ for every operator $K \in \mathcal{K}(X)$, then $s(T;X) = s(T+K;X) \subseteq \sigma(T;X)$ for every $K \in \mathcal{K}(X)$, so

$$s(T;X) \subseteq \bigcap_{K\in\mathcal{K}(X)} \sigma(T+K;X). \tag{3.5.212}$$

In view of the fact that the set in the right-hand side of (3.5.212) is itself a subset of $\sigma(T;X)$ that is invariant under compact perturbation of T, the maximality of $s(T;X)$ implies the we actually have equality in (3.5.212). This establishes (3.5.211).

Our next theorem shows that, in fact, *the Weyl spectrum is equal to the largest subset of the spectrum that is invariant under compact perturbation of the operator in question.* In light of (3.5.211), we may state this result as follows:

Theorem 3.5.15. *Let X be a p-Banach space for some $p \in (0,1]$. Then for each $T \in \mathcal{L}(X)$ one has*

$$\sigma_\omega(T;X) = \bigcap_{K\in\mathcal{K}(X)} \sigma(T+K;X). \tag{3.5.213}$$

Proof. Fix an operator $T \in \mathcal{L}(X)$. Since the Weyl spectrum is contained in the spectrum and is invariant under compact perturbations of T, it is clearly contained in the right-hand side of (3.5.213).

Conversely, the right-hand side of (3.5.213) is contained in the spectrum and, as seen from (3.5.210), its difference with the Weyl spectrum (if there is one) can consist only of Fredholm points of index zero. So we will be done if we can show that no such points lie in the right-hand side of (3.5.213).

Indeed, suppose $\lambda \in \varphi_0(T;X)$. Then $T - \lambda I$ is a Fredholm operator of index zero, and therefore has a compact perturbation $T - \lambda I + K$ that is invertible (according to Theorem 3.3.8). We therefore have $\lambda \notin \sigma(T+K;X)$, hence λ is not in the right-hand side of (3.5.213). □

Equation (3.5.213) is usually taken to be the *definition* of the Weyl spectrum. The equivalence of this and our definition (i.e., Theorem 3.5.15) is due to M. Schechter.

Exercise 3.5.16. Let X be a p-Banach space for some $p \in (0, 1]$, and fix some operator $T \in \mathcal{L}(X)$.

(a) Show that if $\lambda \in \varphi_0(T; X)$ then there is a compact operator K on X with the property that $\lambda \notin \sigma(T + K; X)$.

(b) Assume the space X has enough continuous linear functionals to separate points (cf. Definition 1.6.12), and suppose that $\lambda \in \rho(T; X)$. Show that there is a rank-one operator $K \in \mathcal{L}(X)$ such that $\lambda \in \sigma(T + K; X)$.

Hint: For part (a) use Theorem 3.3.8. As regards part (b), pick some $x_o \in X\setminus\{0\}$ and select $\Lambda \in X^*$ with the property that $\Lambda(x_o) \neq 0$. Then $y_o := (\lambda I - T)(x_o) \in X \setminus \{0\}$, and if for each $x \in X$ we define $K(x) := c\Lambda(x)y_o$ where $c := \Lambda(x_o)^{-1} \in \mathbb{C}$, then $x_o \in \ker(\lambda I - (T + K))$.

3.5.5 Spectral Points of Index Zero

Suppose X is a p-Banach space for some $p \in (0, 1]$. The spectrum of an operator $T \in \mathcal{L}(X)$ naturally decomposes into two pieces. First, there is the Weyl spectrum (cf. (3.5.207), (3.5.209), and (3.5.213)):

$$\begin{aligned}\sigma_\omega(T; X) &= \sigma_e(T; X) \bigsqcup \left(\bigsqcup_{n \neq 0} \varphi_n(T; X) \right) \\ &= \mathbb{C} \setminus \varphi_0(T; X) = \bigcap_{K \in \mathcal{K}(X)} \sigma(T + K; X),\end{aligned} \tag{3.5.214}$$

the part that is unchanged by compact perturbations of T. It can be written as the disjoint union of the essential spectrum and possibly certain "holes" (bounded components of the complement of the essential spectrum; cf. Definition 3.5.8 and Corollary 3.5.9).

What remains, i.e.,

$$\sigma(T; X) \setminus \sigma_\omega(T; X) = \sigma(T; X) \cap \varphi_0(T; X), \tag{3.5.215}$$

is the somewhat mysterious set of spectral points having index zero –a set that is, according to Exercise 3.5.16, extremely *unstable* under compact perturbations of T. Our goal now is to understand the topological nature of this part of the spectrum.

As a start, take a look at the examples we already have in hand, most notably T compact, and the examples in Exercise 3.5.10. In these examples the mystery set is either empty, a discrete set of index-zero eigenvalues clustering only (if anywhere) on the essential spectrum, or an entire component of $\varphi_0(T; X)$. We are going to show that *these are the only possibilities*! The key to our analysis is the following theorem, which asserts, among other things, that the dimension of $\ker(T - \lambda I)$ is *upper semicontinuous* on the essential resolvent. To state it, with $\mathbb{K}$ denoting either $\mathbb{R}$ or $\mathbb{C}$, define

$$B_{\mathbb{K}}(z_0, r) := \{z \in \mathbb{K} : |z - z_0| < r\}, \qquad z_0 \in \mathbb{K}, \quad r > 0. \tag{3.5.216}$$

Theorem 3.5.17 (Punctured Neighborhood). *Let X be a p-Banach space for some $p \in (0, 1]$ and suppose T is a Fredholm operator on X. Then*

$$\begin{aligned} &\exists\, \varepsilon = \varepsilon(T) > 0 \ \textit{ such that for all } \ \lambda \in B_{\mathbb{K}}(0, \varepsilon) \setminus \{0\} \ \textit{ it follows that} \\ &\qquad \dim(\ker(T - \lambda I)) \ \textit{ is a constant } \ \leq \dim\ker T. \end{aligned} \tag{3.5.217}$$

Proof. The argument breaks into two distinct pieces: semicontinuity and constancy. That is, in a first stage we shall show that the function

$$\mathbb{K} \ni \lambda \mapsto \dim(\ker(T - \lambda I)) \in \mathbb{N} \cup \{0\} \tag{3.5.218}$$

is upper semicontinuous at $\lambda = 0$. In particular there exists $\varepsilon = \varepsilon(T) > 0$, depending on T, such that

$$\lambda \in B_{\mathbb{K}}(0, \varepsilon) \implies \dim(\ker(T - \lambda I)) \leq \dim\ker T. \tag{3.5.219}$$

Here we shall follow the arguments from [6, Theorem 3.2.10, p. 40].

In a second stage, possibly shrinking $\varepsilon > 0$, we shall show that

$$\lambda \in B_{\mathbb{K}}(0, \varepsilon) \setminus \{0\} \implies \dim(\ker(T - \lambda I)) \ \text{ is constant.} \tag{3.5.220}$$

Here we follow the arguments from [6, Theorem 3.2.11, pp. 41-42] . Clearly (3.5.219) and (3.5.220) yield the conclusion of Theorem 3.5.17.

To start the semicontinuity argument, we recall from Theorem 3.1.7 the fundamental Fredholm decomposition

$$X = \ker T \oplus Y = M \oplus \operatorname{ran} T. \tag{3.5.221}$$

Then, as in the proof of Atkinson's Theorem (Theorem 3.1.10), using that T is a linear homeomorphism of Y onto $\operatorname{ran} T$, we define the linear mapping $S : X \to X$ by setting

$$S = \begin{cases} (T\big|_{Y \to \operatorname{ran} T})^{-1} & \text{on } \operatorname{ran} T, \\ 0 \ \text{ on } \ M, \end{cases} \tag{3.5.222}$$

which, as before, is continuous on X (because it is continuous on $\operatorname{ran} T$ and M). Also, recall that ST is the natural projection of X onto Y, and (as may be seen from the definition of S) we have $\operatorname{ran} S = Y$.

After this preamble, we claim that the following implication holds

$$|\lambda| < \|S\|^{1/p} \implies \ker(S(T - \lambda I)) \cap Y = \{0\}. \tag{3.5.223}$$

Assuming (3.5.223) for the moment, observe that since

$$\ker(T - \lambda I) \subseteq \ker(S(T - \lambda I)), \tag{3.5.224}$$

for each $\lambda \in \mathbb{K}$ such that $|\lambda| < \|S\|^{-1/p}$ we may write

$$\dim\ker(T - \lambda I) \leq \dim\ker(S(T - \lambda I)) \leq \operatorname{codim} Y = \dim\ker T, \tag{3.5.225}$$

where the second inequality above follows from (3.5.223) and the equality is a consequence of (3.5.221). This shows (3.5.220) with $\varepsilon := \|S\|^{1/p} > 0$.

Turning now to the proof of the claim made in (3.5.223), notice that since ST is a projection of X onto Y, its restriction to Y is the identity operator on Y, denoted by I_Y. Thus for each $\lambda \in \mathbb{K}$ we have

$$S(T - \lambda I)\big|_Y = ST\big|_Y - \lambda S\big|_Y = I_Y - \lambda S\big|_Y = (I - \lambda S)\big|_Y \tag{3.5.226}$$

so, on the one hand,

$$Y \cap \ker(S(T - \lambda I)) = Y \cap \ker(I - \lambda S). \tag{3.5.227}$$

On the other hand,

$$\ker(I - \lambda S) = \{x \in X : x = \lambda S x\} \subseteq \operatorname{ran} S = Y. \tag{3.5.228}$$

Combining (3.5.227) and (3.5.228) we obtain

$$Y \cap \ker(S(T - \lambda I)) = \ker(I - \lambda S). \tag{3.5.229}$$

However, when $\lambda \in \mathbb{K}$ satisfies $|\lambda| < \|S\|^{-1/p}$ we have $\|\lambda S\| < 1$. By Corollary 3.4.5 the operator $I - \lambda S$ is therefore invertible on X and, as such, has trivial kernel. This and (3.5.229) then yield (3.5.223), as desired. This completes the semicontinuity part of the argument.

We shall turn next to the constancy part of the proof. The idea behind this step is to reduce the problem to the semicontinuity result we just proved. For this, we start by using part (*b*) of Corollary 3.2.2 noticing that since T is Fredholm so is T^n, for every $n \in \mathbb{N}$. In particular $\operatorname{ran} T^n$ is closed in X for each $n \in \mathbb{N}$. As a consequence, the subspace

$$R_\infty := \bigcap_{j \in \mathbb{N}} R_j \quad \text{where} \quad R_j := \operatorname{ran}(T^j) \quad \text{for each} \quad j \in \mathbb{N}, \tag{3.5.230}$$

is a closed subspace of X. In particular,

$$R_\infty \text{ is a } p\text{-Banach space when equipped with the } p\text{-norm inherited from } X. \tag{3.5.231}$$

Since T is assumed to be a Fredholm operator, it follows (cf. Definition 3.1.1) that

$$\dim\ker T < +\infty. \tag{3.5.232}$$

Granted this, Theorem 2.2.12 applies and gives that R_∞ is stable under the action of the operator T, i.e.,

$$T(R_\infty) = R_\infty. \tag{3.5.233}$$

With (3.5.233) in hand, we introduce the operator

$$T_\infty : R_\infty \longrightarrow R_\infty \ \text{ given by setting } \ T_\infty := T\Big|_{R_\infty}, \tag{3.5.234}$$

and claim that

$$T_\infty \ \text{ is a Fredholm operator on } R_\infty. \tag{3.5.235}$$

(In this regard the reader should recall that the restriction of a Fredholm operator to an invariant subspace may *not* be Fredholm; see Exercise 3.1.4.)

Since we have already established that $\operatorname{ran} T_\infty = R_\infty$ in (3.5.233), in order to conclude that T_∞ is a Fredholm operator on R_∞, we are left with checking that $\ker T_\infty$ is finite dimensional and topologically complemented in R_∞, and that T_∞ is relatively open. Finite dimensionality is immediate since $\ker T_\infty$ is a subspace of $\ker T$, which is finite dimensional as $T \in \Phi(X)$. As a consequence $\ker T_\infty$ is topologically complemented in $\ker T$. Since $\ker T$ is topologically complemented in X, by Proposition 1.8.23 regarding the transitivity of topological complementation it follows that

$$\ker T_\infty \ \text{ is topologically complemented in } X. \tag{3.5.236}$$

As such, there exists Z subspace of X with the property that

$$X = \ker T_\infty \oplus Z, \tag{3.5.237}$$

and Z is closed in X by Corollary 1.8.18. Going further, using (3.5.237) for each $w \in R_\infty$ there exist $z \in Z$ and $x \in \ker T_\infty$ such that $w = x + z$. Since $x \in \ker T_\infty \subseteq R_\infty$, this further implies that $z \in Z \cap R_\infty$. This shows that

$$R_\infty \subseteq (Z \cap R_\infty) + \ker T_\infty. \tag{3.5.238}$$

Since also $(Z \cap R_\infty) + \ker T_\infty \subseteq R_\infty$ it follows that R_∞ is the algebraic sum of the subspaces $Z \cap R_\infty$ and $\ker T_\infty$. Since $(Z \cap R_\infty) \cap \ker T_\infty \subseteq Z \cap \ker T_\infty = \{0\}$, where the equality follows from (3.5.237), we can conclude that

$$R_\infty = (Z \cap R_\infty) \boxplus \ker T_\infty, \tag{3.5.239}$$

where, as usual, $\boxplus$ denotes algebraic direct sum. Now, since Z is closed in X, we have that $Z \cap R_\infty$ is closed in R_∞. This, in combination with (3.5.239), the fact that $\dim(\ker T_\infty) < \infty$, and Theorem 1.8.36 implies that the summation in (3.5.239) is topological, i.e.,

$$R_\infty = (Z \cap R_\infty) \oplus \ker T_\infty, \tag{3.5.240}$$

Hence, $\ker T_\infty$ is topologically complemented in R_∞, as desired. Finally, since R_∞ is a p-Banach space and T_∞ is onto, the Open Mapping Theorem (recalled in Theorem 1.6.7) applies and gives that T_∞ is relatively open. This completes the proof of the claim (3.5.235).

Having established (3.5.235), we next use the fact that the class of Fredholm operators $\Phi(R_\infty)$ is open in $\mathcal{L}(R_\infty)$ (cf. Corollary 3.4.10) to conclude the existence of a number $\varepsilon > 0$ such that

$$\forall\, \lambda \in B_{\mathbb{K}}(0,\varepsilon) \text{ the operator } T_\infty - \lambda I_\infty \text{ is Fredholm on } R_\infty. \tag{3.5.241}$$

Here I_∞ stands for the identity operator on R_∞. Consider next $\{i(T_\infty; R_\infty)\}$, the singleton of the index of the Fredholm operator $T_\infty \in \Phi(R_\infty)$, which is an open set in $\mathbb{Z}$ (equipped with the discrete topology). Using the continuity of the Fredholm index (cf. Corollary 3.4.13) and eventually shrinking $\varepsilon > 0$ we may conclude that

$$\begin{aligned} \forall\, \lambda \in B_{\mathbb{K}}(0,\varepsilon) \text{ we have } T_\infty - \lambda I_\infty \in \Phi(R_\infty) \\ \text{and } i(T_\infty - \lambda I_\infty; R_\infty) = i(T_\infty; R_\infty). \end{aligned} \tag{3.5.242}$$

Further combining (3.5.242) with the semicontinuity result we have already proved in (3.5.219) (presently employed with X replaced by R_∞, which is a permissible choice thanks to (3.5.231), and T_∞ in place of T, which constitutes a valid choice thanks to (3.5.235)) we can conclude that there exists $\varepsilon > 0$ such that

$$\lambda \in B_{\mathbb{K}}(0,\varepsilon) \implies \begin{cases} T_\infty - \lambda I_\infty \in \Phi(R_\infty), \\ \dim(\ker(T_\infty - \lambda I_\infty) \leq \dim(\ker T_\infty), \\ i(T_\infty - \lambda I_\infty; R_\infty) = i(T_\infty; R_\infty). \end{cases} \tag{3.5.243}$$

However, since T_∞ acting on R_∞ is onto (cf. (3.5.233)) we have

$$i(T_\infty; R_\infty) = \dim(\ker T_\infty). \tag{3.5.244}$$

Thus, for each $\lambda \in B_{\mathbb{K}}(0,\varepsilon)$, we may write (with justifications provided below)

$$\begin{aligned} i(T_\infty; R_\infty) &= i(T_\infty - \lambda I_\infty; R_\infty) \\ &= \dim(\ker(T_\infty - \lambda I_\infty)) - \operatorname{codim}(\ker(T_\infty - \lambda I_\infty)) \\ &\leq \dim(\ker T_\infty) \\ &= i(T_\infty; R_\infty). \end{aligned} \tag{3.5.245}$$

Indeed, the first equality above comes from (3.5.243), the second equality is the definition of the Fredholm index, while the first inequality is obtained by appealing again to (3.5.243), and the last equality is a consequence of (3.5.244). Then (3.5.245) immediately leads to the conclusion that $\operatorname{codim}(\ker(T_\infty - \lambda I_\infty)) = 0$, and thus

$$\lambda \in B_{\mathbb{K}}(0,\varepsilon) \implies \begin{cases} (T_\infty - \lambda I_\infty)(R_\infty) = R_\infty \\ \dim(\ker(T_\infty - \lambda I_\infty)) = i(T_\infty; R_\infty). \end{cases} \tag{3.5.246}$$

Going further, notice that

$$\ker(T - \lambda I) = \ker(T_\infty - \lambda I_\infty) \qquad \forall\, \lambda \in \mathbb{K} \setminus \{0\}. \tag{3.5.247}$$

Indeed, if $\lambda \in \mathbb{K} \setminus \{0\}$ and $x \in \ker(T - \lambda I)$ then $T(x/\lambda) = x$ and iterating this yields $T^n(x/\lambda^n) = x$ showing that $x \in \operatorname{ran} T^n$ for each $n \in \mathbb{N}$. Hence $x \in R_\infty$ and $(T_\infty - \lambda I_\infty)x = (T - \lambda I)x = 0$. This shows that $x \in \ker(T_\infty - \lambda I_\infty)$, completing the proof of the left-to-right inclusion in (3.5.247). Since the reverse inclusion is trivial, the equality (3.5.247) follows. Combining now (3.5.246) and (3.5.247) we may therefore conclude that $\dim(\ker(T - \lambda I))$ is constant on $B_{\mathbb{K}}(0, \varepsilon) \setminus \{0\}$, i.e., (3.5.220) holds.

The proof of Theorem 3.5.17 is therefore complete. □

Exercise 3.5.18. Let X be a p-Banach space for some $p \in (0, 1]$. Prove that the following collections of operators are open in $\mathcal{L}(X)$:

(a) the Fredholm operators that are one-to-one;
(b) the Fredholm operators that map X onto itself.

Hint: For part (a), assume $T \in \Phi(X)$ is injective. Then $Y := \operatorname{ran} T$ is a closed subspace of finite codimension in X. If M is an algebraic complement of Y (which exists by Theorem 1.3.29), then $X = Y \oplus M$ by Theorem 1.8.36. Consequently, the canonical projection $P_Y : X \to X$ of X onto Y along M is a continuous operator (Proposition 1.8.22). Also, $\widetilde{T} : X \to Y$ defined by $\widetilde{T}x := Tx$ for each $x \in X$ is an invertible operator (thanks to Theorem 1.6.7). Suppose now $S \in \mathcal{L}(X)$ is such that $\|T - S\|_{X \to X} < \varepsilon$ for some small $\varepsilon > 0$ to be chosen shortly. If we define $\widetilde{S} : X \to Y$ by setting $\widetilde{S}x := P_Y(Sx)$ for each $x \in X$, then $\widetilde{S}$ is a continuous operator and

$$\|\widetilde{T} - \widetilde{S}\|_{X \to Y} = \|P_Y \circ T - P_Y \circ S\|_{X \to X} \leq \|P_Y\|_{X \to X}\|T - S\|_{X \to X} < \varepsilon\|P_Y\|_{X \to X}. \quad (3.5.248)$$

Since $\widetilde{T} : X \to Y$ is invertible, (3.5.248) forces, if $\varepsilon > 0$ is small enough, the operator $\widetilde{S} : X \to Y$ to be invertible as well. Given that $\widetilde{S}x = P_Y(Sx)$ for each $x \in X$, it follows that S is injective in this case. Use this and Corollary 3.4.10 to conclude the proof of (a).

For part (b), combining Proposition 3.6.38 with Theorem 1.6.7 (cf. also [29, Proposition 1.2.13, p. 18]) one obtains:

$$\{T \in \mathcal{L}(X) : T(X) = X\} \text{ is open in } \mathcal{L}(X). \quad (3.5.249)$$

Use this in concert with Corollary 3.4.10.

Remark 3.5.19. The results of Exercise 3.5.18, when properly rephrased, remain true *without* the requirement that T be Fredholm. For this see Proposition 3.6.35.

Now look back at the examples and exercises that deal with the Fredholm decomposition of the spectrum, and observe that whenever a spectral boundary point is not isolated, it belongs to the essential spectrum. The next result shows that this is no accident!

Theorem 3.5.20 (Putnam's Theorem). *Suppose X is a p-Banach space for some $p \in (0, 1]$. Pick an operator $T \in \mathcal{L}(X)$ and suppose λ_0 is a point of $\partial\sigma(T; X)$ that is not isolated. Then $\lambda \in \sigma_e(T; X)$.*

Proof. Fix $\lambda_0 \in \partial\sigma(T;X)$, and suppose $\lambda_0 \notin \sigma_e(T;X)$. We will show that λ_0 is an isolated point of $\sigma(T;X)$, i.e., there is a disk centered at λ_0 not containing any other point of $\sigma(T;X)$. Since λ_0 lies outside the essential spectrum, it is a Fredholm point of T, so applying Theorem 3.5.17 to $T - \lambda_0 I$ and using the continuity of the index we conclude that for some disk $\Delta = \{\lambda : |\lambda - \lambda_0| < \varepsilon\}$ the index of $T - \lambda I$ is constant, while the dimension of $\ker(T - \lambda I)$ is constant in the punctured disk $\Delta' := \Delta\backslash\{\lambda_0\}$.

Since λ_0 lies on the boundary of $\sigma(T;X)$ it is a limit point of the resolvent set of T, hence Δ' contains points λ for which $T - \lambda I$ is invertible, i.e., of index zero with zero-dimensional null space. Thus both $\dim(\ker(T - \lambda I))$ and $i(T - \lambda I)$ are $\equiv 0$ on Δ', hence $T - \lambda I$ is invertible for every $\lambda \in \Delta'$. Thus λ_0 is the only point of Δ that lies in $\sigma(T;X)$. □

At last, we find ourselves in a position to complete our analysis of the "unstable piece" $\varphi_0(T;X) \cap \sigma(T;X)$ of the spectrum of T.

Theorem 3.5.21. *Let X be a p-Banach space for some $p \in (0,1]$ and fix some operator $T \in \mathcal{L}(X)$. Then for each component G of $\varphi_0(T;X)$ there are only three possibilities:*

(a) $G \cap \sigma(T;X) = \varnothing$.
(b) $G \subset \sigma(T;X)$.
(c) $G \cap \sigma(T;X)$ is a set of isolated points that clusters (if it is an infinite set) only on $\sigma_e(T;X)$.

Proof. Suppose the component G of $\varphi_0(T;X)$ satisfies neither *(a)* nor *(b)*, i.e., it intersects both the spectrum and resolvent of T. Fix a point $\lambda_0 \in \sigma(T;X) \cap G$. Our first task is to show that λ_0 is isolated from the rest of the spectrum.

To this end fix another point $\lambda_1 \in \rho(T;X) \cap G$ (our hypothesis is that there is such a point). Since G is a component of the open set $\varphi_0(T;X)$, it is connected and open, and is therefore arcwise connected. So there is a curve Γ that lies in G and has endpoints λ_0 and λ_1. By the openness of the resolvent set, there is an open disk $\Delta(\lambda_1)$ centered at λ_1 and contained in G so that $T - \lambda I$ is invertible for every $\lambda \in \Delta(\lambda_1)$. For every other point γ of Γ there is, by Theorem 3.5.17, a disk $\Delta(\gamma)$ such that $\dim(\ker(T - \lambda I))$ is constant on the punctured version $\Delta'(\gamma) = \Delta(\gamma)\backslash\{\gamma\}$. By compactness there is a finite subcollection of these disks $\{\Delta(\gamma_j)\}$ where $\{\gamma_j\}$ are consecutive points on Γ and $\gamma_1 = \lambda_1$, $\gamma_n = \lambda_0$.

We have chosen the disk $\Delta(\gamma_1)$ so that $T - \lambda I$ is invertible for every λ therein. Since $\ker(T - \lambda I)$ has constant dimension on $\Delta'(\gamma_2)$, and this punctured disk has nontrivial overlap with $\Delta(\gamma_1)$, we see that $\ker(T - \lambda I)$ has dimension zero for all λ in $\Delta'(\gamma_2)$. Continuing in this manner we eventually obtain the same conclusion for $\Delta'(\gamma_n) = \Delta'(\lambda_0)$. Upon recalling that all the disks in sight lie in G (all points of which have index zero) we see that in particular $T - \lambda I$ has index zero on $\Delta(\lambda_0)$, so it is invertible on $\Delta'(\lambda_0)$, and therefore establishing λ_0 as an isolated spectral point.

To finish the proof, note that since each spectral point of G is isolated, it is in particular a spectral boundary point. Thus by Theorem 3.5.20, any limit point of these has to lie in the essential spectrum. □

3.5.6 Riesz Operators

Having begun our work on operator theory with the study of compact operators, it seems only fitting to end it by introducing a class of operators that is "spectrally indistinguishable" from the compacts. Here we take X to be a *Banach space* so that we can, without further complication, assert that both spectrum and essential spectrum of any $T \in \mathcal{L}(X)$ are nonempty.

Indeed, whenever A is a Banach algebra with unit e (normalized so that $\|e\| = 1$), and $x \in A$, the analyticity of the resolvent map $\rho(x; A) \ni z \mapsto (ze - x)^{-1} \in A$ in concert with Liouville's Theorem prevents the spectrum $\sigma(x; A)$ from being empty. When this is used for $A := \mathcal{L}(X)$ with X a Banach space, we obtain that $\sigma(T; X) \neq \varnothing$ for every $T \in \mathcal{L}(X)$. When the same abstract result is used for the Calkin algebra $A := \mathcal{L}(X)/\mathcal{K}(X)$ with X a Banach space, by relying on (3.5.149) we conclude that $\sigma_e(T; X) \neq \varnothing$ for every $T \in \mathcal{L}(X)$.

Definition 3.5.22. *Given a Banach space X, call $T \in \mathcal{L}(X)$ a* `Riesz operator` *if $\sigma_e(T; X) = \{0\}$.*

Here is our main result for this class of operators:

Theorem 3.5.23. *Let X be a Banach space. Then for any given $T \in \mathcal{L}(X)$, it follows that T is a Riesz operator if and only if $\sigma(T; X)$ consists of the origin, along with an at most countable set of Fredholm points. If there are infinitely many such Fredholm points, they form a sequence tending to zero.*

Proof. If T is a Riesz operator then, thanks to Definition 3.5.22, (3.5.148), and (3.5.160), we have

$$\mathbb{C} \setminus \{0\} = \mathbb{C} \setminus \sigma_e(T; X) = \rho_e(T; X) = \bigsqcup_{n \in \mathbb{Z}} \varphi_n(T; X). \tag{3.5.250}$$

Upon recalling from (3.5.164) that each $\varphi_n(T; X)$ with $n \neq 0$ is a union of bounded components of $\rho_e(T; X)$, we obtain that $\varphi_0(T; X) = \mathbb{C} \setminus \{0\}$. This is a connected set which, by virtue of being unbounded, cannot lie entirely in the spectrum $\sigma(T; X)$. Thus possibility *(b)* of Theorem 3.5.21 does not materialize. This leaves either possibility *(a)*—the spectrum is just the origin, or *(c)*—the non-zero part of the spectrum consists of index-zero isolated points clustering, if there are infinitely many, at the origin. This proves the right-pointing implication in the statement.

In the converse direction, if T has the spectrum indicated in the statement, then since all Fredholm points lie in the essential resolvent it follows that $\sigma_e(T; X) \subseteq \{0\}$. However, it has been noted in the preamble to the present theorem that we cannot possibly have $\sigma_e(T; X) = \varnothing$, since X is a Banach space. Thus, necessarily, $\sigma_e(T; X) = \{0\}$, proving that T is a Riesz operator. □

3.6 General Fredholm Theory: Operators Between Two Spaces

The treatment up to this point in the current chapter has solely focused on the "diagonal" case, i.e., the situation when a linear operator T acts from a linear space X into itself. However, it is of substantial interest to consider the "off-diagonal" case, allowing an operator to act between two (possibly) different linear spaces, say $T : X \to Y$. The goal here is to further extend the Fredholm theory developed so far to the latter, more inclusive, setting. Note that when $T : X \to Y$ with $X \neq Y$, it may not be possible to iterate T, and speak of its powers T^n, if $n \in \mathbb{N}$ has $n \geq 2$. As such, our previous proofs do not adapt to the more general case we presently have in mind. Our approach is to nonetheless use the previously developed theory as a stepping stone and show that this essentially self-improves, from the "diagonal" case to the "off-diagonal" one.

The bulk of the main results are derived in §3.6.1 for linear topological spaces and in §3.6.2 for quasi-Banach spaces, while §3.6.3 contains miscellaneous results, in the direction of a semi-Fredholm theory on quasi-Banach spaces.

3.6.1 Off-Diagonal Theory in Topological Vector Spaces

Given two linear topological spaces X, Y, denote the collection of `continuous` linear operators from X into Y by

$$\mathcal{L}(X \to Y) := \{T : X \to Y \,:\, T \text{ is linear and continuous}\}. \tag{3.6.251}$$

To proceed, we need to extend the notion of compact operators from §2.1.1 to the "off-diagonal" setting considered here. The following is the natural such version of Definition 2.1.1.

Definition 3.6.1. *Given two linear topological spaces X and Y, a linear operator $K : X \to Y$ is said to be* `compact` *if there is a zero-neighborhood V in X such that $K(V)$ is relatively compact in Y (cf. Definition 1.1.33). Denote the set of all compact operators from X into Y by $\mathcal{K}(X \to Y)$; that is,*

$$\mathcal{K}(X \to Y) := \{K : X \to Y \,:\, K \textit{ is linear and compact}\}. \tag{3.6.252}$$

The following result is a generalization of Theorem 2.1.3 to the situation in which operators are allowed to act between different spaces.

Theorem 3.6.2. *Assume X, Y, Z, W are linear topological space. Then the following properties hold:*

(a) One has $\mathcal{K}(X \to Y) \subseteq \mathcal{L}(X \to Y)$. If one additionally assumes that Y is Hausdorff, then $\mathcal{K}(X \to Y)$ becomes a linear subspace of $\mathcal{L}(X \to Y)$.
(b) One has

$$K \circ L \in \mathcal{K}(X \to Z) \textit{ for all } K \in \mathcal{K}(Y \to Z) \textit{ and } L \in \mathcal{L}(X \to Y). \quad (3.6.253)$$

If one additionally assumes W to be Hausdorff, then

$$\begin{aligned} &J \circ K \circ L \textit{ belongs to } \mathcal{K}(X \to W) \textit{ for all} \\ &J \in \mathcal{L}(Z \to W),\ K \in \mathcal{K}(Y \to Z),\ L \in \mathcal{L}(X \to Y). \end{aligned} \quad (3.6.254)$$

(c) If Y is Hausdorff, then every $P \in \mathcal{L}(X \to Y)$ with $\dim P(X) < \infty$ *belongs to $\mathcal{K}(X \to Y)$.*
(d) Every $K \in \mathcal{K}(X \to Y)$ maps bounded subsets of X into relatively compact subsets of Y.

Proof. Denote the vector space topologies on X, Y, Z, W by τ_X, τ_Y, τ_Z, and τ_W, respectively. For the first assertion in part *(a)*, let $K \in \mathcal{K}(X \to Y)$. To show that $K \in \mathcal{L}(X \to Y)$, fix a zero-neighborhood U in Y. Since K is compact, there exists a zero-neighborhood V in X such that $K(V)$ is relatively compact in Y. In particular, $K(V)$ is bounded in Y by Exercise 1.6.41. Hence $K(V)$ is absorbed by U, so there exists $\varepsilon > 0$ such that

$$K(\varepsilon V) = \varepsilon \cdot K(V) \subseteq U. \quad (3.6.255)$$

Because εV is itself a zero-neighborhood in X, it follows from this and (1.1.4) that K is continuous at the origin. By linearity, we conclude that $K \in \mathcal{L}(X \to Y)$.

For the second assertion in part *(a)*, assume in addition that Y is Hausdorff. Since multiplication by non-zero scalars is a homeomorphism on X, it is immediate that $\mathcal{K}(X \to Y)$ is closed under scalar multiplication. It remains to verify stability under addition. Let $K_1, K_2 \in \mathcal{K}(X \to Y)$. Then for each $j = 1, 2$ there exists $V_j \in \mathcal{N}(0; \tau_X)$ such that $K_j(V_j)$ is relatively compact in Y. To proceed, set $V := V_1 \cap V_2 \in \mathcal{N}(0; \tau_X)$ and observe that

$$(K_1 + K_2)V \subseteq K_1(V) + K_2(V) \subseteq K_1(V_1) + K_2(V_2) \subseteq \overline{K_1(V_1)} + \overline{K_2(V_2)}. \quad (3.6.256)$$

Since each $\overline{K_j(V_j)}$ is compact in Y, Exercise 1.6.9 implies that the final set in (3.6.256) is compact. As Y is Hausdorff, Proposition 1.1.34 then yields that the set $(K_1 + K_2)V$ is relatively compact in Y. Hence $K_1 + K_2 \in \mathcal{K}(X \to Y)$.

For the first assertion in part *(b)*, assume $K \in \mathcal{K}(Y \to Z)$ and $L \in \mathcal{L}(X \to Y)$. Then there exists $V \in \mathcal{N}(0; \tau_Y)$ such that $K(V)$ is relatively compact in (Z, τ_Z). Since $L : (X, \tau_X) \to (Y, \tau_Y)$ is continuous and satisfies $L0 = 0$, (1.1.4) ensures the existence of $U \in \mathcal{N}(0; \tau_X)$ with $L(U) \subseteq V$. Consequently,

$$(K \circ L)(U) = K(L(U)) \subseteq K(V), \quad (3.6.257)$$

which is relatively compact in (Z, τ_Z) by Proposition 1.1.35. As $K \circ L$ is linear, Definition 3.6.1 shows that $K \circ L \in \mathcal{K}(X \to Z)$.

For the second assertion in part *(b)*, assume further that (W, τ_W) is Hausdorff. Given what has already been established, (3.6.254) will follow once we verify that $J \circ K$ belongs to $\mathcal{K}(X \to W)$ whenever $K \in \mathcal{K}(X \to Z)$ and $J \in \mathcal{L}(Z \to W)$. By Definition 3.6.1, there exists $V \in \mathcal{N}(0; \tau_X)$ such that $K(V)$ is relatively compact in

(Z, τ_Z). Since $J : (Z, \tau_Z) \to (W, \tau_W)$ is continuous, Exercise 1.1.36 implies that

$$(J \circ K)(V) = J(K(V)) \tag{3.6.258}$$

is relatively compact set in (W, τ_W). Given that $J \circ K$ is linear, Definition 3.6.1 yields $J \circ K \in \mathcal{K}(X \to W)$.

We now address the claim in part *(c)*. Assume (Y, τ_Y) is Hausdorff and suppose $P \in \mathcal{L}(X \to Y)$ is such that $M := P(X)$ is a finite-dimensional subspace of Y. By Proposition 1.6.8, the space $(M, \tau_Y|_M)$ is a Hausdorff topological vector space. Hence, by Theorem 1.8.31, there exists a linear homeomorphism $\Phi : M \to \mathbb{C}^n$, where $n := \dim M$. Let $B(0, 1)$ denote the open unit ball in $\mathbb{C}^n$, and set $V := \Phi^{-1}(B(0, 1))$. Since $B(0, 1)$ is a relatively compact zero-neighborhood in $\mathbb{C}^n$ and Φ^{-1} is a linear homeomorphism, V is a relatively compact zero-neighborhood in $(M, \tau_Y|_M)$. As M is closed in (Y, τ_Y) by Theorem 1.8.1*(b)*, Proposition 1.1.35 implies that V is relatively compact in (Y, τ_Y). Consider the linear operator

$$\begin{aligned} R : (X, \tau) &\longrightarrow (M, \tau_Y\big|_M) \text{ defined by} \\ Rx &:= Px \text{ for each } x \in X. \end{aligned} \tag{3.6.259}$$

Since P is continuous, Exercise 1.1.11 shows that R is continuous as well. Using (1.1.4), we obtain a zero-neighborhood O in (X, τ_X) such that $P(O) = R(O) \subseteq V$. Thus $P(O)$ is relatively compact in (Y, τ_Y) by Proposition 1.1.35 and the relative compactness of V. Hence $P \in \mathcal{K}(X \to Y)$, as wanted.

Finally, for the claim in part *(d)*, let $K \in \mathcal{K}(X \to Y)$ and let B be a bounded subset of (X, τ_X). Since K is compact, there exists $V \in \mathcal{N}(0; \tau_X)$ such that $K(V)$ is relatively compact in (Y, τ_Y). As V absorbs B, there exists $\varepsilon > 0$ such that $\varepsilon B \subseteq V$. Hence

$$\varepsilon K(B) = K(\varepsilon B) \subseteq K(V), \tag{3.6.260}$$

and by Proposition 1.1.35, the set $\varepsilon K(B)$ is relatively compact in (Y, τ_Y). Because multiplication by non-zero scalars is a homeomorphism on Y, it follows that $K(B)$ is relatively compact in (Y, τ_Y). This completes the proof of Theorem 3.6.2. □

We now introduce the class of bounded operators acting between two linear topological spaces.

Definition 3.6.3. *Suppose X and Y are two linear topological spaces. A linear operator $T : X \to Y$ is called* bounded *if there exists a zero-neighborhood U in X such that $T(U)$ is a bounded subset of Y. Denote the collection of all bounded operators from X to Y by $\mathcal{B}(X \to Y)$; that is,*

$$\mathcal{B}(X \to Y) := \{T : X \longrightarrow Y : \ T \text{ linear and bounded}\}. \tag{3.6.261}$$

On occasions when it is desirable to emphasize the underlying topologies τ_X on X and τ_Y on Y, write $\mathcal{B}((X, \tau_X) \to (Y, \tau_Y))$ in place of $\mathcal{B}(X \to Y)$.

Our next goal is to establish an "off-diagonal" analogue of Theorem 2.1.14.

Theorem 3.6.4. *Let X, Y be linear topological spaces. Recall $\mathcal{L}(X \to Y)$ from* (3.6.251)*, together with $\mathcal{K}(X \to Y)$ from* (3.6.252)*, and $\mathcal{B}(X \to Y)$ from* (3.6.261)*. Then*

$$\mathcal{K}(X \to Y) \subseteq \mathcal{B}(X \to Y) \subseteq \mathcal{L}(X \to Y), \tag{3.6.262}$$

and

$$\mathcal{B}(X \to Y) = \mathcal{L}(X \to Y) \quad \text{whenever } Y \text{ is locally bounded.} \tag{3.6.263}$$

Moreover, if $(X, \|\cdot\|_X)$ and $(Y, \|\cdot\|_Y)$ are actually quasi-normed spaces and if $T : X \to Y$ is linear, then T is bounded in the sense of Exercise 1.4.23 if and only if $T \in \mathcal{B}((X, \tau_{\|\cdot\|_X}) \to (Y, \tau_{\|\cdot\|_Y}))$. In particular, if X and Y are quasi-normed spaces then $\mathcal{B}(X \to Y)$ and $\mathcal{L}(X \to Y)$ coincide.

Proof. The first inclusion in (3.6.262) follows directly from the definitions together with item *(c)* of Theorem 1.6.40.

To establish the second inclusion in (3.6.262), let $T \in \mathcal{B}(X \to Y)$ and let W be a zero-neighborhood in Y. Since T is bounded, there exists a zero-neighborhood U in X such that $T(U)$ is bounded in Y. Hence $T(U)$ is absorbed by W, so there exists $\varepsilon > 0$ such that $\varepsilon T(U) \subseteq W$. Thus $T(\varepsilon U) \subseteq W$, and since εU is again a zero-neighborhood in X, item *(1)* of Definition 1.1.7 implies that T is continuous at the origin. Therefore $T \in \mathcal{L}(X \to Y)$.

For the next step, assume that Y is locally bounded. Then Y contains a zero-neighborhood V which is bounded. Let $T \in \mathcal{L}(X \to Y)$. Since T is continuous at $0 \in X$, item *(1)* of Definition 1.1.7 yields a zero-neighborhood U in X such that $T(U) \subseteq V$. By Proposition 1.6.39, the set $T(U)$ is bounded in Y, and hence Definition 3.6.3 shows that $T \in \mathcal{B}(X \to Y)$. Thus $\mathcal{L}(X \to Y) \subseteq \mathcal{B}(X \to Y)$ whenever Y is locally bounded, and together with (3.6.262) this proves (3.6.263).

Finally, the remaining assertions in the statement follow by combining (3.6.263), (1.11.598), and Proposition 1.11.3. □

We now turn to the task of providing an appropriate definition for the notion of Fredholm operator acting between to (possibly) different linear topological spaces.

Definition 3.6.5. *Given two Hausdorff linear topological spaces X, Y, a linear continuous mapping $T : X \to Y$ is said to be a* `Fredholm operator` *(or, simply,* `Fredholm`*) provided it verifies the following three axioms:*

(F1) $\operatorname{ran} T := \operatorname{ran}(T : X \to Y) := T(X)$ *is a closed subspace of Y, of finite codimension in Y.*

(F2) $\ker T := \ker(T : X \to Y) := \{x \in X : Tx = 0\}$ *is finite dimensional and topologically complemented in X.*

(F3) T is a relatively open map (which means that $T(O)$ is a relatively open subset of $\operatorname{ran}(T : X \to Y)$ *whenever O is open in X).*

Set

$$\Phi(X \to Y) := \{T : X \to Y : T \ \text{Fredholm}\} \tag{3.6.264}$$

and consider the index function

$$\begin{gathered} i : \Phi(X \to Y) \longrightarrow \mathbb{Z} \ \textit{defined by} \\ i(T) := \dim(\ker T) - \operatorname{codim}(\operatorname{ran} T) \ \textit{for each} \ \ T \in \Phi(X \to Y). \end{gathered} \tag{3.6.265}$$

If we wish to stress the spaces on which the operator T is considered, we write $i(T : X \to Y)$ in place of the less revealing symbol $i(T)$.

In relation to the rather general definition given above, it is reassuring to note that in a more restrictive setting we can get away with assuming less, namely,

if X and Y are actually Banach spaces, then $T \in \mathcal{L}(X \to Y)$ is Fredholm in the sense of Definition 3.6.5 if and only if T has a finite codimensional range, and a finite-dimensional kernel. (3.6.266)

Specifically, we make the following important remark:

Remark 3.6.6. Regarding condition *(F1)* in Definition 3.6.5, the reader is alerted to the fact that if X, Y are quasi-Banach spaces and $T \in \mathcal{L}(X \to Y)$ has a range of finite codimension in Y, then necessarily $\operatorname{ran}(T : X \to Y)$ is a closed subspace of Y (see, e.g., [29, Lemma 1.2.7, p. 13]).

Regarding condition *(F2)* in Definition 3.6.5, recall that a finite-dimensional subspace of a Hausdorff linear topological space is always closed, but it need not necessarily be topologically complemented. On the other hand, if X has enough continuous linear functionals to separate points (cf. Definition 1.6.12; this is the case if, e.g., X is a normed space, or more generally, X is a Hausdorff locally convex topological vector space – cf. Theorem 1.12.2) then every finite-dimensional subspace *is* topologically complemented (cf. item *(c)* in Remark 1.8.30 and Corollary 1.12.3).

Regarding condition *(F3)* in Definition 3.6.5, we wish to note that for quasi-Banach spaces the Open Mapping Theorem (recorded in Theorem 1.11.5) provides this relative openness property once we know that the range of T is closed.

Some basic examples of "off-diagonal" Fredholm operators are offered by the next three propositions.

Proposition 3.6.7. *Let X be a Hausdorff linear topological space, and let $Y \subseteq X$ be a closed subspace of finite codimension in X. Then the canonical inclusion*

$$\begin{gathered} \iota : Y \hookrightarrow X \ \ \textit{is a Fredholm operator,} \\ \textit{with index} \ \ i(\iota) = -\dim(X/Y). \end{gathered} \tag{3.6.267}$$

Also, if M is any algebraic complement of Y in X then the projection

$$\begin{gathered} P : X \to Y \ \ \textit{given by} \ \ P(y+m) := y \ \ \textit{for all} \ y \in Y \ \textit{and} \ m \in M \\ \textit{is a Fredholm operator with index} \ \ i(P) = \dim(X/Y). \end{gathered} \tag{3.6.268}$$

Proof. The claims in (3.6.267) are seen straight from Definition 3.6.5, bearing in mind Proposition 1.6.8, while the claims in (3.6.268) are implied by Definition 3.6.5, Theorem 1.8.36, Proposition 1.8.22 (cf. also Exercise 1.1.11), and Proposition 1.8.25. □

Proposition 3.6.8. *Let X be a Hausdorff topological vector space, and assume F is a finite-dimensional linear subspace of X which is topologically complemented in X. Then the quotient map $\pi : X \to X/F$ is a Fredholm operator with index $i(\pi) = \dim F$.*

Proof. The desired claims follow from item *(b)* in Theorem 1.8.1, item *(1)* of Lemma 1.6.10, and definitions. □

Here is the last installment of examples of "off-diagonal" Fredholm operators, advertised earlier.

Proposition 3.6.9. *Let X, Y be two Hausdorff linear topological spaces. Also, given two operators $T \in \mathcal{L}(X \to Y)$ and $S \in \mathcal{L}(Y \to X)$, consider $R : X \times Y \to X \times Y$ defined as*

$$R(x,y) := \begin{pmatrix} 0 & S \\ T & 0 \end{pmatrix} \begin{pmatrix} x \\ y \end{pmatrix} \in X \times Y \ \textit{ for each } \ (x,y) \in X \times Y. \tag{3.6.269}$$

Then

$$\operatorname{ran} R = (\operatorname{ran} S) \times (\operatorname{ran} T) \ \textit{ and } \ \ker R = (\ker T) \times (\ker S). \tag{3.6.270}$$

Also,

$$R \textit{ is relatively open if and only if } T \textit{ and } S \textit{ are relatively open.} \tag{3.6.271}$$

Finally,

$$R \textit{ is Fredholm if and only if } T \textit{ and } S \textit{ are Fredholm} \tag{3.6.272}$$

and, if any of these eventualities materializes,

$$i(R) = i(T) + i(S). \tag{3.6.273}$$

Proof. The claims in (3.6.270) are clear from definitions. Since for each $V \subseteq X$ and $W \subseteq Y$ we have $R(V \times W) = (SW) \times (TV)$, we conclude from Definition 1.1.15 that the left-pointing implication in (3.6.271) holds. To prove the right-pointing implication in (3.6.271), pick an open set $U \subseteq X$. Fix an arbitrary $u \in U$ and note that $(0, Tu) = R(u, 0) \in R(U \times Y)$. Since the current working hypothesis guarantees that $R(U \times Y)$ is relatively open in $\operatorname{ran} R = (\operatorname{ran} S) \times (\operatorname{ran} T)$, it follows that there exists an open set O in $X \times Y$ such that $R(U \times Y) = O \cap ((\operatorname{ran} S) \times (\operatorname{ran} T))$. In particular, $(0, Tu) \in O$ which means that we may find an open set V in X and an open set W in Y such that $(0, Tu) \in V \times W \subseteq O$. In particular,

$$(V \times W) \cap ((\operatorname{ran} S) \times (\operatorname{ran} T)) \subseteq R(U \times Y) = (\operatorname{ran} S) \times (TU), \tag{3.6.274}$$

hence

$$(V \cap \operatorname{ran} S) \times (W \cap \operatorname{ran} T) \subseteq (\operatorname{ran} S) \times (TU) \tag{3.6.275}$$

which further implies that

$$Tu \in W \cap \operatorname{ran} T \subseteq T(U). \tag{3.6.276}$$

This ultimately proves that TU is a relatively open subset of $\operatorname{ran} T$, thus T is a relatively open map. In a similar fashion, S is a relatively open map, completing the proof of (3.6.271).

To justify the right-pointing implication in (3.6.272), assume R is a Fredholm operator on $X \times Y$. From the implication *(a)*$\Rightarrow$*(d)* in Theorem 3.1.10 we see that there exists

$$Q := \begin{pmatrix} Q_{11} & Q_{12} \\ Q_{21} & Q_{22} \end{pmatrix} \in \mathcal{L}(X \times Y) \tag{3.6.277}$$

such that

$$K := I_{X\times Y} - RQ \in \mathcal{K}(X \times Y) \text{ and } \widetilde{K} := I_{X\times Y} - QR \in \mathcal{K}(X \times Y). \tag{3.6.278}$$

In matrix formalism, writing $K = (K_{ij})_{1\leq i,j\leq 2}$ and $\widetilde{K} = (\widetilde{K}_{ij})_{1\leq i,j\leq 2}$, it follows from item *(b)* in Theorem 3.6.2 and (3.6.278) that

$$I_X - S\,Q_{21} = K_{11} \in \mathcal{K}(X), \quad I_Y - Q_{21}S = \widetilde{K}_{22} \in \mathcal{K}(Y), \tag{3.6.279}$$

$$I_Y - T\,Q_{12} = K_{22} \in \mathcal{K}(Y), \quad I_X - Q_{12}T = \widetilde{K}_{11} \in \mathcal{K}(X). \tag{3.6.280}$$

Since $Q_{21}S$ is equal to $I_Y - \widetilde{K}_{22}$, itself a compact perturbation of the identity on Y, we may invoke Riesz Theory (cf. item *(a)* in Theorem 2.3.6) and conclude that $\ker(Q_{21}S)$ is both finite dimensional and topologically complemented in Y. Since $\ker S$ is contained in $\ker(Q_{21}S)$, it follows that $\ker S$ is finite dimensional, and also topologically complemented in $\ker(Q_{21}S)$ (by finite dimensionality). Keeping in mind that $\ker(Q_{21}S)$ is topologically complemented in Y, Proposition 1.8.23 applies and ultimately proves that

$$\ker S \text{ is finite dimensional and topologically complemented in } Y. \tag{3.6.281}$$

Focusing now on $\operatorname{ran} S$, we start by observing that since $S\,Q_{21}$ is equal to $I_X - K_{11}$ which is a compact perturbation of the identity on X, item *(a)* in Theorem 2.3.6 guarantees that $\operatorname{ran}(S\,Q_{21})$ is closed and of finite codimension in X. Since $\operatorname{ran} S \supseteq \operatorname{ran}(S\,Q_{21})$ we deduce that $\operatorname{ran} S$ has finite codimension in X, by Exercise 1.3.32. Appealing now to Proposition 1.3.46, it follows that $\operatorname{ran}(S\,Q_{21})$ has finite codimension in $\operatorname{ran} S$. The latter, combined with the fact that $\operatorname{ran}(S\,Q_{21})$ is closed in Y, allows us to apply Theorem 1.8.35 to conclude that $\operatorname{ran} S$ is also closed in X. In summary:

$$\operatorname{ran} S \text{ is a closed subspace of finite codimension in } X. \tag{3.6.282}$$

A similar type of argument, now using the second line in (3.6.279), leads to the conclusion that

$$\begin{array}{l}\ker T \text{ is finite dimensional and topologically complemented in} \\ X, \text{ while } \operatorname{ran} T \text{ is a closed subspace of finite codimension in } Y.\end{array} \tag{3.6.283}$$

Let us also note that the present working assumption (that R is Fredholm) and (3.6.271) imply that

$$T \text{ and } S \text{ are relatively open.} \tag{3.6.284}$$

At this stage, from (3.6.281)-(3.6.284) and Definition 3.6.5 we obtain that S and T are Fredholm operators. This establishes the right-pointing implication in (3.6.272).

To justify the left-pointing implication in (3.6.272), assume next that S and T are Fredholm operators. Then they are relatively open and, as such, (3.6.271) implies that

$$\text{the operator } R \text{ is relatively open.} \tag{3.6.285}$$

Moreover, since the ranges of S and T are closed, from the first equality in (3.6.270) and Exercise 1.1.17 we see that

$$\text{the operator } R \text{ has a closed range.} \tag{3.6.286}$$

In addition, given that the ranges of S and T are finite codimensional, we may rely on the first equality in (3.6.270) and Proposition 1.3.42 to conclude that

$$\begin{array}{c}\text{the range of the operator } R \text{ has finite codimension in } X \times Y \\ \text{and, in fact, } \operatorname{codim} \operatorname{ran} R = \operatorname{codim} \operatorname{ran} S + \operatorname{codim} \operatorname{ran} T.\end{array} \tag{3.6.287}$$

Finally, since S and T have finite dimensional kernels which are topologically complemented, we deduce from the second equality in (3.6.270), Proposition 1.8.26, and Exercise 1.3.33 that

$$\begin{array}{c}\text{the kernel of } R \text{ is finite dimensional,} \\ \text{as well as topologically complemented in } X \times Y, \\ \text{and one has } \dim \ker R = \dim \ker S + \dim \ker T.\end{array} \tag{3.6.288}$$

On account of (3.6.285)-(3.6.288) and Definition 3.6.5 we then see that R is a Fredholm operator on $X \times Y$ and (3.6.273) holds. This completes the proof of Proposition 3.6.9. □

We continue by presenting the "off-diagonal" version of Theorem 3.1.7.

Theorem 3.6.10 (Fundamental Fredholm Decomposition). *Let X, Y be two Hausdorff linear topological spaces and consider a Fredholm operator $T : X \to Y$. Then there exist subspaces Z of X and M of Y, where Z is closed and of finite codimension, and M is finite dimensional (hence closed), such that:*

(a) $X = Z \oplus \ker T$, *and*
(b) $Y = \operatorname{ran} T \oplus M$,

where T maps Z homeomorphically onto $\operatorname{ran} T$.

Proof. This is mostly a restatement of Definition 3.6.5. Since $\operatorname{ran} T$ is closed and has finite codimension, it is topologically complemented in Y (cf. Theorem 1.8.36), so M can be taken to be any topological complement of $\operatorname{ran} T$. Similarly we can take Z to be any topological complement of $\ker T$ in X (which exists by *(F2)* in Definition 3.6.5). These subspaces are automatically closed by Corollary 1.8.18.

It is clear from the decomposition *(a)* that T maps Z one-to-one and onto $\operatorname{ran} T$. To complete the proof, it remains to show that $T : Z \to \operatorname{ran} T$ is open. To this end, let U be an open subset of Z. Then the decomposition *(a)* together with Proposition 1.8.13 imply that $U + \ker T$ is an open subset of X, and *(F3)* in Definition 3.6.5 guarantees that $T(U) = T(U + \ker T)$ is open in $\operatorname{ran} T$. □

An important consequence of Theorem 3.6.10 is given below.

Proposition 3.6.11. *Let X, Y be two Hausdorff linear topological spaces and consider a Fredholm operator $T : X \to Y$. In addition, let the spaces Z, M be associated with T as in Theorem 3.6.10. Then*

$$\begin{aligned} &T_{\text{inv}} : Y \longrightarrow X \ \ \textit{defined by setting} \\ &T_{\text{inv}}(a+m) := (T|_Z : Z \to \operatorname{ran} T)^{-1}(a) \\ &\textit{for each} \ \ a \in \operatorname{ran} T \ \ \textit{and} \ \ m \in M \end{aligned} \tag{3.6.289}$$

is a Fredholm operator. Furthermore,

$$\begin{aligned} &\widetilde{T} : X \times Y \longrightarrow X \times Y \ \ \textit{defined as} \\ &\widetilde{T}(x,y) := (T_{\text{inv}}y,\, Tx) \ \ \textit{for each} \ \ (x,y) \in X \times Y \end{aligned} \tag{3.6.290}$$

is also a Fredholm operator.

Proof. Note that $\ker T_{\text{inv}} = M$ is a finite-dimensional subspace of Y which is topologically complemented in Y, and that $\operatorname{ran} T_{\text{inv}} = Z$ is a closed linear subspace of X, of finite codimension in Y (cf. Theorem 3.1.7). Moreover, whenever V is a relatively open set in $\operatorname{ran} T$ and W is a relatively open set in M, it follows that $T_{\text{inv}}(V + W) = T_{\text{inv}}(V)$ and $T_{\text{inv}}(V)$ is the image of V under the inverse of the linear homeomorphism $T : Z \to \operatorname{ran} T$. As a result, $T_{\text{inv}}(V + W)$ is a relatively open subset of $Z = \operatorname{ran} T_{\text{inv}}$. Since Proposition 1.8.13 guarantees that the sets of the form $V + W$, where V is relatively open in $\operatorname{ran} T$ and W is relatively open in M, form a basis for the topology of $\operatorname{ran} T \oplus M = X$, we conclude that T_{inv} is indeed a relatively open map. In view of Definition 3.6.5, we therefore see that T_{inv} is Fredholm.

Consider next the claim that (3.6.290) is a Fredholm operator. To justify it, start by observing that since both $\ker T$ and $\ker T_{\text{inv}}$ are finite-dimensional (as kernels of Fredholm operators), from (1.3.156) we deduce that $\ker \widetilde{T} = (\ker T) \times (\ker T_{\text{inv}})$ is a finite-dimensional subspace of $X \times Y$. Also, note that $\ker T$ is topologically complemented in X since T is a Fredholm operator and, as remarked earlier, $\ker T_{\text{inv}}$ is topologically complemented in Y. Granted these, we may invoke Proposition 1.8.26 to conclude that $\ker \widetilde{T}$ is topologically complemented in $X \times Y$. In addition, $\operatorname{ran} \widetilde{T} = (\operatorname{ran} T) \times (\operatorname{ran} T_{\text{inv}})$ is a closed space of finite codimension in $X \times Y$.

Indeed, this is a consequence of the fact that T and T_{inv} are Fredholm operators, together with Exercise 1.1.17, and Proposition 1.3.42.

Finally, since both T and T_{inv} are relatively open mappings, it follows from (3.6.290) (see also (3.6.271)) that so is $\widetilde{T}$. Having established these properties, Definition 3.6.5 guarantees that $\widetilde{T}$ is Fredholm. □

In turn, Proposition 3.6.11 becomes an ingredient in the proof of the "off-diagonal" version of Atkinson's Theorem (compare with Theorem 3.1.10). Recall that we agree to denote by I_X the identity operator on a given vector space X.

Theorem 3.6.12 (Atkinson's Theorem). *Let X, Y be two Hausdorff linear topological spaces and suppose $T \in \mathcal{L}(X \to Y)$. Then the following statements are equivalent:*

(a) The operator T is a Fredholm operator (cf. Definition 3.6.5).
(b) The operator T is invertible modulo operators with finite-dimensional ranges, in the sense that there exists $S \in \mathcal{L}(Y \to X)$ such that both $I_X - ST$ and $I_Y - TS$ have finite-dimensional ranges.
(c) The operator T is invertible modulo compact operators in the sense that there exists some $S \in \mathcal{L}(Y \to X)$ such that both $I_X - ST$ and $I_Y - TS$ are compact.
(d) There exist $S_1, S_2 \in \mathcal{L}(Y \to X)$ such that both $I_X - S_1T$ and $I_Y - TS_2$ are compact.

Moreover, if the operator $T \in \mathcal{L}(X \to Y)$ is Fredholm then any operators $S, S_1, S_2 \in \mathcal{L}(Y \to X)$ doing the job in items (b)-(d) above are also Fredholm.

Proof. Assume $T \in \mathcal{L}(X \to Y)$ is Fredholm. Then $\widetilde{T} \in \mathcal{L}(X \times Y \to X \times Y)$ associated with T as in Proposition 3.6.11 is also Fredholm. For this, we may invoke Theorem 3.1.10 and conclude that there exists a Fredholm operator $\widetilde{S} \in \mathcal{L}(X \times Y \to X \times Y)$ with the property that $I_{X\times Y} - \widetilde{S}\widetilde{T}$ and $I_{X\times Y} - \widetilde{T}\widetilde{S}$ have finite-dimensional ranges. Using matrix formalism, let us express the action of the operator $\widetilde{S}$ on each $(x, y) \in X \times Y$ as

$$\widetilde{S}(x, y) = \begin{pmatrix} S_{11} & S_{12} \\ S_{21} & S_{22} \end{pmatrix} \begin{pmatrix} x \\ y \end{pmatrix} \in X \times Y, \tag{3.6.291}$$

where

$$\begin{aligned} S_{11} &\in \mathcal{L}(X \to X), & S_{12} &\in \mathcal{L}(Y \to X), \\ S_{21} &\in \mathcal{L}(X \to Y), & S_{22} &\in \mathcal{L}(Y \to Y). \end{aligned} \tag{3.6.292}$$

It is also natural to identify $\widetilde{T}$ with a 2×2 matrix, namely

$$\widetilde{T} = \begin{pmatrix} 0 & T_{\text{inv}} \\ T & 0 \end{pmatrix}. \tag{3.6.293}$$

In this formalism we may then write

$$I_{X\times Y} - \widetilde{S}\widetilde{T} = \begin{pmatrix} I_X & 0 \\ 0 & I_Y \end{pmatrix} - \begin{pmatrix} S_{11} & S_{12} \\ S_{21} & S_{22} \end{pmatrix}\begin{pmatrix} 0 & T_{\rm inv} \\ T & 0 \end{pmatrix}$$
$$= \begin{pmatrix} I_X - S_{12}T & -S_{11}T_{\rm inv} \\ -S_{22}T & I_Y - S_{21}T_{\rm inv} \end{pmatrix} \tag{3.6.294}$$

and

$$I_{X\times Y} - \widetilde{T}\widetilde{S} = \begin{pmatrix} I_X & 0 \\ 0 & I_Y \end{pmatrix} - \begin{pmatrix} 0 & T_{\rm inv} \\ T & 0 \end{pmatrix}\begin{pmatrix} S_{11} & S_{12} \\ S_{21} & S_{22} \end{pmatrix}$$
$$= \begin{pmatrix} I_X - T_{\rm inv}S_{21} & -T_{\rm inv}S_{22} \\ -TS_{11} & I_Y - TS_{12} \end{pmatrix}. \tag{3.6.295}$$

In view of the fact that the property of having a finite-dimensional range for operators in $\mathcal{L}(X\times Y \to X\times Y)$, naturally identified as 2×2 matrices, is equivalent to the property that each entry has a finite-dimensional range, from (3.6.294)-(3.6.295) we conclude that

$$\begin{aligned} C_{11} &:= I_X - S_{12}T, \quad C_{12} := -S_{11}T_{\rm inv} \\ C_{21} &:= -S_{22}T, \quad C_{22} := I_Y - S_{21}T_{\rm inv} \end{aligned} \tag{3.6.296}$$

as well as

$$\begin{aligned} K_{11} &:= I_X - T_{\rm inv}S_{21}, \quad K_{12} := -T_{\rm inv}S_{22} \\ K_{21} &:= -TS_{11}, \quad K_{22} := I_Y - TS_{12} \end{aligned} \tag{3.6.297}$$

are continuous linear operators with finite-dimensional ranges. As a result, we see that $S := S_{12} \in \mathcal{L}(Y \to X)$ does the job in item *(b)*. Hence *(a)*$\Rightarrow$*(b)*. The fact that *(b)*$\Rightarrow$*(c)* is seen from item *(c)* in Theorem 2.1.3.

Going further, to justify the implication *(c)*$\Rightarrow$*(a)*, let us assume that there exists some $S \in \mathcal{L}(Y \to X)$ such that $C_X := I_X - ST$ and $C_Y := I_Y - TS$ are compact operators. Since ST is a compact perturbation of the identity on X, the Riesz Theory (cf. item *(a)* in Theorem 2.3.6) gives that $\ker(ST)$ is both finite dimensional and topologically complemented in X. Consequently, the same is true of $\ker T$, because it is a subspace of $\ker(ST)$, hence complemented therein by finite dimensionality, and therefore, using Proposition 1.8.23, it is topologically complemented in X. Thus,

$$\ker T \text{ is finite dimensional and topologically complemented in } X. \tag{3.6.298}$$

Next, since $I_Y - TS$ is a compact operator on Y, it follows that TS is a compact perturbation of the identity on Y. According to item *(a)* in Theorem 2.3.6, $\operatorname{ran}(TS)$ is closed and of finite codimension in Y. But $\operatorname{ran} T \supseteq \operatorname{ran}(TS)$, so $\operatorname{ran} T$ also has finite codimension in Y, by Exercise 1.3.32. Invoking now Proposition 1.3.46, it follows that $\operatorname{ran}(TS)$ has finite codimension in $\operatorname{ran} T$. The latter, combined with the fact that $\operatorname{ran}(TS)$ is closed in Y and Theorem 1.8.35, implies $\operatorname{ran} T$ is also closed in Y. In conclusion:

$$\operatorname{ran} T \text{ is a closed subspace of finite codimension in } Y. \tag{3.6.299}$$

The same type of argument but now emphasizing S in place of T (hence, with the roles of X and Y reversed) leads to the conclusion that

$$\begin{array}{l}\ker S \text{ is finite dimensional and topologically complemented in} \\ Y, \text{ while } \operatorname{ran} S \text{ is a closed subspace of finite codimension in } X.\end{array} \tag{3.6.300}$$

To conclude that T, S are Fredholm operators, there remains to show that T, S are relatively open mappings. To this end, consider the operator $R : X \times Y \to X \times Y$ acting on each $(x, y) \in X \times Y$ as

$$R(x, y) := \begin{pmatrix} 0 & S \\ T & 0 \end{pmatrix} \begin{pmatrix} x \\ y \end{pmatrix} \in X \times Y. \tag{3.6.301}$$

Then $R \in \mathcal{L}(X \times Y \to X \times Y)$ and

$$\begin{aligned} I_{X\times Y} - R^2 &= \begin{pmatrix} I_X & 0 \\ 0 & I_Y \end{pmatrix} - \begin{pmatrix} 0 & S \\ T & 0 \end{pmatrix}^2 \\ &= \begin{pmatrix} I_X - ST & 0 \\ 0 & I_Y - TS \end{pmatrix} \in \mathcal{K}(X \times Y \to X \times Y). \end{aligned} \tag{3.6.302}$$

Granted this, Theorem 3.1.10 applies and gives that R is a Fredholm operator. In particular, R is a relatively open map. From this and (3.6.271) we then conclude that both T and S are relatively open maps. In concert with (3.6.298)-(3.6.300) this ultimately shows that T, S are Fredholm operators. The above argument justifies the implication *(c)* $\Rightarrow$ *(a)*.

Since at this point we have shown that *(a)*-*(c)* are all equivalent, there remains to include item *(d)* in this circle of equivalences. Trivially, *(c)* $\Rightarrow$ *(d)*, so there remains to prove that *(d)* $\Rightarrow$ *(c)*. With this in mind, assume there exist $S_1, S_2 \in \mathcal{L}(Y \to X)$ such that

$$K_1 := I_X - S_1 T \in \mathcal{K}(X \to X) \quad \text{and} \quad K_2 := I_Y - T S_2 \in \mathcal{K}(Y \to Y) \tag{3.6.303}$$

Then

$$\begin{aligned} K_1 S_2 &= S_2 - (S_1 T) S_2 = S_2 - S_1 (T S_2) \\ &= S_2 - S_1 (I_Y - K_2) = S_2 - S_1 + S_1 K_2 \end{aligned} \tag{3.6.304}$$

which further implies

$$S_1 - S_2 = K := S_1 K_2 - K_1 S_2 \in \mathcal{K}(Y \to X), \tag{3.6.305}$$

where the membership comes from item *(b)* in Theorem 3.6.2. As a result,

$$\begin{aligned} I_Y - TS_1 &= I_Y - T(S_2 + K) \\ &= (I_Y - TS_2) - TK = K_2 - TK \in \mathcal{K}(Y), \end{aligned} \tag{3.6.306}$$

again by item *(b)* in Theorem 3.6.2. Hence, we have found $S_1 \in \mathcal{L}(Y \to X)$ with the property that $I_X - S_1T \in \mathcal{K}(X)$ and $I_Y - TS_1 \in \mathcal{K}(Y)$, finishing the justification of *(d)* ⇒ *(c)*.

There remains to prove the very last claim in the statement of the theorem. First, thanks to the current implication *(c)* ⇒ *(a)* and item *(c)* in Theorem 3.6.2, any operator $S \in \mathcal{L}(Y \to X)$ doing the job in either item *(b)* or item *(c)* is necessarily Fredholm. Assume now $S_1, S_2 \in \mathcal{L}(Y \to X)$ are as in items *(d)*. Reasoning as in (3.6.303)-(3.6.305) with the role of S_2 played by S from item *(c)*, we see that

$$S_1 - S \in \mathcal{K}(Y \to X). \tag{3.6.307}$$

Since $S \in \mathcal{L}(Y \to X)$ is already known to be Fredholm, from the current item *(c)* we know that that there exists some $R \in \mathcal{L}(X \to Y)$ such that both $I_Y - SR$ and $I_X - RS$ are compact. In view of (3.6.307) and item *(c)* in Theorem 3.6.2 we then conclude that both $I_Y - S_1R$ and $I_X - RS_1$ are compact, hence $S_1 \in \mathcal{L}(Y \to X)$ is actually Fredholm. In a similar fashion, $S_2 \in \mathcal{L}(Y \to X)$ is also seen to be Fredholm, and this completes the proof of Theorem 3.6.12. □

The version of Atkinson's Theorem worked out above is very versatile, and it finds itself at the core of the Fredholm theory. A modest but useful application (which should be compared with Lemma 1.6.11) is as follows.

Proposition 3.6.13. *Let X, Y be two Hausdorff linear topological spaces and consider a Fredholm operator $T : X \to Y$. If E is a finite-dimensional linear subspace of Y then $T^{-1}(E)$ is a finite-dimensional linear subspace of X.*

Proof. From the equivalence *(a)* ⇔ *(b)* in Theorem 3.6.12 we know that there exists some operator $S \in \mathcal{L}(Y \to X)$ such that $Q := I_X - ST$ has a finite-dimensional range. Then for every $x \in T^{-1}(E)$ we have $Qx = x - S(Tx)$, hence $x \in \operatorname{ran} Q + S(E)$. This goes to show that $T^{-1}(E)$ is contained in the finite-dimensional space $\operatorname{ran} Q + S(E)$, so it must be finite dimensional itself. □

The stage is now set for proving the following fundamental result.

Theorem 3.6.14. *Let X, Y, Z be three Hausdorff linear topological spaces. Then the following properties hold.*

(a) If $T : X \to Y$ is a Fredholm operator then so is $T + K$ for every $K \in \mathcal{K}(X \to Y)$, and one has $i(T) = i(T + K)$. In other words, both the class of Fredholm operators as well as the index are stable under compact perturbations.
(b) If $T : X \to Y$ and $S : Y \to Z$ are Fredholm operators, then so is $ST : X \to Z$.

Proof. Let $\widetilde{T} \in \mathcal{L}(X \times Y \to X \times Y)$ be associated with T as in (3.6.290). From Proposition 3.6.11 we know that $\widetilde{T}$ is a Fredholm operator. In the matrix formalism employed in the proof of Theorem 3.6.12 this is naturally identified with a 2×2 matrix, namely

$$\widetilde{T} = \begin{pmatrix} 0 & T_{\mathrm{inv}} \\ T & 0 \end{pmatrix}. \tag{3.6.308}$$

Let us also introduce

$$\widetilde{K} := \begin{pmatrix} 0 & 0 \\ K & 0 \end{pmatrix} \tag{3.6.309}$$

and observe that $\widetilde{K} \in \mathcal{K}(X \times Y \to X \times Y)$. Corollary 3.2.2 then guarantees that

$$\widetilde{T} + \widetilde{K} = \begin{pmatrix} 0 & T_{\mathrm{inv}} \\ T + K & 0 \end{pmatrix} \tag{3.6.310}$$

is a Fredholm operator on $X \times Y$. With this in hand, we conclude from (3.6.272) that $T + K : X \to Y$ is also a Fredholm operator. In addition, based on (3.6.273) and Corollary 3.3.16 we may write

$$i(T) + i(T_{\mathrm{inv}}) = i(\widetilde{T}) = i\,(\widetilde{T} + \widetilde{K}) = i(T + K) + i(T_{\mathrm{inv}}) \tag{3.6.311}$$

from which we conclude that $i(T) = i(T + K)$. This takes care of item *(a)*.

To deal with the claim in item *(b)*, assume $T : X \to Y$ and $S : Y \to Z$ are Fredholm operators. Then the operators

$$\widetilde{T} = \begin{pmatrix} 0 & T_{\mathrm{inv}} \\ T & 0 \end{pmatrix} \in \mathcal{L}(X \times Y \to X \times Y), \tag{3.6.312}$$

and

$$\widetilde{S} = \begin{pmatrix} 0 & S_{\mathrm{inv}} \\ S & 0 \end{pmatrix} \in \mathcal{L}(Y \times Z \to Y \times Z) \tag{3.6.313}$$

associated with T and S as in Proposition 3.6.11 are both Fredholm. To proceed, consider

$$\begin{gathered} \widetilde{\widetilde{T}} : X \times Y \times Z \longrightarrow X \times Y \times Z \ \text{ defined as} \\ \widetilde{\widetilde{T}}(x, y, z) := (\widetilde{T}(x, y)\,,\, z) \ \text{ for each } \ (x, y, z) \in X \times Y \times Z, \end{gathered} \tag{3.6.314}$$

and

$$\begin{gathered} \widetilde{\widetilde{S}} : X \times Y \times Z \longrightarrow X \times Y \times Z \ \text{ defined as} \\ \widetilde{\widetilde{S}}(x, y, z) := (x\,,\, \widetilde{S}(y, z)) \ \text{ for each } \ (x, y, z) \in X \times Y \times Z. \end{gathered} \tag{3.6.315}$$

Since $\widetilde{T}$ and $\widetilde{S}$ are Fredholm, the above definitions imply that both $\widetilde{\widetilde{T}}$ and $\widetilde{\widetilde{S}}$ are Fredholm operators from $X \times Y \times Z$ into itself. As such, Corollary 3.2.2 applies and gives that $\widetilde{\widetilde{S}} \circ \widetilde{\widetilde{T}}$ is a Fredholm operator on $X \times Y \times Z$. In the matrix formalism, we have

$$\widetilde{\widetilde{T}} = \begin{pmatrix} 0 & T_{\text{inv}} & 0 \\ T & 0 & 0 \\ 0 & 0 & I_Z \end{pmatrix} \tag{3.6.316}$$

and

$$\widetilde{\widetilde{S}} = \begin{pmatrix} I_X & 0 & 0 \\ 0 & 0 & S_{\text{inv}} \\ 0 & S & 0 \end{pmatrix} \tag{3.6.317}$$

hence

$$\widetilde{\widetilde{S}} \circ \widetilde{\widetilde{T}} = \begin{pmatrix} 0 & T_{\text{inv}} & 0 \\ 0 & 0 & S_{\text{inv}} \\ ST & 0 & 0 \end{pmatrix} \text{ is a Fredholm operator on } X \times Y \times Z. \tag{3.6.318}$$

Granted this, Atkinson's Theorem in the format recorded Theorem 3.1.10 applies and gives an operator

$$Q = \begin{pmatrix} Q_{11} & Q_{12} & Q_{13} \\ Q_{21} & Q_{22} & Q_{23} \\ Q_{31} & Q_{32} & Q_{33} \end{pmatrix} \in \mathcal{L}(X \times Y \times Z) \tag{3.6.319}$$

with the property that

$$\begin{gathered} K := I_{X\times Y\times Z} - Q(\widetilde{\widetilde{S}} \circ \widetilde{\widetilde{T}}) \text{ and } \widetilde{K} := I_{X\times Y\times Z} - (\widetilde{\widetilde{S}} \circ \widetilde{\widetilde{T}})Q \\ \text{are both compact operators on } X \times Y \times Z. \end{gathered} \tag{3.6.320}$$

Unraveling notation, we deduce from the equality in (3.6.318), the format of Q in (3.6.319), and the definitions in (3.6.320) that

$$STQ_{13} = I_Z - \widetilde{K}_{13} \text{ and } Q_{13}ST = I_X - K_{13}, \tag{3.6.321}$$

where $K_{13} \in \mathcal{L}(Z \to X)$ and $\widetilde{K}_{13} \in \mathcal{L}(Z \to X)$ are the $(1, 3)$-entries in the 3×3 matrices K and $\widetilde{K}$, respectively. Since the latter are compact operators on $X \times Y \times Z$, we may rely on item *(b)* in Theorem 3.6.2 to conclude that

$$K_{13} \in \mathcal{K}(Z \to X) \text{ and } \widetilde{K}_{13} \in \mathcal{K}(Z \to X). \tag{3.6.322}$$

Thus, we have found an operator $Q_{13} \in \mathcal{L}(Z \to X)$ doing the job in (3.6.321)-(3.6.322). With this in hand, we may call upon the brand of Atkinson Theorem stated in Theorem 3.6.12 and conclude that $ST : X \to Z$ is indeed a Fredholm operator.

The formalism developed above will be used again later. This is why we chose to include this proof, even though a shorter argument – based entirely on Theorem 3.6.12– may be employed to arrive at the conclusion that ST is a Fredholm operator whenever S and T are Fredholm operators. This alternative, more economical, proof goes as follows. Since T and S are Fredholm, Theorem 3.6.12 guarantees the existence of

$$\begin{aligned} &Q \in \mathcal{L}(Y \to X) \text{ such that both } I_X - QT \text{ and } I_Y - TQ \text{ are compact,} \\ &\text{and } R \in \mathcal{L}(Z \to Y) \text{ such that both } I_Z - SR \text{ and } I_Y - RS \text{ are compact.} \end{aligned} \tag{3.6.323}$$

Then, based on repeated applications of (3.6.323) and item *(b)* in Theorem 3.6.2, we may write

$$\begin{aligned} (ST)(QR) &= S(TQ)R = S(I_Y + \text{compact})R \\ &= SR + \text{compact} = I_Z + \text{compact}, \end{aligned} \tag{3.6.324}$$

and

$$\begin{aligned} (QR)(ST) &= Q(RS)T = Q(I_Y + \text{compact})T \\ &= QT + \text{compact} = I_X + \text{compact}. \end{aligned} \tag{3.6.325}$$

Once (3.6.324)-(3.6.325) have been established, the implication *(d)* ⇒ *(a)* in Theorem 3.6.12 gives that ST is a Fredholm operator. This completes the proof of Theorem 3.6.14. □

Corollary 3.6.15. *Let X, Y, Z be three Hausdorff linear topological spaces and let $T \in \mathcal{L}(X \to Y)$ and $S \in \mathcal{L}(Y \to Z)$ be such that ST is a Fredholm operator. Then T is Fredholm if and only if S is Fredholm.*

Proof. To fix ideas, suppose first that T is a Fredholm operator. Theorem 3.6.12 then ensures the existence of a Fredholm operator $Q : Y \to X$ with the property that $I_Y - TQ$ is compact on Y. Since ST is, by assumption, a Fredholm operator, item *(b)* in Theorem 3.6.14 guarantees that $(ST)Q$ is Fredholm. However, $(ST)Q = S(TQ) = S(I_Y + \text{compact}) = S + \text{compact}$, by item *(b)* in Theorem 3.6.2, which ultimately shows that S is a compact perturbation of a Fredholm operator. Having established this, item *(a)* in Theorem 3.6.14 applies and gives that S itself is a Fredholm operator, as wanted. Finally, in the case when we start with the assumption that S is a Fredholm operator, the same type of argument shows that T is a Fredholm operator. □

It turns out that the additivity property of the index is robust enough so that we have the following "off-diagonal" version of Theorem 3.3.15.

Theorem 3.6.16 (Addition). *Let X, Y, Z be three Hausdorff linear topological spaces and suppose $T : X \to Y$ and $S : Y \to Z$ are Fredholm operators. Then the operator $ST : X \to Z$ is also Fredholm and*

$$i(ST) = i(S) + i(T). \tag{3.6.326}$$

Proof. The fact that $ST : X \to Z$ is Fredholm has already been established in Theorem 3.6.14. Granted this, we may invoke the additivity property of the index for algebraic Fredholm operators established in Theorem 3.3.26 to conclude that (3.6.326) holds. An alternative proof of the index formula claimed in (3.6.326) may be given, by reducing matters to the "diagonal" case, treated in Theorem 3.3.15. This is done by recycling definitions, notation, and results from the proof of Theorem 3.6.14. For one thing, $\widetilde{\widetilde{T}}$ and $\widetilde{\widetilde{S}}$ from (3.6.314)-(3.6.315) are Fredholm operators from $X \times Y \times Z$ into itself. As such, Theorem 3.3.15 applies and gives that, on the one hand,

$$i(\widetilde{\widetilde{S}} \circ \widetilde{\widetilde{T}}) = i(\widetilde{\widetilde{S}}) + i(\widetilde{\widetilde{T}}). \tag{3.6.327}$$

On the other hand, from the equality in (3.6.318) we see that

$$\operatorname{ran}(\widetilde{\widetilde{S}} \circ \widetilde{\widetilde{T}}) = \operatorname{ran}(T_{\text{inv}}) \times \operatorname{ran}(S_{\text{inv}}) \times \operatorname{ran}(ST) \tag{3.6.328}$$

and

$$\ker(\widetilde{\widetilde{S}} \circ \widetilde{\widetilde{T}}) = \ker(T_{\text{inv}}) \times \ker(S_{\text{inv}}) \times \ker(ST). \tag{3.6.329}$$

In turn, (3.6.328)-(3.6.329) imply (see Exercise 1.3.33 and Proposition 1.3.42)

$$i(\widetilde{\widetilde{S}} \circ \widetilde{\widetilde{T}}) = i(T_{\text{inv}}) + i(S_{\text{inv}}) + i(ST), \tag{3.6.330}$$

while from (3.6.316)-(3.6.317) we see, in a similar manner, that

$$i(\widetilde{\widetilde{S}}) = i(S) + i(S_{\text{inv}}) \quad \text{and} \quad i(\widetilde{\widetilde{T}}) = i(T) + i(T_{\text{inv}}). \tag{3.6.331}$$

At this stage, (3.6.326) is obtained by combining (3.6.327) with (3.6.330)-(3.6.331). □

Even when dealing with the "diagonal" case, the extra flexibility acquired by considering "off-diagonal" operators comes in handy. Here is an example of this sort.

Proposition 3.6.17. *Suppose X is a Hausdorff linear topological space, and suppose $T \in \mathcal{L}(X)$ is a Fredholm operator. Assume V is a finite-dimensional subspace of X which is topologically complemented in X, and such that $T(V) \subseteq V$. Then*

$$[T] : \frac{X}{V} \longrightarrow \frac{X}{V}, \qquad [T](x + V) := (Tx) + V, \quad \forall\, x \in X, \tag{3.6.332}$$

is a Fredholm operator on X/V and $i([T]) = i(T)$.

Proof. Consider the quotient map $\pi : X \to X/V$. Then (3.6.332) implies

$$[T] \circ \pi = \pi \circ T. \tag{3.6.333}$$

From assumptions, Proposition 3.6.8, and item *(b)* in Theorem 3.6.14 we know that $\pi \circ T$ is Fredholm. Hence, so is $[T] \circ \pi$, thanks to (3.6.333). With this in hand, from

Corollary 3.6.15 and Proposition 3.6.8 we conclude that $[T]$ is a Fredholm operator on X/V. Finally, from (3.6.333), Proposition 3.6.8, and Theorem 3.6.16 we see that $i([T]) + i(\pi) = i(\pi) + i(T)$, from which the desired formula $i([T]) = i(T)$ follows. □

The flexibility gained by considering "off-diagonal" operators in the present section allows us to establish the following useful result.

Theorem 3.6.18. *Suppose X_1, X_2 are two Hausdorff topological vector space and, for $j \in \{1, 2\}$, consider $Y_j \subseteq X_j$ closed subspace of finite codimension in X_j. In addition, assume that $T \in \mathcal{L}(X_1 \to X_2)$ is a Fredholm operator such that $T(Y_1) \subseteq Y_2$ and set*

$$S : Y_1 \to Y_2, \quad Sx := Tx, \quad \forall\, x \in Y_1. \tag{3.6.334}$$

Then S is a Fredholm operator and $i(S) = \dim(X_2/Y_2) + i(T) - \dim(X_1/Y_1)$. In particular, if Y_1, Y_2 have equal codimensions, then $i(S) = i(T)$.

Proof. Denote by $\iota_1 : Y_1 \hookrightarrow X_1$ the canonical embedding of Y_1 into X_1, and bring in a projection P_2 of X_2 onto Y_2 (defined much as in (3.6.268)). Then (3.6.334) may be simply recast as $S = P_2 \circ T \circ \iota_1$. In concert with Proposition 3.6.7 and Theorem 3.6.16 (applied twice), this allows us to conclude that S is a Fredholm operator and

$$i(S) = i(P_2) + i(T) + i(\iota_1) = \dim(X_2/Y_2) + i(T) - \dim(X_1/Y_1), \tag{3.6.335}$$

as wanted. □

Recall from Exercise 3.1.4 that, generally speaking, the restriction of a Fredholm operator to an invariant subspace need not be Fredholm. In the next two corollaries we present positives result of this flavor.

Corollary 3.6.19. *Let X be a Hausdorff topological vector space and let $F \subseteq X$ be a closed subspace of finite codimension in X. Assume that $T \in \mathcal{L}(X)$ is a Fredholm operator on X such that $T(F) \subseteq F$ and set*

$$T\big|_F : F \to F, \quad (T\big|_F)(x) := Tx, \quad \forall\, x \in F. \tag{3.6.336}$$

Then the operator $T\big|_F$ is a Fredholm operator on F and $i(T\big|_F) = i(T)$.

Proof. This is an immediate consequence of Theorem 3.6.18 used with $X_j := X$ and $Y_j := F$ for $j = 1, 2$. □

Here is the second corollary mentioned above.

Corollary 3.6.20. *Suppose X and Y are two Hausdorff linear topological spaces, and assume that $T \in \mathcal{L}(X \to Y)$ is a Fredholm operator. Let W be a closed linear subspace of finite codimension in Y, and define $V := T^{-1}W$. Then V is a closed linear subspace of finite codimension in X, and the operator*

$$\begin{aligned} &T|_V : V \to W \ \text{ is Fredholm, with} \\ &i(T|_V) = \dim(Y/W) + i(T) - \dim(X/V). \end{aligned} \tag{3.6.337}$$

Proof. Since T is linear and continuous, Lemma 1.6.11 guarantees that $V = T^{-1}W$ is a closed linear subspace of finite codimension in X. Granted this, we may invoke Theorem 3.6.18 with $X_1 := X$, $X_2 := Y$, $Y_1 := V$, $Y_2 := W$, and the desired result follows. □

We bring this subsection to a close by presenting a fundamental result concerning "off-diagonal" Fredholm operators acting between p-Banach spaces.

Theorem 3.6.21. *If X, Y are p-Banach spaces for some $p \in (0, 1]$, then the following hold:*

(a) The set $\Phi(X \to Y)$ is open in $\mathcal{L}(X \to Y)$.
(b) The index function

$$i : \Phi(X \to Y) \longrightarrow \mathbb{Z} \tag{3.6.338}$$

is continuous on $\Phi(X \to Y)$, where $\mathbb{Z}$ is equipped with the discrete topology.
(c) For each $n \in \mathbb{Z}$, the set $\Phi_n(X \to Y)$ of Fredholm operators from X to Y of index n is both open and closed in $\Phi(X \to Y)$, i.e., is a clopen subset of $\Phi(X \to Y)$.

Proof. Recall from Exercise 1.4.3 that $X \times Y$ becomes a p-Banach space when equipped with the product p-norm defined in (1.4.211). To proceed in earnest, fix some $T \in \Phi(X \to Y)$. Then the operator

$$\widetilde{T} = \begin{pmatrix} 0 & T_{\text{inv}} \\ T & 0 \end{pmatrix} \in \mathcal{L}(X \times Y \to X \times Y) \tag{3.6.339}$$

associated with T as in Proposition 3.6.11 is Fredholm. Corollary 3.4.13 then guarantees the existence of some small $\varepsilon > 0$ such that the ball in $\mathcal{L}(X \times Y \to X \times Y)$ centered at $\widetilde{T}$ and having radius ε consists of Fredholm operators on $X \times Y$ of the same index as $\widetilde{T}$.

Suppose now $S \in \mathcal{L}(X \to Y)$ has the property that $\|T - S\|_{X\to Y} < \varepsilon$. If we consider

$$R := \begin{pmatrix} 0 & T_{\text{inv}} \\ S & 0 \end{pmatrix} \in \mathcal{L}(X \times Y \to X \times Y) \tag{3.6.340}$$

then

$$\|\widetilde{T} - R\|_{X\times Y\to X\times Y} = \|T - S\|_{X\to Y} < \varepsilon. \tag{3.6.341}$$

Consequently, R is a Fredholm operator on $X \times Y$ of the same index as $\widetilde{T}$. In concert with Proposition 3.6.9, this implies that S is a Fredholm operator and

$$i(T_{\text{inv}}) + i(S) = i(R) = i(\widetilde{T}) = i(T_{\text{inv}}) + i(T). \tag{3.6.342}$$

This forces $i(S) = i(T)$. Altogether, the argument so far proves that the quality of being Fredholm and the value of the index are invariant under small norm-perturbations. From this, we conclude that the claims in items *(a)-(b)* hold and that for each $n \in \mathbb{Z}$ the set $\Phi_n(X \to Y)$ is open in $\mathcal{L}(X \to Y)$.

Finally, the continuity of the index function (3.6.338) also implies that, for each $n \in \mathbb{Z}$, the set $\Phi_n(X \to Y) = i^{-1}(\{n\})$ is a relatively closed subset of $\Phi(X \to Y)$. □

3.6.2 Fredholm Theory in Quasi-Banach Spaces

In this subsection, the goal is to present a coherent brand of Fredholm theory specifically adapted to quasi-Banach spaces. We begin by making a number of definitions and remarks.

For any two given quasi-normed spaces $(X, \|\cdot\|_X)$ and $(Y, \|\cdot\|_Y)$, the collection of continuous linear operators from $(X, \tau_{\|\cdot\|_X})$ into $(Y, \tau_{\|\cdot\|_Y})$ originally introduced in (3.6.251) now becomes

$$\begin{aligned}\mathcal{L}(X \to Y) &= \{T : X \to Y \,:\, T \text{ is linear and continuous}\} \\ &= \{T : X \to Y \,:\, T \text{ is linear and bounded}\} \\ &= \Big\{T : X \to Y \,:\, T \text{ is linear and } \|T\|_{X\to Y} < \infty\Big\} \end{aligned} \tag{3.6.343}$$

where the operator quasi-norm has been introduced in (1.4.241). This follows from Theorem 1.11.1 and Theorem 2.1.14. Moreover,

$$\mathcal{L}(X \to Y) \ni T \mapsto \|T\|_{X\to Y} \in [0, \infty) \tag{3.6.344}$$

is a quasi-norm on $\mathcal{L}(X \to Y)$ (cf. Exercise 1.4.23), and in [29, Lemma 1.2.11, p. 16] it has been shown that

$$\Big(\mathcal{L}(X \to Y),\, \|\cdot\|_{X\to Y}\Big) \text{ is a quasi-Banach space if } Y \text{ is complete.} \tag{3.6.345}$$

For further use, we recall below the following useful result

$$\begin{aligned}&\text{whenever } X, Y \text{ are quasi-Banach spaces and } T \in \mathcal{L}(X \to Y) \\ &\text{has a range of finite codimension in } Y, \text{ it necessarily follows} \\ &\text{that ran}\,(T : X \to Y) \text{ is a closed subspace of } Y.\end{aligned} \tag{3.6.346}$$

(see, e.g., [29, Lemma 1.2.7, p. 13]).

Pressing on, given a quasi-normed space $(X, \|\cdot\|)$, one typically abbreviates $\mathcal{L}(X \to \mathbb{C})$ simply as X^*, a space referred to as the (topological) `dual` of X. Thus,

$$X^* := \Big\{\Lambda : X \to \mathbb{C} \text{ linear} : \|\Lambda\|_{X^*} := \sup_{x\in X\setminus\{0\}} \frac{|\Lambda x|}{\|x\|} < \infty\Big\}. \tag{3.6.347}$$

Then, according to (3.6.345),

$$(X^*,\, \|\cdot\|_{X^*}) \text{ is a quasi-Banach space.} \tag{3.6.348}$$

As opposed to genuine Banach spaces which have an abundance of continuous linear functionals, the dual of a quasi-normed space (or even quasi-Banach space) may be quite poor (think $L^p([0, 1])$ equipped with the usual L^p quasi-norm, when $0 < p < 1$; cf. Exercise 7 in §1.9).

This justifies making the following definition (itself, a more specialized version of Definition 1.6.12):

Definition 3.6.22. *Let X be a quasi-normed space. Call X* `dual-rich` *if there are enough continuous linear functionals on X to separate points i.e., for each $x \in X\backslash\{0\}$ there exists $\Lambda \in X^*$ so that $\Lambda(x) \neq 0$.*

It is not always the case that a finite-dimensional subspace of a Hausdorff linear topological space X is necessarily topologically complemented. However, as pointed out in part *(c)* of Remark 1.8.30 this does happen if X is dual-rich. As such, we have the following result:

Proposition 3.6.23. *Any finite-dimensional subspace F of a dual-rich quasi-normed space X is topologically complemented.*

Moving on, with $\mathcal{K}(X \to Y)$ denoting the set of linear compact operators from X into Y, defined as in (3.6.252), we have the following result.

Theorem 3.6.24. *Assume $(X, \|\cdot\|_X)$, $(Y, \|\cdot\|_Y)$, $(Z, \|\cdot\|_Z)$, and $(W, \|\cdot\|_W)$ are quasi-normed space. Then the following properties hold:*

(a) $\mathcal{K}(X \to Y)$ is a subspace of $\mathcal{L}(X \to Y)$, and if Y is complete then $\mathcal{K}(X \to Y)$ is a closed subspace of the quasi-Banach space $\big(\mathcal{L}(X \to Y), \|\cdot\|_{X\to Y}\big)$.
(b) One has

$$\begin{aligned} &J \circ K \circ L \in \mathcal{K}(X \to W) \ \text{ for all} \\ &J \in \mathcal{L}(Z \to W),\ K \in \mathcal{K}(Y \to Z),\ L \in \mathcal{L}(X \to Y). \end{aligned} \tag{3.6.349}$$

(c) Every finite-rank operator $T \in \mathcal{L}(X \to Y)$ belongs to $\mathcal{K}(X \to Y)$.
(d) Every $K \in \mathcal{K}(X \to Y)$ maps bounded subsets of X into relatively compact subsets of Y. In fact, a linear map $K : X \to Y$ belongs to $\mathcal{K}(X \to Y)$ if and only if

$$\begin{aligned} &\text{for each sequence } \{x_j\}_{j\in\mathbb{N}} \subseteq X \text{ with } \sup_{j\in\mathbb{N}} \|x_j\|_X < \infty \text{ there} \\ &\text{exists a subsequence } \{x_{j_k}\}_{k\in\mathbb{N}} \text{ so that } \{Kx_{j_k}\}_{k\in\mathbb{N}} \text{ converges in } Y. \end{aligned} \tag{3.6.350}$$

Proof. All claims are established by reasoning as in the proofs of Theorem 3.6.2, Proposition 2.1.5, and Proposition 2.1.2. □

The class of Fredholm operators acting between quasi-Banach spaces is introduced next.

Definition 3.6.25. *Suppose X and Y are two quasi-Banach spaces. Call an operator $T \in \mathcal{L}(X \to Y)$* `Fredholm` *if:*

(1) T has a range of finite codimension in Y;
(2) $\ker T$ is finite dimensional and topologically complemented in X.

Introduce $\Phi(X, Y) := \{T \in \mathcal{L}(X \to Y) : T \text{ Fredholm}\}$ and define the index function

$$\begin{aligned} i : \Phi(X, Y) \longrightarrow \mathbb{Z} \ &\text{by setting, for each } T \in \Phi(X, Y), \\ i(T) &:= \dim(\ker T) - \operatorname{codim}(\operatorname{ran} T). \end{aligned} \tag{3.6.351}$$

If one wishes to stress the spaces on which the operator T is considered, write $i(T : X \to Y)$ in place of just $i(T)$.

It is apparent from the above definition and the Open Mapping Theorem (in the version recorded in Theorem 1.6.7) that:

$$\text{for a Fredholm operator of index zero acting between two quasi-Banach spaces, being bijective with a continuous inverse is equivalent to mere injectivity.} \tag{3.6.352}$$

This is one of the most striking –and useful– features of Fredholm theory: it breaks the notoriously difficult problem of proving that an operator is a linear homeomorphism into a collection of far more manageable subtasks!

From Definition 3.6.25 and (3.6.346) we see that:

Proposition 3.6.26. *Suppose X and Y are two quasi-Banach spaces. Then every operator $T \in \Phi(X \to Y)$ has closed range.*

In view of Proposition 3.6.26 and the last part of Remark 3.6.6, it is reassuring to note that Definition 3.6.25 is a particular case of that in Definition 3.6.5.

Also, based on Definition 3.6.25, Proposition 3.6.26, and Proposition 3.6.23 we obtain:

Proposition 3.6.27. *Assuming X is a dual-rich quasi-Banach space, Y is a quasi-Banach space, and $T \in \mathcal{L}(X \to Y)$, then T is Fredholm if and only if T has finite codimensional range and* $\ker T$ *is finite dimensional.*

Specializing Theorem 3.6.12 to the class of quasi-Banach spaces yields the following result.

Theorem 3.6.28 (Atkinson's Theorem for quasi-Banach spaces). *Assume X, Y are two quasi-Banach spaces, and suppose $T \in \mathcal{L}(X \to Y)$. Then the following statements are equivalent:*

(a) The operator T is a Fredholm operator (in the sense of Definition 3.6.25).
(b) The operator T is invertible modulo operators with finite-dimensional ranges, in the sense that there exists $S \in \mathcal{L}(Y \to X)$ such that both $I_X - ST$ and $I_Y - TS$ have finite-dimensional ranges.
(c) The operator T is invertible modulo compact operators in the sense that there exists some $S \in \mathcal{L}(Y \to X)$ such that both $I_X - ST$ and $I_Y - TS$ are compact.
(d) There exist $S_1, S_2 \in \mathcal{L}(Y \to X)$ such that both $I_X - S_1T$ and $I_Y - TS_2$ are compact.

In addition, if the operator $T \in \mathcal{L}(X \to Y)$ is Fredholm then any operators $S, S_1, S_2 \in \mathcal{L}(Y \to X)$ doing the job in items (b)-(d) above are also Fredholm.

Combining Theorem 3.6.14 with Theorem 3.6.16 specialized to quasi-Banach spaces we see that, in this setting, both the class of Fredholm operators as well as the index are stable under compact perturbations. Moreover, the class of Fredholm operators is stable under composition and the index satisfies the multiplication-to-addition law. Specifically, we have the following result:

Theorem 3.6.29. *Let X, Y, Z be three quasi-Banach spaces. Then the following properties hold.*

(a) If $T : X \to Y$ is a Fredholm operator then so is $T + K$ for every $K \in \mathcal{K}(X \to Y)$, and one has $i(T) = i(T + K)$.
(b) If $T : X \to Y$ and $S : Y \to Z$ are Fredholm operators, then so is $ST : X \to Z$ and

$$i(ST) = i(S) + i(T). \tag{3.6.353}$$

Corollary 3.6.30. *Suppose X, Y, Z are three quasi-Banach spaces and consider two operators, $T \in \mathcal{L}(X \to Y)$ and $S \in \mathcal{L}(Y \to Z)$, with the property that ST is a Fredholm operator. Then T is Fredholm if and only if S is Fredholm.*

Proof. This is readily verified from Theorem 3.6.28 and Theorem 3.6.29. □

Exercise 3.6.31. Let $\mathcal{X}$, $\mathcal{Y}$, $\mathcal{Z}$, $\mathcal{W}$ be quasi-Banach spaces and consider the commutative diagram:

$$\begin{array}{ccc} \mathcal{X} & \longrightarrow & \mathcal{Y} \\ \downarrow & & \downarrow \\ \mathcal{Z} & \longrightarrow & \mathcal{W} \end{array} \tag{3.6.354}$$

where all arrows are linear and continuous operators. Show that if three of the four arrows are Fredholm operators then so is the fourth one.

Hint: Observe that if three of the four arrows are Fredholm operators, then two "composable" arrows are Fredholm operators. Use this, the first part of item *(b)* in Theorem 3.6.29, and the commutativity of the diagram, to conclude that the composition of the remaining two "composable" arrows is Fredholm. Since one the two operators involved in this composition is Fredholm, use Corollary 3.6.30 to obtain that the other operator involved in this composition is Fredholm.

Our next major result in this subsection deals with the topological properties of the Fredholm index and of the "off-diagonal" Fredholm operators acting between quasi-Banach spaces.

Theorem 3.6.32. *If X, Y are quasi-Banach spaces, then the following hold:*

(a) The set $\Phi(X \to Y)$ is open in $\mathcal{L}(X \to Y)$.
(b) The index function

$$i : \Phi(X \to Y) \longrightarrow \mathbb{Z} \tag{3.6.355}$$

is continuous on $\Phi(X \to Y)$, where $\mathbb{Z}$ is equipped with the discrete topology.
(c) For each $n \in \mathbb{Z}$, the set $\Phi_n(X \to Y)$ of Fredholm operators from X to Y of index n is both open and closed in $\Phi(X \to Y)$, i.e., is a clopen subset of $\Phi(X \to Y)$.

Proof. Thanks to Theorem 1.11.1 and Exercise 1.4.9, there is no loss of generality in assuming that X, Y are both p-Banach spaces for a common $p \in (0, 1]$. Assuming this is the case, Theorem 3.6.21 applies and gives all desired results. □

Comment 3.6.33 *Parenthetically, we wish to observe another route for arriving at the conclusions in items (a)-(b) of Theorem 3.6.32 if the corresponding results are already known in the diagonal case (i.e., when $X = Y$).*

Specifically, assume $T \in \Phi(X \to Y)$ has been given and suppose $T_1 \in \mathcal{L}(X \to Y)$ is such that $\|T - T_1\|_{X \to Y} < \varepsilon$, for some sufficiently small $\varepsilon > 0$. Associate with T an operator $S \in \Phi(Y \to X)$ as in item (c) of Theorem 3.6.28. Then the operator $R_1 : X \times Y \to X \times Y$ defined as

$$R_1(x, y) := \begin{pmatrix} 0 & S \\ T_1 & 0 \end{pmatrix} \begin{pmatrix} x \\ y \end{pmatrix} \in X \times Y \text{ for each } (x, y) \in X \times Y, \tag{3.6.356}$$

belongs to $\mathcal{L}(X \times Y \to X \times Y)$, while Proposition 3.6.9 tells that $R : X \times Y \to X \times Y$ defined as

$$R(x, y) := \begin{pmatrix} 0 & S \\ T & 0 \end{pmatrix} \begin{pmatrix} x \\ y \end{pmatrix} \in X \times Y \text{ for each } (x, y) \in X \times Y, \tag{3.6.357}$$

belongs to $\Phi(X \times Y \to X \times Y)$ and has

$$i(R) = i(T) + i(S). \tag{3.6.358}$$

Since, by design, $\|R - R_1\|_{X \times Y \to X \times Y} = \|T - T_1\|_{X \to Y} < \varepsilon$, and since both R, R_1 act from the same quasi-Banach space to itself (namely $X \times Y$), our working hypotheses guarantee that R_1 belongs to $\Phi(X \times Y \to X \times Y)$ and

$$i(R_1) = i(R) \tag{3.6.359}$$

Granted these, Proposition 3.6.9 further tells us that $T_1 \in \Phi(X \to Y)$ and

$$i(R_1) = i(T_1) + i(S). \tag{3.6.360}$$

Combining (3.6.358)-(3.6.360) *also yields $i(T_1) = i(T)$, as wanted.*

We conclude with an abstract regularity result (compare with Proposition 3.3.22).

Proposition 3.6.34. *Assume X_j and Y_j, with $j = 1, 2$, are two pairs of quasi-Banach spaces such that the inclusions $X_1 \hookrightarrow X_2$, $Y_1 \hookrightarrow Y_2$ are well defined, continuous, and the second one has dense range. If $T \in \Phi(X_1, Y_1) \cap \Phi(X_2, Y_2)$ is such that $i\,(T : X_1 \to Y_1) = i\,(T : X_2 \to Y_2)$ then*

$$\ker(T : X_1 \to Y_1) = \ker(T : X_2 \to Y_2). \tag{3.6.361}$$

Proof. Since TX_1 has finite codimension in Y_1, there exists a finite-dimensional subspace M of Y_1 such that $TX_1 \boxplus M = Y_1$ (direct, non-orthogonal sum). We claim that $TX_2 + M = Y_2$. To prove the claim, observe that $Y_1 = TX_1 + M \subseteq TX_2 + M$. Hence, since Y_1 is densely embedded into Y_2, so is $TX_2 + M$. Moreover, because TX_2 is closed and M is finite dimensional, Theorem 1.8.35 implies that $TX_2 + M$

is closed in Y_2. Combining these results, the claim follows. Going further, by using the claim we obtain that

$$\dim\Big(\frac{Y_1}{TX_1}\Big) = \dim M \geq \dim\Big(\frac{Y_2}{TX_2}\Big) \tag{3.6.362}$$

which, in turn, implies that $\operatorname{codim}\operatorname{ran}(T : X_1 \to Y_1) \geq \operatorname{codim}\operatorname{ran}(T : X_2 \to Y_2)$. The latter inequality together with the fact that the index of T is the same when acting from X_j onto Y_j for $j = 1$ and $j = 2$ ultimately allow us to conclude that $\dim\ker(T : X_1 \to Y_1) \geq \dim\ker(T : X_2 \to Y_2)$. The reversed inequality is obvious, thus the conclusion follows. □

3.6.3 Odds and Ends

Here we collect a number of miscellaneous results, related to *semi-Fredholm* theory on quasi-Banach spaces. First, we record a result proved in [23]; cf. Proposition 7.8 on p. 132, and Proposition 7.9 on p. 134. To state it, given two quasi-Banach spaces, we let $G_1(X, Y)$ denote the set of "isomorphic embeddings" of X into Y (aka, linear continuous operators from X into Y that are isomorphisms onto their images; cf. Definition 1.4.24), and $G_2(X, Y)$ the set of linear continuous open mappings of X into Y.

Proposition 3.6.35. *For any two quasi-Banach spaces X and Y, the set $G_j(X, Y)$ defined above is open in $\mathcal{L}(X \to Y)$, for $j = 1, 2$, and $G_1(X, Y) \cap G_2(X, Y)$ is both closed and open in either of $G_1(X, Y)$, $G_2(X, Y)$.*

The result below is contained in [23, Lemma 4.11, p. 74].

Proposition 3.6.36. *Suppose that X, Y are two quasi-Banach spaces. Then $A + K$ has closed range for every $A \in G_1(X, Y)$ and $K \in \mathcal{K}(X \to Y)$.*

We wish to complement the above proposition with the following result.

Proposition 3.6.37. *Let X, Y be two quasi-Banach spaces, and let $T \in \mathcal{L}(X \to Y)$ be such that* $\operatorname{ran} T$ *has finite codimension in Y while* $\ker T$ *is topologically complemented in X. Then for every $K \in \mathcal{K}(X \to Y)$ it follows that* $\operatorname{ran}(T + K)$ *is a closed subspace of finite codimension in Y.*

Proof. In a first stage, we shall work under the stronger assumption that

$$T : X \longrightarrow Y \text{ is actually surjective.} \tag{3.6.363}$$

Let Z be a topological complement of $\ker T$ in X. Thus, $Z \oplus \ker T = X$ and Corollary 1.8.18 ensures that Z is closed in X. In particular, Z is itself a quasi-Banach space when equipped with the quasi-norm inherited from X. Also, the restriction of T to Z, namely

$$T|_Z : Z \longrightarrow Y, \tag{3.6.364}$$

is linear, bounded, and bijective. Since both Z and Y are quasi-Banach spaces, the Open Mapping Theorem (recorded in Theorem 1.11.5) ensures this operator has a linear continuous inverse, $(T|_Z)^{-1} : Y \longrightarrow Z$. If $\iota : Z \hookrightarrow X$ is the canonical inclusion, we now define

$$R := \iota \circ (T|_Z)^{-1} : Y \longrightarrow X. \tag{3.6.365}$$

Then $R \in \mathcal{L}(Y \to X)$. Also, for each $y \in Y$ there exists a unique $z \in Z$ such that $Tz = y$, which further implies $Ry = (T|_Z)^{-1}y = z$, so $(T \circ R)y = T(Ry) = Tz = y$. This proves that $T \circ R = I_Y$, the identity on Y, hence R is an inverse from the right for T. Having picked some $K \in \mathcal{K}(X \to Y)$, it follows that $C := KR \in \mathcal{K}(Y \to Y)$ and $I_Y + C = (T + K) \circ R$. In particular,

$$\operatorname{ran}(I_Y + C) \subseteq \operatorname{ran}(T + K). \tag{3.6.366}$$

But $C \in \mathcal{K}(Y \to Y)$ hence $\operatorname{ran}(I_Y + C)$ is a closed subspace of finite codimension in Y by Theorem 2.1.22. In view of this, we conclude from (3.6.366) and Theorem 1.8.38 that $\operatorname{ran}(T + K)$ is also a closed subspace of finite codimension in Y.

Moving on, we dispense with the stronger hypothesis adopted in (3.6.363), by reverting to the original assumptions made in the statement of the current proposition. Since we are assuming that $\operatorname{ran} T$ has finite codimension in Y, there exists a finite-dimensional subspace M of Y with the property that $M \boxplus \operatorname{ran} T = Y$. From Theorem 1.8.36 we then see that this algebraic sum is actually topological, i.e., $M \oplus \operatorname{ran} T = Y$. Fix $K \in \mathcal{K}(X \to Y)$ and introduce

$$\begin{aligned} &\widetilde{T}, \widetilde{K} : X \times M \longrightarrow Y \ \text{ defined for each } \ (x, y) \in X \times M \\ &\quad \text{by } \ \widetilde{T}(x, y) := Tx + y \ \text{ and } \ \widetilde{K}(x, y) := Kx. \end{aligned} \tag{3.6.367}$$

Then $\widetilde{T} \in \mathcal{L}(X \times M \to Y)$ is surjective, and $\ker \widetilde{T} = (\ker T) \times \{0\}$ is topologically complemented in the quasi-Banach space $X \times M$ (this follows from Proposition 1.8.26; alternatively, use that $Z \times M$ is an algebraic complement of $(\ker T) \times \{0\}$ in $X \times M$, and invoke Proposition 1.8.19, keeping in mind (1.11.601)). Also, $\widetilde{K}$ belongs to $\mathcal{K}(X \times M \to Y)$. As such, from what we have established in the first part of the proof we see that

$$\widetilde{Y} := \operatorname{ran}(\widetilde{T} + \widetilde{K}) \ \text{ is a closed subspace of finite codimension in } \ Y. \tag{3.6.368}$$

In particular, $\widetilde{Y}$ becomes a quasi-Banach spaces when equipped with the quasi-norm inherited from Y. In addition,

$$\begin{aligned} \widetilde{Y} &= \operatorname{ran}(\widetilde{T} + \widetilde{K}) = \{(Tx + y) + Kx : \ x \in X, \ y \in M\} \\ &= \operatorname{ran}(T + K) + M, \end{aligned} \tag{3.6.369}$$

from which we conclude that $T + K : X \to \widetilde{Y}$ is a linear continuous operator of finite codimensional range (cf. Proposition 1.3.31). In light of this, (3.6.346), (3.6.368), and Exercise 1.3.45 we then see that ran $(T+K)$ is indeed a closed subspace of finite codimension in Y. □

Consider next two quasi-Banach spaces $(X, \|\cdot\|_X)$, $(Y, \|\cdot\|_Y)$ and let $T : X \to Y$ be a linear, bounded operator. Define $\kappa(T;\, X, Y)$ to be $+\infty$ if T is not surjective and, otherwise, as the smallest constant with the property that for each $y \in Y$ there exists $x \in X$ so that $Tx = y$ and $\|x\|_X \leq \kappa(T;\, X, Y)\|y\|_Y$. Note that, by the Open Mapping Theorem (recorded in Theorem 1.11.5),

$$\kappa(T;\, X, Y) \text{ is finite if and only if } T \text{ maps } X \text{ onto } Y. \tag{3.6.370}$$

We also let $\eta(T;\, X, Y)$ be the largest constant ≥ 0, so that $\eta(T;\, X, Y)\|x\|_X \leq \|Tx\|_Y$ for each $x \in X$. Once again by virtue of the Open Mapping Theorem (cf. Theorem 1.11.5)

$$\eta(T;\, X, Y) > 0 \text{ if and only if } T \text{ is injective with closed range.} \tag{3.6.371}$$

The result below has been proved in [22].

Proposition 3.6.38. *Suppose that* $(X, \|\cdot\|_X)$, $(Y, \|\cdot\|_Y)$ *are two quasi-normed spaces such that* X *is complete. Also, suppose that* $T : X \to Y$ *is a linear, bounded operator for which the following property is true: there exist* $0 < C_0 < \infty$ *and* $0 < a < 1$ *such that for each* y *in the unit sphere of* Y *one can find* $x \in X$ *with* $\|x\|_X \leq C_0$ *and* $\|y - Tx\|_Y \leq a$.

Then T *is onto and* $\kappa(T;\, X, Y) \leq C_1$ *for some* C_1 *depending exclusively on* C_0, *the quasi-norm constant of* X *and* a.

Other categories of semi-Fredholm operators, related to the class of Fredholm operators introduced in Definition 3.6.25, are introduced below.

Definition 3.6.39. *If* X *and* Y *are two quasi-Banach spaces, set*

$$\begin{aligned}\Phi_+(X, Y) := \{T \in \mathcal{L}(X \to Y) :\ & T \text{ has closed range, and a finite-dimensional} \\ & \text{kernel which is topologically complemented in } X\}\end{aligned} \tag{3.6.372}$$

and

$$\Phi_-(X, Y) := \{T \in \mathcal{L}(X \to Y) : T \text{ has a closed range of finite codimensional}\}. \tag{3.6.373}$$

The set of semi-Fredholm operators *is then defined as* $\Phi_-(X, Y) \cup \Phi_+(X, Y)$. *The index function* (3.6.351) *can then be extended to the set of all semi-Fredholm operators by setting*

$$\begin{aligned} i : \Phi_-(X, Y) \cup \Phi_+(X, Y) &\longrightarrow \mathbb{Z} \cup \{\pm\infty\}, \\ i(T) &:= \dim(\ker T) - \operatorname{codim}(\operatorname{ran} T). \end{aligned} \tag{3.6.374}$$

Clearly, $\Phi(X, Y)$ (the class of Fredholm operators from X into Y) may then be characterized as follows:

$$\Phi(X,Y) = \Phi_-(X,Y) \cap \Phi_+(X,Y). \tag{3.6.375}$$

Thanks to (3.6.346), the demand of "having closed range" is superfluous (hence, may be omitted) in the definition of $\Phi_-(X,Y)$ (cf. (3.6.373)).

The last result in this section, further summarizing various properties of semi-Fredholm operators, is stated next.

Theorem 3.6.40. *Let X and Y be two quasi-Banach spaces and let $T \in \mathcal{L}(X \to Y)$. Then the following assertions hold.*

(1) $T \in \Phi_+(X,Y)$ if and only if T has a kernel which is topologically complemented in X and T is bounded from below up to addition of compact operators. The latter property means that there exist a quasi-Banach space Z, a compact operator $K : X \to Z$, and a constant $C \in (0,\infty)$ such that

$$\|x\|_X \leq C\|Tx\|_Y + \|Kx\|_Z \quad \textit{for all} \quad x \in X. \tag{3.6.376}$$

(2) The set $\Phi_-(X,Y)$ is open in $\mathcal{L}(X \to Y)$ and $\Phi_-(X,Y)$ is stable under addition of continuous finite-rank operators. Moreover, if $T \in \Phi_-(X,Y)$ and $S \in \Phi_-(Y,Z)$ then $ST \in \Phi_-(X,Z)$.

Proof. Let us consider *(1)*. In one direction, if T is bounded from below, up to addition of a compact operator K, introduce $A := (T,K) : X \to Y \oplus Z$ (with the latter space equipped with the natural quasi-norm $\|(y,z)\|_{Y\oplus Z} := \|y\|_Y + \|z\|_Z$). Then (3.6.376) amounts to $\eta(A; X, Y\oplus Z) > 0$, i.e., $A \in G_1(X, Y\oplus Z)$ (in the terminology of Proposition 3.6.35). Since $(0,-K) \in \mathcal{K}(X \to Y \oplus Z)$, Proposition 3.6.36 then gives that $(T,0) = A + (0,-K)$ has closed range. Thus, T has closed range, as desired. To show that $N := \ker T$, which is a closed subspace of X, is finite dimensional, it suffices to check that its unit ball is sequentially relatively compact (here, *(i)* $\Leftrightarrow$ *(ii)* in Proposition 2.1.4 is used). To this end, fix an arbitrary sequence $\{x_j\}_j$ of vectors in X with $\|x_j\|_X \leq 1$ and $Tx_j = 0$. Without loss of generality, it can be assumed that $\{Kx_j\}_j$ converges in Z. Writing (3.6.376) for $x = x_j - x_k$, then proves that $\{x_j\}_j$ is Cauchy, hence, convergent in X. This concludes the proof of the fact that, for an operator in $\mathcal{L}(X \to Y)$ whose kernel is topologically complemented in X, being bounded from below up to addition of compact operators entails membership to $\Phi_+(X,Y)$.

Conversely, if $T \in \Phi_+(X,Y)$ and Z is a topological complement of $W := \ker T$ (which, by Corollary 1.8.18, means that Z is closed in X), define $K : X = W\oplus Z \to W$ by $K(x,y) := x$ for each $x \in W = \ker T$ and $y \in Z$. Note that $K \in \mathcal{K}(X \to W)$, since K has finite rank. Then, since $T : Z \to \operatorname{ran} T$ is an isomorphism, for each $x \in X$ with $x = x_o + y$, where $x_o \in Z$ and $y \in \ker T$, we may write

$$\|x\|_X \leq C(\|x_o\|_X + \|y\|_X) \leq C(\|Tx_o\|_Y + \|Kx\|_X) = C(\|Tx\|_Y + \|Kx\|_W), \tag{3.6.377}$$

using the quasi-triangle inequality on X and Proposition 1.4.25. Thus, (3.6.376) follows. This completes the treatment of item *(1)*.

Moving on, consider the claims made in item *(2)*. Let $T \in \Phi_-(X,Y)$. Then there exists $M \subseteq Y$ such that $Y = TX \boxplus M$ and $\dim M < \infty$. Define $\tilde{T} : X \oplus M \to Y$

as $\tilde{T}(x,m) := Tx + m$ for each $x \in X$ and each $m \in M$. Then $\tilde{T}$ is onto, and, as a result, $C_o := \kappa(\tilde{T}; X \oplus M, Y) < \infty$ by (3.6.370). Let $R \in \mathcal{L}(X \to Y)$ be such that $C_o\|R\|_{X\to Y} < \frac{1}{2}$. Define $\tilde{R} : X \oplus M \to Y$ by setting $\tilde{R}(x,m) := Rx$ for each $x \in X$ and each $m \in M$. As a consequence, we have $C_o\|\tilde{R}\|_{X\oplus M\to Y} < \frac{1}{2}$. Then from the definition of $\kappa(\tilde{T}; X\oplus M, Y)$, for any $y \in Y$ with $\|y\|_Y \leq 1$, there exists $(x,m) \in X\oplus M$ such that $\tilde{T}(x,m) = y$ and $\|(x,m)\|_{X\oplus M} \leq C_o$. Then

$$\|y - (\tilde{T} + \tilde{R})(x,m)\|_Y \leq \|\tilde{R}\|_{X\oplus M\to Y}\|(x,m)\|_{X\oplus M} < \tfrac{1}{2}, \tag{3.6.378}$$

and so it follows from Proposition 3.6.38 that $\tilde{T} + \tilde{R}$ is onto. Then

$$Y = \operatorname{ran}(\tilde{T} + \tilde{R}) = \{Tx + m + Rx : x \in X, m \in M\} = (T + R)X + M, \tag{3.6.379}$$

and so the range of $T + R$ has finite codimension in Y (by Proposition 1.3.31). From (3.6.346) we then conclude that $T + R$ has closed range, and so $T + R \in \Phi_-(X, Y)$. Therefore $\Phi_-(X, Y)$ is open in $\mathcal{L}(X \to Y)$.

To see that the space $\Phi_-(X, Y)$ is stable under addition of compact operators, let $T \in \Phi_-(X, Y)$ and let $K \in \mathcal{L}(X \to Y)$ have finite rank. The goal is to show that $T + K \in \Phi_-(X, Y)$. First we will treat the case when T is onto. In this scenario,

$$Y \supseteq \operatorname{ran}(T + K) + \operatorname{ran} K \supseteq \operatorname{ran} T = Y, \tag{3.6.380}$$

hence $\operatorname{ran}(T+K)+\operatorname{ran} K = Y$. Based on this, the fact that $\operatorname{ran} K$ is finite dimensional (since K has finite rank), and Proposition 1.3.31, we conclude that $T + K$ has finite codimensional range. Granted this, (3.6.346) applies and gives that the range of $T + K$ is also closed. This establishes that $T + K \in \Phi_-(X, Y)$ under the assumption that T is onto.

Next, we consider the general case (i.e., drop the auxiliary assumption that T is onto). Let $M \subseteq Y$ be such that $Y = TX \oplus M$ and $\dim M < \infty$. Define the operators $\tilde{T}, \tilde{K} : X \oplus M \to Y$ by setting, for each $x \in X$ and $y \in M$,

$$\tilde{T}(x,y) := Tx + y \quad \text{and} \quad \tilde{K}(x,y) := Kx. \tag{3.6.381}$$

Since $\tilde{T}$ is onto and $\tilde{K}$ is compact, using the previous case, we know that $\tilde{T} + \tilde{K}$ has closed range of finite codimension in Y. Then since

$$\operatorname{ran}(\tilde{T} + \tilde{K}) = \{(Tx + y) + Kx : x \in X, y \in M\} = \operatorname{ran}(T + K) + M, \tag{3.6.382}$$

it follows that the range of $T+K$ has finite codimension in $\operatorname{ran}(\tilde{T}+\tilde{K})$. Then the range of $T + K$ also has finite codimension in Y (by Exercise 1.3.45). Based on (3.6.346) we then conclude that the range of $T+K$ is also closed, and hence $T+K \in \Phi_-(X, Y)$.

Pressing on, let $T \in \Phi_-(X, Y)$ and $S \in \Phi_-(Y, Z)$, hence $ST \in \mathcal{L}(X \to Z)$. Also, there exist $M \subseteq Y$ and $N \subseteq Z$ finite-dimensional subspaces such that $\operatorname{ran} T + M = Y$ and $\operatorname{ran} S + N = Z$. This permits us to write

$$Z = S(Y) + N = S(\operatorname{ran} T) + S(M) + N = \operatorname{ran}(ST) + (S(M) + N), \tag{3.6.383}$$

where $\dim(S(M)+N) < +\infty$. By Proposition 1.3.31, this shows that ran (ST) has finite codimension in Z. Bearing (3.6.346) in mind, we ultimately conclude that ST belongs to $\Phi_-(X,Z)$, finishing the proof of *(2)*. □

Open Problem: *Fully develop the theory of semi-Fredholm operators on quasi-Banach spaces, including establishing the continuity of the index.*

References

1. Alvarado, R., Mitrea, M.: Hardy spaces on Ahlfors-regular quasi metric spaces. A sharp theory. Lecture Notes in Mathematics, vol. 2142, Springer (2015)
2. Aoki, T.: Locally bounded topological spaces. Proc. Japan Acad. Tokyo. 18, 588–594 (1942)
3. Bleecker, D.D., Booss-Bavnbek, B.: Index Theory with Applications to Mathematics and Physics. International Press, Boston (2013)
4. Böttcher, A., Karlovich, Y.I.: Carleson Curves, Muckenhoupt Weights, and Toeplitz Operators. Progress in Mathematics, vol. 154, Birkhäuser (1997)
5. Brown, R.: The Neumann problem on Lipschitz domains in Hardy spaces of order less than one. Pacific J. Math. 171(2), 389–407 (1995)
6. Caradus, S.R., Pfaffenberger, W.E., Yood, B.: Calkin Algebras and Algebras of Operators on Banach Spaces. Marcel Dekker (1974)
7. Coifman, R.R., Weiss, G.: Extensions of Hardy spaces and their use in analysis. Bull. Amer. Math. Soc. 83(4), 569–645 (1977)
8. Conway, J. B.: A Course in Functional Analysis. 2nd edition. Springer (1990)
9. Dahlberg, B., Kenig, c.: Hardy spaces and the L^p-Neumann problem for Laplace's equation in a Lipschitz domain. Ann. of Math. 125, 437–465 (1987)
10. Cater, S.: Note on a theorem of Day. Bulletin of the AMS. 69, 638–640 (1962)
11. Duren, P.L., Romberg, B.W., Shields, A.L.: Linear functionals on H^p spaces with $0 < p < 1$. J. Reine und Angew. Math. 238, 32–60 (1969)
12. Fabes, E., Mendez, O., Mitrea, M.: Boundary layers on Sobolev-Besov spaces and Poisson's equation for the Laplacian in Lipschitz domains. J. Funct. Anal. 159(2), 323–368 (1998)
13. Fefferman, C., Stein, E.M.: H^p spaces of several variables. Acta Math. 129 (3-4), 137–193 (1972)
14. Folland, G.B.: Real Analysis, Modern Techniques and Their Applications. 2-nd edition. John Wiley and Sons, Inc., New York (1999)
15. Frazier, M., Jawerth, B.: A discrete transform and decompositions of distribution spaces. J. Funct. Anal. 93(1), 34–170 (1990)
16. Gohberg, I., Goldberg, S., Kaashoek,M.A.: Classes of Linear Operators, vol. I. Operator Theory: Advances and Applications, vol. 49, Birkhäuser, Basel (1990)
17. Gohberg, I., Krupnik, N.: One-Dimensional Linear Singular Integral Equations, vol. I-II. Operator Theory: Advances and Applications, vol. 53-54, Birkhäuser Verlag Basel (1992)
18. Goldberg, D.: A local version of real Hardy spaces. Duke Math. J. 46(1), 27–42 (1979)
19. Gramsch, B.: Ein Schema zur Theorie Fredholmscher Endomorphismen und eine Anwendung auf die Idealkette der Hilberträume. Math. Ann. 171, 263–272 (1967)
20. Horváth, J.: Topological Vector Spaces and Distributions, vol. II. Addison–Wesley (1966)
21. Hyers, D.H.: Locally bounded linear topological spaces. Revista Ci. 41, 555–574 (1939)
22. Kalton, N., Mitrea, M.: Stability of Fredholm properties on interpolation scales of quasi-Banach spaces and applications. Transactions of the Amer. Math. Soc. 350, 3837–3901 (1998)

D. Mitrea et al., *Fredholm Theory in Topological Vector Spaces*, Springer Monographs in Mathematics, https://doi.org/10.1007/978-3-032-24738-4

23. Kalton, N., Peck, N.T., Roberts, J.: An F-Space Sampler. London Mat. Soc. Lecture Notes, vol. 89, Cambridge University Press (1984)
24. Kato, T.: Perturbation Theory for Linear Operators. Classics in Mathematics, vol. 132. Springer (1995)
25. Macías, R.A., Segovia, C.: A Decomposition into Atoms of Distributions on Spaces of Homogeneous Type. Adv. in Math. 33(3), 271–309 (1979)
26. Mendez, O., Mitrea, M.: The Banach envelope of Besov and Triebel-Lizorkin spaces and applications. J. Four. Anal. Appl. 6, 503–531 (2000)
27. Mikhlin, S.G., Prössdorf, S.: Singular Integral Operators. Springer Verlag, Berlin, 1986.
28. Mitrea, D., Mitrea, I., Mitrea, M.: Geometric Harmonic Analysis I: A Sharp Divergence Theorem with Nontangential Pointwise Traces. Developments in Mathematics, vol. 72, Springer Nature, Switzerland (2022)
29. D. Mitrea, I. Mitrea, and M. Mitrea, Geometric Harmonic Analysis II: Function Spaces Measuring Size and Smoothness on Rough Sets. Developments in Mathematics, vol. 73, Springer Nature, Switzerland (2022)
30. Mitrea, D., Mitrea, I., Mitrea, M.: Geometric Harmonic Analysis III: Integral Representations, Calderón-Zygmund Theory, Fatou Theorems, and Applications to Scattering. Developments in Mathematics, vol. 74, Springer Nature, Switzerland (2023)
31. Mitrea, D., Mitrea, I., Mitrea, M.: Geometric Harmonic Analysis IV: Boundary Layer Potentials in Uniformly Rectifiable Domains, and Applications to Complex Analysis. Developments in Mathematics, vol. 75, Springer Nature, Switzerland (2023)
32. Mitrea, D., Mitrea, I., Mitrea, M.: Geometric Harmonic Analysis V: Fredholm Theory and Finer Estimates for Integral Operators, with Applications to Boundary Problems. Developments in Mathematics, vol. 76, Springer Nature, Switzerland (2023)
33. Mitrea, D., Mitrea, I., Mitrea, M., Monniaux, S.: Groupoid Metrization Theory: With Applications to Analysis on Quasi-Metric Spaces and Functional Analysis. Applied and Numerical Harmonic Analysis, Birkhäuser (2013)
34. Mukherjee, A.: Atiyah-Singer Index Theorem. An Introduction. Hindustan Book Agency, New Delhi (2013)
35. Müller, V.: Spectral Theory of Linear Operators and Spectral Systems in Banach Algebras. 2-nd edition. Operator Theory: Advances and Applications, Birkhäuser (2007)
36. Pietsch, A.: Operator Ideals. North-Holland, 1980.
37. Pietsch, A.: Eigenvalues and s-Numbers. Cambridge University Press (1987)
38. Rolewicz, S.: On a certain class of linear metric spaces. Bull. Acad. Polon. Sci. Sér. Sci. Math. Astronom. Phys., 5, 471–473 (1957)
39. Rudin, W.: Functional Analysis. 2-nd edition. International Series in Pure and Applied Mathematics, McGraw-Hill, Inc. (1991)
40. Schechter, M.: Principles of Functional Analysis. 2-nd edition. Graduate Studies in Math., vol. 36, Amer. Math. Soc., Providence, RI (2002)
41. Shapiro, J.H.: Examples of proper, closed, weakly dense subspaces in non-locally convex F-spaces. Israel J. Math. 7, 369–380 (1969)
42. Shapiro, J.: Algebraic Fredholm Theory. Open Math Notes, AMS, Reference # OMN:202412.111451, December 19 (2024)
43. Waelbroeck, L.: Topological Vector Spaces and Algebras. Springer (1971)
44. Wengenroth, J.: Derived Functors in Functional Analysis. Lecture Notes in Mathematics, vol. 1810, Springer (2003)
45. Williamson, J.H.: Compact linear operators in linear topological spaces. J. London Math. Soc. 29, 149–156 (1954)
46. Yosida, K.: Functional Analysis. 6th edition, Springer (1980)

Author Index

D. Mitrea et al., *Fredholm Theory in Topological Vector Spaces*, Springer Monographs in Mathematics, https://doi.org/10.1007/978-3-032-24738-4

Subject Index

D. Mitrea et al., *Fredholm Theory in Topological Vector Spaces*, Springer Monographs in Mathematics, https://doi.org/10.1007/978-3-032-24738-4

Theorem Index

D. Mitrea et al., *Fredholm Theory in Topological Vector Spaces*, Springer Monographs in Mathematics, https://doi.org/10.1007/978-3-032-24738-4

H

L

M

O

P

R

S

T

Z

Symbol Index

Symbols

D. Mitrea et al., *Fredholm Theory in Topological Vector Spaces*, Springer Monographs in Mathematics, https://doi.org/10.1007/978-3-032-24738-4